PLANT REPRODUCTION

SECOND EDITION

T. PULLAIAH
Department of Botany
Sri Krishnadevaraya University
Anantapur 515 003, AP, India

K. LAKSHMINARAYANA
&
B. HANUMANTHA RAO
Department of Botany
Andhra University
Visakhapatnam 530 003, AP, India

Published by:

SCIENTIFIC PUBLISHERS (INDIA)
5A, New Pali Road, P.O. Box 91
Jodhpur 342 001, India
E-mail: info@scientificpub.com
Website: www.scientificpub.com

ISBN: 978-93-89061-83-3
eISBN: 978-93-89061-85-7

Printed in India

PREFACE TO SECOND EDITION

We wish to express our grateful thanks to the Teachers of various colleges and universities within and outside the State of Andhra Pradesh for their overwhelming support given to our text book which is evident by complete sale out of the first edition. In this connection, we would like to thank all our Teacher colleagues who gave their valuable suggestions for the improvement of the text. In this edition, which is retitled as Plant Reproduction basing on the UGC Model Syllabus, most of the chapters have been thoroughly revised. The revised edition includes Practical methods. Some molecular aspects have also been included.

We again request all our Teacher, colleagues and student friends to extend their whole hearted support as was done earlier and make this book also a grand and useful edition.

T. Pullaiah
K. Lakshminarayana
B. Hanumantha Rao

ABOUT THE AUTHORS

Prof. T. Pullaiah obtained his M. Sc. (1973) and Ph.D. (1976) degrees in Botany from Andhra University. He was a Post-doctoral Fellow at Moscow State University, Russia, during 1976-1978. He travelled widely in Europe and visited Universities and Botanic Gardens in about 17 countries. He joined Sri Krishnadevaraya University as Lecturer in 1979 and became Professor in 1993. He held several positions in the University as Dean, Faculty of Life Sciences; Head, Department of Botany; Chairman, Board of Studies in Botany; Head, Department of Sericulture; Co-ordinator and Chairman, Board of Studies in Biotechnology; Principal, Sri Krishnadevaraya University College. He retired from active Service in 2011 and was selected by UGC as UGC-BSR Faculty Fellow. He has published 73 books, 330 research papers and 35 popular articles. Under his guidance, 54 students obtained their Ph. D. degrees and 34 students their M. Phil. Degrees. He is the recipient of Best Teacher Award from the Government of Andhra Pradesh, Prof. P. Maheshwari Gold Medal and Dr. G.Panigrahi Memorial Award of Indian Botanical Society and Prof. Y. D. Tiagi Gold Medal of the Indian Association for Angiosperm Taxonomy. He was President of Indian Association for Angiosperm Taxonomy (2013) and President of the Indian Botanical Society (2014). He was a member of Species Survival Commission of International Union for Conservation of Nature and Natural Resources (IUCN).

Prof. K. Lakshminaryana took his M.Sc. (with distinction) and Ph.D. degrees from Andhra University in 1974 and 1978 respectively. He is recipient of National Associateship of UGC in 1985. He has more than 34 years of teaching experience. He has completed five research projects and published more than 80 research papers in National and international journals. He is the editorial board member of the journal Flora and Fauna. His research work is cited in several international standard books. He is an active member of various academic and research associations and actively participated in several conferences/seminars/symposia. Under his guidance 10 students obtained their Ph.D. degrees and 15 students obtained their M.Phil. degrees. He is the Editor of some of the Books in Telugu and English for Undergraduate and Postgraduate students.

Prof. B. Hanumantha Rao obtained his M.Sc. (Gold Medalist) and Ph.D. Degrees from Andhra University. He is a former Professor in the Department of Botany, Andhra University, Visakhapatnam, with many years of Teaching and Research experience. Prof. Rao published more than 80 Research Papers in Journals of National and International repute. Several students obtained their Ph.D. and M.Phil. Degrees under his guidance. He is the Editor of some of the Books in Telugu and English for Undergraduate and Postgraduate students. He is associated with several National and International Professional Bodies.

CONTENTS

	Preface	*iii*
CHAPTER - I	HISTORY OF EMBRYOLOGY OF ANGIOSPERMS	1
CHAPTER - II	MICROSPORANGIUM AND MICROSPOROGENESIS	9
CHAPTER - III	MALE GAMETOPHYTE	26
CHAPTER - IV	PALYNOLOGY	41
CHAPTER - V	OVULE AND MEGASPOROGENESIS	58
CHAPTER - VI	THE FEMALE GAMETOPHYTE	81
CHAPTER - VII	FLORAL CHARACTERS, POLLINATION MECHANISMS AND VECTORS — *A. Janaki Bai*	117
CHAPTER - VIII	FERTILIZATION	132
CHAPTER - IX	INCOMPATIBILITY	159
CHAPTER - X	ENDOSPERM	179
CHAPTER - XI	EMBRYO	204
CHAPTER - XII	POLYEMBRYONY	229
CHAPTER - XIII	APOMIXIS	242
CHAPTER - XIV	EMBRYOLOGY IN RELATION TO TAXONOMY	264
CHAPTER - XV	EXPERIMENTAL EMBRYOLOGY	276
CHAPTER - XVI	PRACTICALS	306
	• SUBJECT AND PLANT INDEX	313

CHAPTER – I

HISTORY OF EMBRYOLOGY OF ANGIOSPERMS

Nehemiah Grew (1682) for the first time highlighted the fact that stamens are the male sex organs of a flower and the pollen grains in some way influenced the ovary to produce the fruit. Rudolph Camerarius (1694) was the first person to say that plants do have sexuality in them. He noticed that a female mulberry tree, near which no male plant was growing, formed only abortive seeds. He also experimented with some female plants of *Mercurialis annua, Ricinus* and maize. He concluded that anthers are the male sex organs of the plant and the ovary with style and stigma, the female sex organs. The significance of pollination in seed-set and the importance of insects in pollination was recognized by Kolreuter (1761).

Johansen (1950), Soueges and Crete (1963) defined embryology as the study of the development of embryo. However, Schnarf (1929), Fagerlind, Maheshwari (1950) and Johri (1992) are of the opinion that embryology is not only the study of the development of embryo but also the events that lead to the development of embryo. In this book embryology is taken in its wider sense to include development of anther, male gametophyte, ovule and female gametophyte, fertilization, development of embryo and endosperm, apomixis and polyembryony.

History of embryology of Angiosperms can be broadly divided into three phases: Descriptive embryology, Comparative embryology and Experimental embryology.

Descriptive Embryology

In this phase dissections, free hand sections and then microtome sections of anthers and ovules were examined under the microscope and considerable amount of embryological information was collected on the development and organization of the gametophytes, endosperm and embryo.

An Italian mathematician and astronomer Giovanni Battista Amici (1824), while examining the stigma of *Portulaca oleracea* saw the pollen grain adhering to a stigmatic hair. He kept it under observation and, to his surprise the pollen grain split open followed by emergence of tube or "gut" which grew along the side of the hair and entered the tissues of the stigma. Repeating his observation in 1830, Amici confirmed that, the pollen tube grew bit by bit through stigma and style and came in contact with the ovule.

Schleiden (who with Schwann propounded the cell theory in 1839) in 1830 made a detailed observation on the ovule. He confirmed Amici's statement that the pollen tubes made their way from the stigma to the ovule, entering the latter through the micropyle. Schleiden (1837) followed its growth in many plants and established its occurrence as a general phenomenon in angiosperms. However, he considered the embryo sac to be only an incubator or a sort of hatching place for the embryo which is said to originate from the tip of the pollen tube. This idea really countered the existence of any kind of sexual process.

Amici (1842) opposed the views of Schleiden and tried to show that the embryo developed from a portion of the ovule which was already in existence and was fertilized by the fluid in the tube. In 1847 Amici provided a conclusive evidence to show that in *Orchis* the 'germinal vesicle' (now called as egg) was already present inside the embryo sac before the arrival of the pollen tube, and that this vesicle which gave rise to the embryo, stimulated no doubt by the presence of the pollen tube. Support for Amici's views came from Hofmeister (1849) who described his observations on 38 species belonging to 19 genera and showed that without exception the embryo originates from a preexisting cell in the embryo sac and not from the pollen tube. Radlkoefer (1856) also supported the views of Amici.

Hofmeister (1848) also figured the division stages leading to formation of microspore tetrads in *Tradescantia.* Later he concluded that the pollen tube contained some liquid substance which diffused out and caused the egg to develop into an embryo.

Hanstein (1870) gave a detailed description of the development of embryo in *Capsella* and *Alisma.* Famintzin (1879) confirmed these observations, while Treub (1879) described the embryos of several orchids. Guignard (1881) gave a full account of massive suspensors of the Leguminosae.

Strasburger (1878) demonstrated for the first time that in *Hosta, Nothoscordum, Citrus* and *Mangifera,* the nucellar cells lying close to the apex of the embryo sac become richly protoplasmic and divide to form adventive embryos.

Strasburger (1879) described for the first time development of embryo sac in *Polygonum divaricatum.* Elfving (1879) published some excellent illustrations of the nuclei contained in pollen grains. Strasburger (1884), on the basis of his studies on the embryo sac of *Monotropa hypopitys,* further showed that the pollen tube discharges its nuclei into the embryo sac and one of the nuclei fuses with the egg. This phenomenon is known as syngamy.

Hegelmaier (1885, 1886) gave a comparative account of the histology of the endosperm in many dicotyledons. Treub (1882) discovered chalazogamy in *Casuarina.* Nawaschin (1894, 1897) found chalazogamy in *Betula* and *Juglans,* while Benson (1894) reported its occurrence in several members of the Amentiferae.

The fate of the second male gamete discharged by the pollen tube was not known so far. Nawaschin (1898), Guignard (1899) working independently on identical material, *i.e., Lilium martagon* and *Fritillaria tenella* showed that one male gamete fused with the egg (syngamy), and the other male gamete fused with the polar nuclei (triple fusion). The phenomenon, called double fertilization, was later found to be universal in angiosperms. Juel (1900) and Murbeck (1901) working on *Antennaria* and *Alchemilla* described parthenogenesis.

By 1900 enough information was available to summarize the embryological information which was done by Coulter and Chamberlain (1903) in their book "*Morphology of Angiosperms*". This was followed 26 years later by "*Embryologie der Angiospermen*" by Schnarf (1929).

In thirties and forties of 20th Century India took lead in the field of embryology and the name of P. Maheshwari stands foremost. His book "*An Introduction to the Embryology of*

Angiosperms" (P. Maheshwari, 1950) was considered as Old Testament while "*Recent Advances in the Embryology of Angiosperms*" (P. Maheshwari, ed. 1963) was considered as New Testament. Later Johri (1984) has brought out an edited volume "*Embryology of Angiosperms*".

In the second half of the twentieth century histochemistry, electron microscopy, phase contrast microscopy, fluorescence microscopy, autoradiography and several other modern techniques have been employed to understand fine structure and functional aspects of anther and pollen, ovule and embryo sac, fertilization, endosperm and embryo.

Cass (1973) had shown that the two sperms produced by a male gametophyte remain together for considerable time. Later several embryologists demonstrated that the vegetative nucleus also remains in contact with the sperm cells and the 3 cells move through the pollen tube as a Male Germ Unit. In 1973, Cass also reported separation of sperm cells from the vegetative cell cytoplasm of barley pollen. However, Russell (1986) was the first to obtain an enriched fraction of purified sperm cells. Since then several protocols have been described for the isolation of large numbers of live sperms by bursting of pollen or pollen tubes in a variety of angiosperms. Techniques have also been developed to isolate live embryo sacs and egg cells (Hoshino *et al.*, 2006).

Comparative Embryology

In the beginning of twentieth century embryologists found that embryological characters are useful in understanding the relationships of taxonomic groups. The role of embryology in taxonomy was first brought to prominence by the German embryologist Schnarf (1931) in his book "*Vergleichende Embryologie der Angiospermen*". Later on Davis (1966) brought out a very useful book "*Systematic Embryology of Angiosperms*". Yakovlev and Batygina (1981–90) have brought out in 5 volumes "*Comparative embryology of Flowering Plants*" in Russian language. Johri *et al.* (1992) in their book "*Comparative Embryology of Angiosperms*" discussed the role of embryology in taxonomy.

To quote a few examples, the genus *Paeonia* differs from all other angiosperms in having a free nuclear embryo development.

Similarly, pseudoembryo sac and absence of endosperm development characterize the family Podostemaceae. Endosperm in gymnosperms is a pre-fertilization phenomenon, while in angiosperms it is post-fertilization character. Similarly, the family Onagraceae is characterized by Oenothera type of embryo sac development. The family Cyperaceae is characterized by the formation of pseudomonad, while all other angiosperms show the formation of pollen tetrad.

Experimental Embryology

The beginnings of experimental embryology can be traced to the observations of Gaertner (1949), Massart (1902) and Fitting (1909) who treated the ovaries of certain plants with *Lycopodium* spores, dead pollinia and aqueous extracts of pollen, respectively and still obtained some swelling of the ovary. New light was thereby thrown on the role of pollen and it was suspected that besides carrying sperm nuclei the pollen tube also contributes to the growth of the fruit by supplying auxins. Yet, some workers pointed out, the auxin in pollen grains and pollen tubes is generally too insufficient to account for the tremendous increase in its quantity as shown by analyses of ovaries before and after fertilization. Part of the auxin, no doubt comes from the seeds themselves (Luckwill, 1957), but in his experiments on *Nicotiana* Lund (1956) showed that the growing pollen tubes secrete a substance (probably an enzyme) which possesses the ability to convert free tryptophan into IAA. The formation of auxins is first observed in the style, and then in the ovary, being correlated with the gradual advance of the pollen tube. The ovaries of naturally parthenocarpic plants, either contain a sufficiently high concentration of auxins to enable their development into fruits, or perhaps auxins are transported to them from other parts of the plant. In seeded fruits, on the other hand, a part of the auxin comes from the pollen tube and a part from the ovules themselves after they have been fertilized. The production of seedless fruits by application of hormones and the augmentation of fruit set in greenhouse tomatoes are examples of the application of this knowledge in horticulture. Recently several workers have found that gibberellic acid is even more effective than IAA in inducing parthenocarpy in tomatoes.

It is now clear that pollen performs two separate and independent functions. First, it provides the male gamete for the double fertilization, this gives the stimulus for the development of the embryo and endosperm. Secondly, it either releases hormones which cause the ovary wall to grow into the pericarp, or certain chemicals which lead to the production of hormones.

Since 1960 embryology has become an experimental science having intimate contacts with physiology and genetics. The main objectives of experimental embryology are (1) to understand the factors that control embryological processes and (2) modify these factors.

Anther culture

Guha and Maheshwari (1964) cultured anthers of *Datura innoxia* on a basal medium supplemented with coconut milk, kinetin or grape juice. Numerous embryoids developed inside the anther which grew into plantlets. Guha and Maheshwari (1966) confirmed that these plantlets are haploid. Haploids are very useful in plant breeding as they can be diploidized by treating with colchicine and homozygous diploids for all genes can be obtained. This technique has been used in a large number of species to obtain haploids from anthers and pollen grains.

Embryo culture

The first systematic attempt to culture embryos *in vitro* was made by Hannig (1904) who cultured embryos of *Cochlearia* and *Raphanus.* Dietrich (1925) found that mature embryos when cultured on medium showed normal grcwth, while immature embryos grew directly into the seedlings skipping the stages of embryo development. Laibach (1925, 1928) cultured embryos of interspecific crosses of *Linum* demonstrating the practical application of this technique.

Until 1993 it was not possible to culture isolated pre-globular zygotic embryos. Therefore, in the crosses where abortion occurred at an early stage embryo rescue was possible by *in-ovulo* embryo culture. In 1993, Liu *et al.* achieved complete embryogenesis in the cultures of excised 8-celled proembryos of mustard by using a rich nutrient medium and a modified culture technique. Kranz and

Lorz (1993) and Holm *et al.* (1994) raised full fertile plants from *in vitro* and *in vivo* formed zygotes of maize and barley, respectively, by supplementing the culture medium with nurse cells.

Nucellus culture

Rangaswamy (1961) and Sabharwal (1963) cultured nucellar tissue of polyembryonate *Citrus* species excised from post pollinated stage which formed a callus. Numerous embryoids differentiated from this callus which developed into seedlings. Rangan *et al.* (1968) cultured callus of monoembryonate *Citrus* species and obtained seedlings via callus.

Ovule culture

The first attempt to isolate ovules and culture them under aseptic conditions was made by White (1932) in *Antirrhinum majus.* Maheshwari and Lal (1961) excised ovules of *Papaver somniferum* 6 days after pollination when they contained 2-celled proembryo and few endosperm nuclei and cultured on Nitsch medium. The growth of embryos was initially slower, but after the globular stage they grew rapidly. Poddubnaja-Arnoldi (1959, 1960) grew ovules from pollinated ovaries of certain orchids on simply sucrose solution(10%), Sachar and Kapoor (1959) successfully cultured ovules of *Zephyranthes.*

Ovary culture

Nitsch (1951) attempted the culture of whole ovaries excised after pollination. Fruits were obtained in tomato and gherkin, but these were much smaller than those *in vitro* and the seeds were usually abortive. Maheshwari and Lal (1961) cultured ovaries of *Iberis amara* excised from flowers one day after pollination. Healthy fruits were obtained with the addition of vitamins of the Nitsch Medium.

References

Bhojwani, S.S., S.P. Bhatnagar and P.K.Dantu 2015. *The Embryology of Angiosperms.* 6th edition. Vikas Publishing House, New Delhi.

Bhojwani, S.S. and W.Y.Soh (eds.) 2001. Current Trends in the embryology of Angiosperms. Kluwer Academic Publishers, Dordrecht.

Davis, G.L. 1966. *Systematic Embryology of Angiosperms,* John Wiley, New York.

Johri, B.M. (ed.) 1984. *Embryology of Angiosperms.* Springer-Verlag, Berlin.

Johri, B.M., K.B. Ambegaokar and P.S. Srivastava 1992. *Comparative Embryology of Angiosperms.* Springer-Verlag, Berlin.

Maheshwari, P. 1950. *An Introduction to the Embryology of Angiosperms.* McGraw Hill, New Delhi.

Maheshwari, P. (ed.) 1963. *Recent advances in the embryology of Angiosperms.* International Society of Plant Morphologists, Delhi.

Pandey, A.K. 1997. *Introduction to Embryology of Angiosperms.* CBS Publishers & Distributors, New Delhi.

Ramawat, K.G., Jean-Michel Mérillon, K. R. Shivanna 2014. *Reproductive Biology of Plants.* CRC Press.

Schnarf, K. 1929. Embryologie der Angiospermen. Berlin.

Schnarf, K. 1931. *Vergleichende Embryologie der Angiospermen.* Berlin.

Yakovlev, M.S. and B. Batygina 1981-1990. *Comparative Embryology of Flowering Plants.* Vols. 1-5. Nauka, St. Pertersberg.

CHAPTER – II

MICROSPORANGIUM AND MICROSPOROGENESIS

The male gametes, also known as sperms, are formed in the male gametophyte, which is called as the pollen grain. The pollen grains are formed within the microsporangium. In majority of the Angiosperms, anthers are tetrasporangiate (Fig. 2.1). But in Malvaceae, Adoxaceae, Circaestraceae, Epacridaceae and Restionaceae anthers are bisporangiate. In Bixaceae anthers are 8-sporangiate. Both bisporangiate and tetrasporangiate anthers are seen in the families Asteraceae, Araceae, Asclepiadaceae, Celastraceae, Lauraceae, Lemnaceae, Moringaceae and Piperaceae. Only one microsporangium per anther has been recorded in *Arceuthobium minutissimum*.

In the young stage, the anther in cross section consists of an oval mass of meristematic cells surrounded by an epidermis. When a four lobed contour is formed, archesporium gets differentiated hypodermally at the four corners of the anther. The archesporial cells are larger in size, slightly radially elongated and have prominent nuclei. The extent of archesporium varies from species to species. Either a single archesporial cell may be seen in each lobe in a cross section of the anther as in *Boerhavia diffusa* (P. Maheshwari, 1929), *B. erecta* (Pullaiah and Sarada, 1986), *Elephantopus scaber*, *Blumea malabarica, Zinnia angustifolia* (Pullaiah, 1978, 1979, 1981) or several cells as in *Urginea* (Cooper, 1937), and *Basella rubra* (Maheswari Devi and Pullaiah, 1975). In longitudinal section also the row may comprise of only one cell as in *Enaulus* (Kausik, 1941), or two cells as in *Boerhavia diffusa* (P. Maheshwari, 1929), *B. erecta* (Pullaiah and Sarada, 1986) or several cells as in *Basella rubra* (Maheswari Devi and Pullaiah, 1975). The archesporial cells undergo periclinal divisions forming primary parietal layer and a primary sporogenous cells.

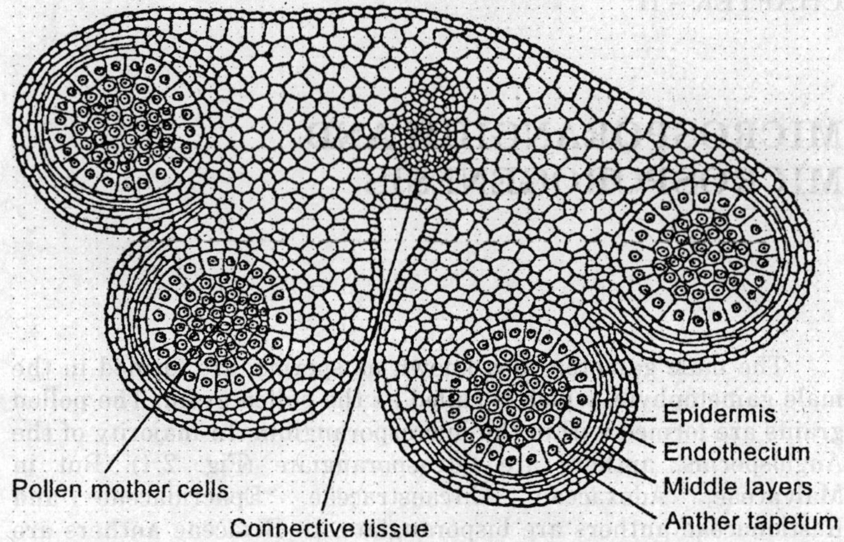

Fig. 2.1. Transverse section of tetrasporangiate anther to show its various tissues.

Anther wall Development

The primary parietal layer undergoes periclinal and anticlinal divisions forming the anther wall. On the basis of the formation of middle layers, Davis (1966) classified the development of anther wall into four types - Basic, Dicotyledonous, Monocotyledonous and Reduced.

(a) Basic type

The primary parietal layer undergoes periclinal division forming outer and inner secondary parietal layers. Both the layers undergo periclinal divisions and the one towards the epidermis differentiates into endothecium followed by a middle layer. The other secondary parietal layer produces another middle layer and the tapetum *e.g.*, Anacardiaceae, Rhamnaceae, Tiliaceae, Vitaceae and Winteraceae. It is considered as the most primitive type.

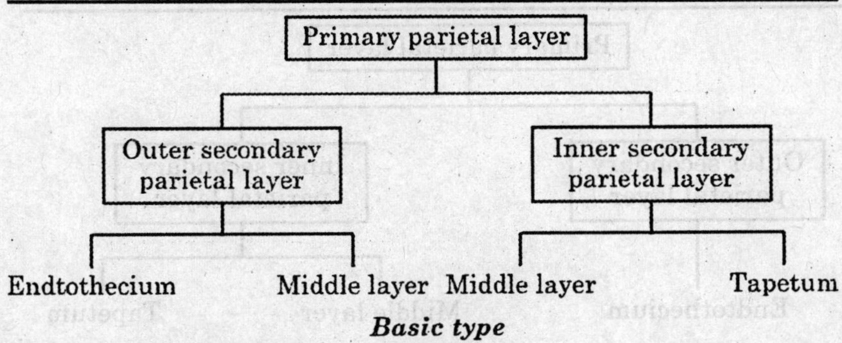

Basic type

(b) Dicotyledonous type

The primary parietal layer undergoes periclinal division forming two secondary parietal layers. Of these the outer parietal layers divides periclinally to give rise to endothecium and middle layer, while the inner layer functions directly as the tapetum *e.g.*, Asteraceae, Asclepiadaceae, Annonaceae, Bignoniaceae, Convolvulaceae, Apiaceae, Solanaceae, Rubiaceae and Acanthaceae.

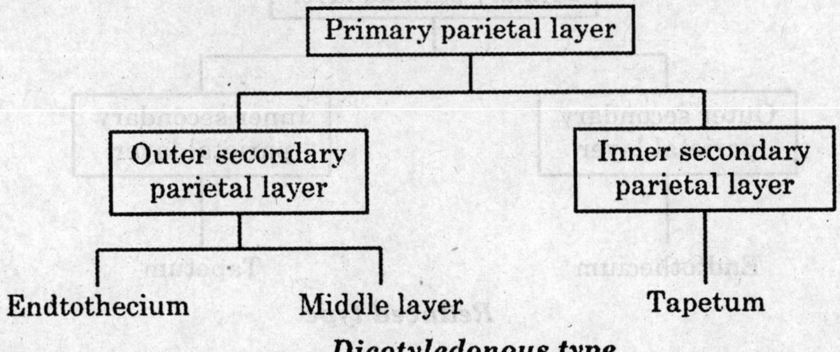

Dicotyledonous type

(c) Monocotyledonous type

The primary parietal layer undergoes periclinal division forming two secondary parietal layers. Of these, outer layer directly functions as the endothecium, while the inner one undergoes periclinal division forming middle layer and tapetum *e.g.*, Liliaceae, Hydrocharitaceae, Basellaceae, Chenopodiaceae, Amaranthaceae, Orchidaceae, Poaceae, Cyperaceae and Brassicaceae.

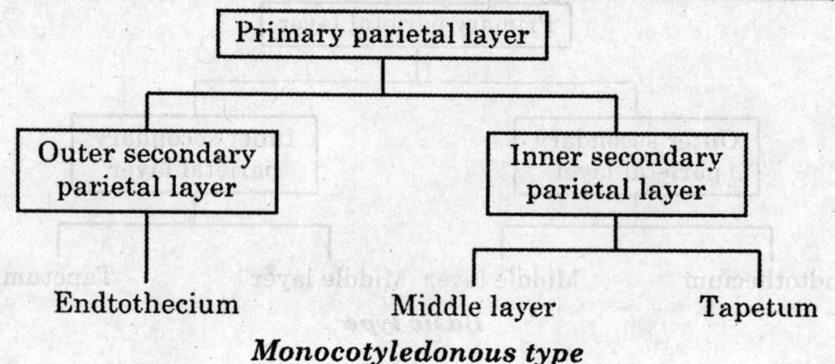

Monocotyledonous type

(d) Reduced type

The primary parietal layer undergoes periclinal division forming two secondary layers. The outer functions directly as endothecium, while the inner functions as tapetum. The middle layer is absent *e.g.*, Lemnaceae and Najadaceae.

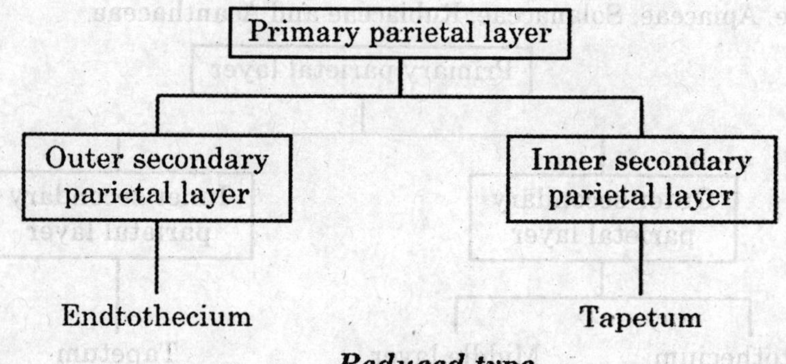

Reduced type

Anther wall structure

The anther wall, at the pollen mother cell stage, comprises of the epidermis, endothecium, middle layer(s) and tapetum (Fig. 2.1).

(a) Epidermis

Epidermis is single-layered and is the external covering of the sporangium. The epidermis at first keeps pace with the developing internal tissues by anticlinal divisions, but soon, due to the pressure exerted by the developing cells inside the sporangium its cells become radially flattened and stretched both tangentially. The epidermis performs its usual protective function.

(b) Endothecium

The layer of cells lying immediately beneath the epidermis is the endothecium. Its cells enlarge and develop fibrous thickenings on their inner tangential walls at maturity. These bands extend radially outward and upwards terminating near the outer tangential wall. In addition, varying number of layers of connective tissue develop fibrous bands and simulate endothecium as in *Swertia angustifolia, Canscora decurrens* (Maheswari Devi and Lakshminarayana, 1977) and *Capparis decidua* (Hanumantha Rao and Prakasa Rao, 1977). The thickenings are usually absent in the region of dehiscence. De Fossard (1969) reported that endothecial thickenings contain mostly α-cellulose. Fibrous thickenings are absent in the cleistogamous flowers and in some members of Dipterocarpaceae, Ericaceae and Hydrocharitaceae. The presence of fibrous bands, differential expansion of the outer and inner tangential walls and hygroscopic nature of the endothecial cells help in the dehiscence of anthers at maturity.

(c) Middle layer(s)

The initial number of middle layers formed in any anther depends on the type of wall formation, two such layers are formed in the basic type, a single layer in both dicotyledonous and monocotyledonous types, and none in the reduced type. In some families the middle layer may undergo further periclinal divisions forming two-many middle layers. In 190 families for which information is available, a single middle layer is characteristic of only 67 families, whereas more than one such layer is found in 113 families. As the cells of the middle layers lack the ability to divide anticlinally; the tissue cannot adjust itself to the pressure exerted by the multiplying and expanding sporogenous cells within each sporangium. The cells of the middle layers ultimately become flattened and crushed at the time of meiotic divisions in the pollen mother cells. In many species the cells of these middle layers are storage centres for starch and other reserves which get mobilized during later development of pollen.

(d) Anther tapetum

The tapetum is the innermost layer of anther wall and surrounds the sporogenous tissue completely. The tapetum is composed of single layer of cells which are full of dense cytoplasm and prominent nuclei. The tapetum on the outer side is derived from the parietal layer, while towards the connective side it develops from the connective tissue with the result it surrounds the sporogenous cells completely. All the cells of the anther tapetum are similar in size and morphology in most of the plants. But in some families like Lamiaceae, the cells of anther tapetum are dimorphic. The cells towards the interior of the anther (connective tapetum) are larger in size, whereas the cells on the outer side of the anther (parietal tapetum) are smaller in size, such dimorphic anther tapetum has been reported in *Salvia mellifera* (Carton and Stuart, 1936), *Plectranthus rugosus*, *Brunella vulgaris*, *Coleus amboinicus* and *Alectra thomsonii* (Fig. 2.2) (Vijayaraghavan and Ratnaparki, 1973). A multiseriate tapetum was reported in *Oxystelma esculentum* (Maheswari Devi and Lakshminarayana, 1977), *Mahonia* and *Arum italicum*. In *Mahonia aquifolium* anther tapetum is 3-layered while in *Arum italicum* it is 2-layered.

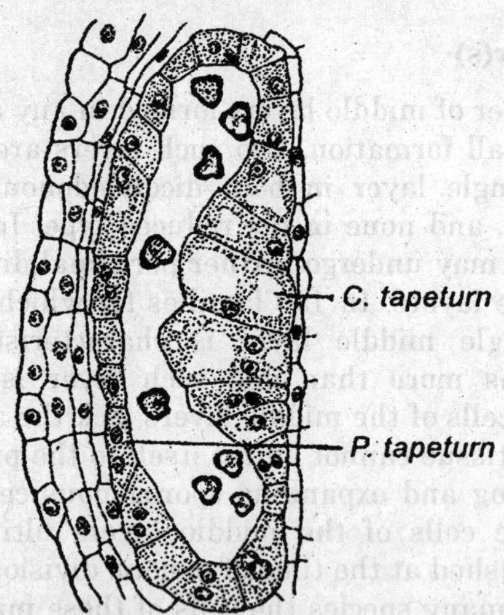

Fig. 2.2. Dimorphic tapeturn in *Alectra thomsonii* (Vijayaraghavan and Ratnaparki, 1973).

In angiosperms, the anther tapetum is of two types: *Glandular* or *secretory* and *Amoeboid* or *periplasmodial*.

Pacini *et al.* (1985) reviewed the form and function of tapetum. According to them Glandular tapetum is characteristic of 175 families, 88 percent of which are dicots. An exclusively amoeboid tapetum has been found in 32 families, of which only 14 are dicots. Both tapetal types are present in 12 families. Glandular tapetum is seen in families like Cucurbitaceae, Amaranthaceae, Chenopodiaceae, Fabaceae and Brassicaceae, while amoeboid tapetum is found in families, such as Asteraceae, Alismataceae etc.

Secretory (Glandular or Parietal or Cellular)

In this type, tapetal cells remain intact *in situ* till the maturation of pollen grains and provide nutrition. A tapetal membrane develops on the inner tangential surface of tapetal cells.

Ultra structure: Echlin and Godwin (1968) have described in detail the events which took place in the tapetal cells of *Helleborus foetidus*. At the sporogenous cells stage the tapetal cells possess mitochondria, plastids, a number of spherical bodies called as ***pro-ubisch bodies***, and ***dictyosomes***. The pro-ubisch bodies are formed from lipid droplets arising probably from mitochondria in tapetal cells. At this stage, the cell walls are relatively thin. Before the onset of meiosis the tapetal cell walls become thick and the cytoplasm becomes more dense due to the increased number of ribosomes and pro-ubisch bodies. During meiosis, there is further increase in the number of pro-ubisch bodies and the size of nuclei. The pro-ubisch bodies, although distributed throughout the cytoplasm, appear to be greater in the region of the cells near the anther cavity. At the tetrad stage, the number of ribosomes further increases and that of microtubules and plastids decreases. At this stage the pro-ubisch bodies, whose number has also increased further appear now to be surrounded by a zone of ribosomes which radiate from them like the spokes of a wheel. The limiting membrane of the pro-ubisch bodies is discontinuous at those places which correspond to the insertion of ribosome rays. The dictyosomes are now found to be associated with a large number of vesicles at the periphery. The dictyosomes and vesicles are never seen associated with pro-ubisch bodies. The tapetal cell membrane at this stage appears highly convoluted.

Soon after the microspores have separated from one another, the ribosomes surrounding the pro-ubisch bodies become more prominent and the inner region of the cytoplasm contain abundant dictyosomes. The pro-ubisch bodies eventually pass through the tapetal cell membranes facing the locule without creating discontinuities in the latter. Once the pro-ubisch bodies have been extruded from the tapetal cytoplasm into the space between the membranc and the cell wall, they rapidly get coated with sporopollenin, and are now called ***ubisch bodies***. The inner wall of the tapetal cells now appears considerably thinner, and a space appears between the wall and the cell membrane. Pro-ubisch bodies continue to be formed within the tapetal cytoplasm and extruded from the cells and sporopollenin is deposited on them. The deposition is initially on the side nearest the locule. All the tapetal cells have by this time disappeared and the tapetal cytoplasm has completely disorganised.

Ubisch bodies

Ubisch bodies are spheroidal structures and they are only a few microns in diameter. The ubisch bodies may remain individually separate or during the course of development, fuse to form larger aggregations or compound bodies. Heslop-Harrison (1968) called these bodies as ***orbicules*** instead of ubisch bodies. These are produced only by the secretory tapetum and not by periplasmodial tapetum. The functions of ubisch bodies are not yet clearly understood. P. Maheshwari (1950) is of the opinion that ubisch granules are involved in external thickening of the exine. Heslop-Harrison and Dickinson (1969) discarded the hypothesis and opined that orbicules are concerned in the transport mechanism of sporopollenin.

Tapetal membrane

Banerjee (1967) observed acetolysis resistant, intricately patterned tapetal membrane in some grasses. Structurally, the tapetal membrane comprises of three layers - fenestrated, reticular and orbicular. Bhandari and Kishori (1969) reported that, in *Nigella damascena*, the original tapetal cell walls disintegrate and, instead, a new membrane resistant to acetolysis becomes organised along the thecal face of tapetal cells. On this, membrane are studded the ubisch granules. In some members of Asteraceae, which have plasmodial tapetum, Heslop-Harrison (1969) reported an acetolysis resistant extra tapetal membrane, in the form of a complete sac surrounding mature pollen.

The tapetal membrane originates from the secretions of the tapetal cells. The extra-tapetal membrane probably acts as a culture sac enclosing the developing microspores.

B. Periplasmodial (Invasive or Amoeboid or Nuclear type)

In this type, the walls of the tapetal cells break down, protoplasts enlarge, enter the locule and fuse to form a coenocytic plasmodium.

Ultrastructure: Echlin (1971) reviewed the structural and functional aspects of tapetum during microsporogenesis. Of the three studies on the plasmodial tapetum - *Tradescantia bracteata, Helianthus annuus* and in *Mahonia aquifolium,* the first one gives considerable information. In *Tradescantia* and *Helianthus* the tapetum is uniseriate, while in *Mahonia* it has three layers which release their cellular contents into the locule one after another. In *Tradescantia,* the tapetal cells, preceding meiosis, show prominent population of organelles. The plastids have dense matrix, prolamellar bodies and different inclusions like plastoglobuli, large polysaccharide granules, and bodies of moderate electron density and granular appearance. Raphides, without any connection to membranous structure, accumulate and grow abundantly. Styloid-like crystals have been observed in *Helianthus annuus.* The peripheral tapetal cells become extremely vacuolate and large dictyosomes are formed, which produce vesicles. The vesicles presumably release hydrolytic enzymes. A wave of lysis ensures from the tapetal cells inward, bringing about dissolution of largely pectic cell walls. This allows the periplasmodial tapetum to invade the locule. It affects the lysis of walls of microspore mother cells which are finally surrounded by the invading tapetal protoplasts. As soon as the tapetal plasmalemma breaks down along the radial surface, the cell contents become confluent carrying the mitochondria, plastids, nuclei and raphides inwards and surrounds the meiocytes.

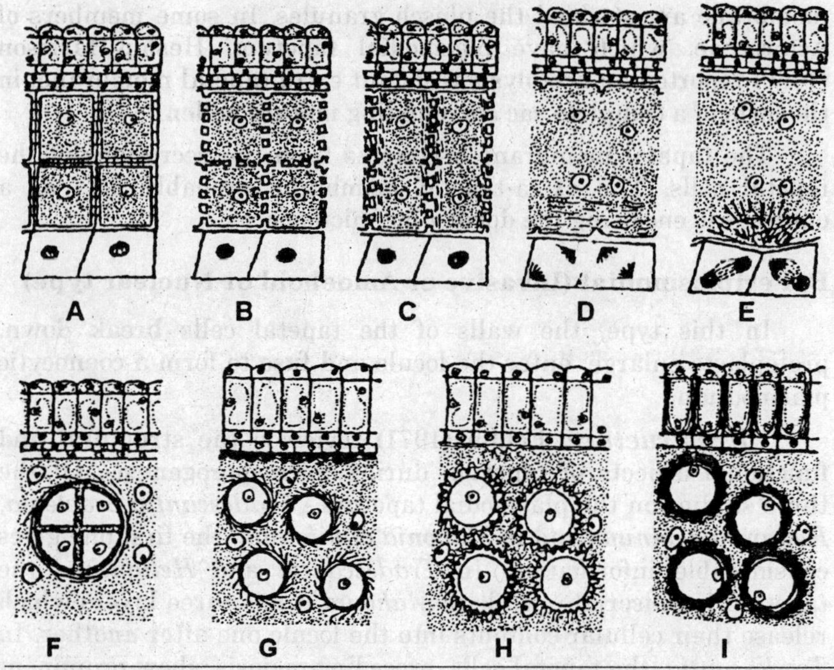

Fig. 2.3. Amoeboid tapetum in *Arum italicum* (semi-diagrammatic). **A.** Premeiotic phase, **B.** Leptotene stage, **C.** Pachytene stage, **D.** Meiosis I-Anaphase stage, **E.** Meiosis I-Telophase stage, **F.** Tetrad stage, **G-H.** Microspore stage, **I.** Uninucleate pollen grain with intine and exine (after Pacini and Juniper, 1983).

Pacini and Juniper (1983) studied ultrastructural details of the development of periplasmodial tapetum in *Arum italicum*. The anther wall comprises the epidermis, fibrous endothecium, three middle layers, and two layered tapetum. At the premeiotic phase of microspore mother cells, the tapetal cells are connected with each other by plasmodesmata, but these are absent in the tapetal walls facing the middle layer and the microspore mother cells (Fig. 2.3A). During the meiotic divisions of pollen mother cells, the tapetum shows conspicuous changes. At leptotene (Fig. 2.3B), the plasmodesmata in the tapetal cells widen to form cytomictic channels, while the tangential walls disappear at pachytene (Fig. 2.3C). During meiosis I anaphase (Fig. 2.3D), the radial walls of tapetal cells also disappear and microtubules lie parallel to the inner radial walls of pollen mother cells. The tapetal cytoplasm starts to intrude between the meiocytes and microtubules appear at that region (Fig. 2.3E). Fig. 2.3D and E show the formation of multinucleate periplasmodium in which the tetrads, microspores and uninucleate pollen grains are embedded (Fig. 2.3F-I). Microtubules surround the tetrad facing the callose wall and the microspore. The tapetal plasmalemma facing the microspores retracts from the exine surface leaving roughly cone-shaped spaces (Fig. 2.3H). Microtubules no longer lie parallel to the exine surface. By the localized retraction of the plasma membrane, the spaces left on the microspore surface and tapetal periphery become filled with a material forming 'spines'; the intine also develops.

Cytology of Anther tapetum

The tapetal cells are uninucleate. Simultaneously, with the onset of the meiosis, the tapetal cells exhibit significant cytological variations.

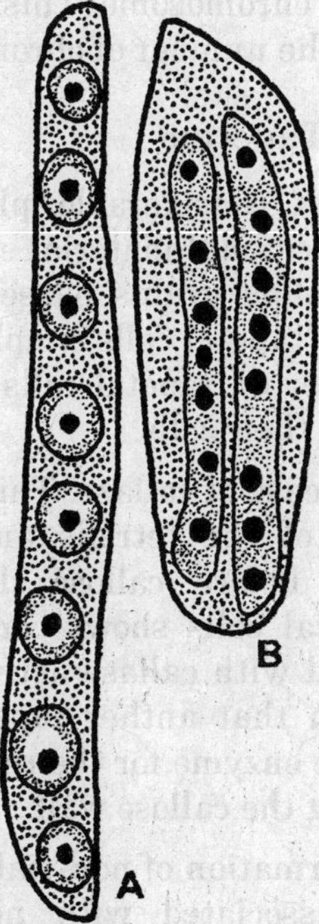

Fig. 2.4. Anther tapetal cells in *Eupatorium odoratum*. **A.** Multinucleate tapetal cell. **B.** Tapetal Cell with polyploid nuclei

1. **Multinucleate condition**: The single nucleus may undergo one or more mitotic divisions resulting in two to four nuclei. Sometimes the number goes up to 16. In *Hepatica acutiloba* tapetal cells contain 13 nuclei. Sometimes the nuclei fuse forming polyploid nuclei. In *Eupatorium odoratum* nuclear divisions and fusions result in multinucleate and polyploid condition (Fig. 2.4) (Maheswari Devi and Pullaiah, 1977).
2. **Endomitosis**: Here the nucleolus and the nuclear membrane remain intact and there is no spindle formation. The chromosomes contract and split longitudinally, but all of them remain within the same nucleus, which becomes tetraploid. Further such endomitosis may results in a large polyploid nucleus. In the uninucleate tapetum of *Cucurbita pepo,* nuclei of four sizes corresponding to their degree of ploidy (2n, 4n, 8n and 16n) are formed as a result of endomitosis.
3. **Polyteny**: This refers to an increase in the number of chromonemata per chromosome. This mode of DNA increases, but does not alter the number of chromosomes per nucleus.

Functions of Anther Tapetum

1. The tapetum is of considerable physiological significance because all the nutritional material entering the sporogenous cells/microspores/pollen grains passes or originates from it. During meiosis the tapetal cells simply transport nutrients to the meiocytes. Later on tapetal cells themselves supply food materials stored in their cells.
2. Production and release of callase: This enzyme depolymerises the callose walls of the tetrads and the microspores are released. Shortly before callose degradation starts, the cytoplasm of tapetal cells shows certain vesicles which are probably associated with callase activity. Stieglitz and Stern (1973) have shown that anther tapetum is involved in the synthesis of callase enzyme for the release of microspores in a tetrad by degrading the callose wall.
3. **Male sterility:** Formation of non-viable pollen grains in male sterile lines is associated with non-functional or faulty development of tapetum. In cytoplasmic male sterile *Petunia* sterility was found to be due to faulty timing of callase

appearance. Whereas in fertile plants callase appears only at tetrad stage, in sterile plants strong callase activity is detected during prophase - I. Precocious release of callase by tapetum is responsible for male sterility in *Petunia*.

4. Tapetum plays an important role in pollen wall formation. The pro-ubisch bodies, which are produced in the secretory tapetal cells, migrate to the cell surface where they get coated with sporopollenin and are released into the anther locule. The sporopollenin coated ubisch bodies are involved in the external thickening of the exine.

5. Pollen kitt substances and tryphine are transferred from tapetal cells to pollen grains. Pollen kitt substances are a mixture of hydrophobic lipids and carotenoids. Tryphine is a mixture of hydrophilic substances including protein. Both these substances are synthesized in a special population of plastids in the tapetal cells. The carotenoid containing lipid granules spread over the exine and help in pollination.

6. Pollen wall contains proteins derived from male gametophyte as well as the tapetal cells. These proteins are responsible for the incompatibility. The tapetum synthesizes certain substances which are released into the locule during maturation of pollen grains. These proteins enter the exine of pollen grain, and are stored in the region of bacula. These wall bound proteins are extra-cellular and capable of recognizing the compatibility of stigma.

Microsporogenesis

The primary sporogenous cells give rise to the microspore mother cells. In some plants the sporogenous cells function directly as pollen mother cells, *e.g.*, *Boerhavia diffusa* (Maheshwari, 1929), *B. erecta* (Pullaiah and Sarada, 1986) and *Arnica alpina* (Engell, 1979). In some others sporogenous cells, undergo only transverse divisions resulting in a single row of pollen mother cells, *e.g.*, *Elephantopus scaber, Vernonia elaegnifolia* (Pullaiah, 1979). In still others divisions in sporogenous cells occur in all planes resulting in a moderate mass of pollen mother cells, *e.g.*, *Basella rubra* (Maheswari Devi and Pullaiah, 1975).

In Mimosaceae and Annonaceae sterile transverse septa are seen in anther lobes. Septate microsporangia also occur in Balsaminaceae, Convolvulaceae, Gentianaceae, Plumbaginaceae and Rhizophoraceae (Lersten, 1971).

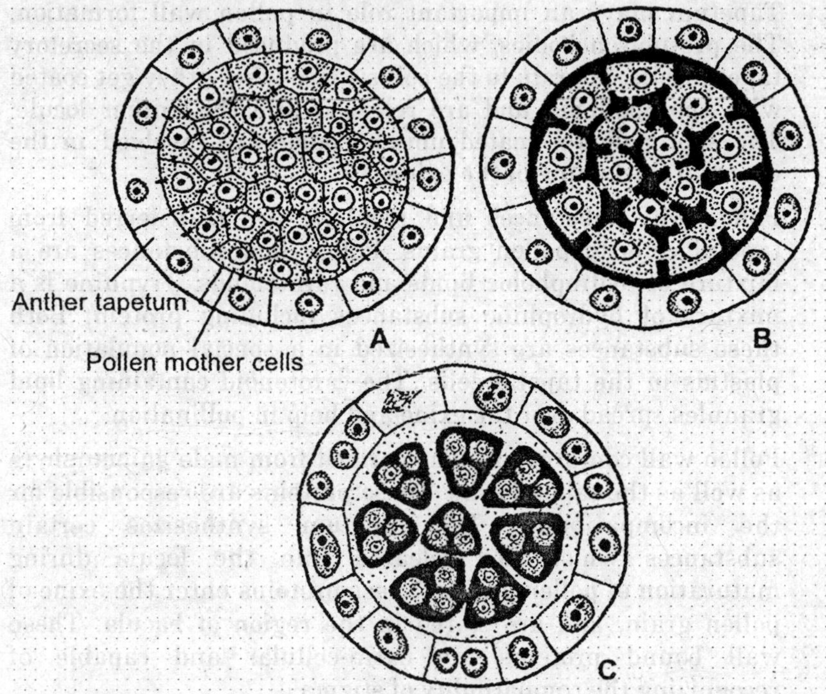

Fig. 2.5. Diagrams to show cellular interconnections in developing anther. **A.** Preleptotene stage, **B.** Meiotic prophase-I stage, **C.** Tetrad stage

The microspore (pollen) mother cells undergo meiosis to form microspores. Before they enter meiosis the pollen mother cells are meristematic cells. The pollen mother cells in an anther locule are interconnected by plasmodesmata (Fig. 2.5A). Plasmodesmatal connections are also known at this stage between tapetal cells and pollen mother cells. When pollen mother cells enter meiosis plasmodesmatal connections between pollen mother cells and tapetal cells are lost due to the deposition of callose (β-1,3-glucan) wall. Pollen mother cells also get enclosed in callose wall but large cytoplasmic channels exist between pollen mother cells (Fig. 2.5B). These channels are 1-2 μm in diameter. Due to the cytoplasmic channels a close synchrony in pollen mother cells is maintained during meiosis.

At the end of meiotic prophase (sometimes at the end of meiosis I) the callose walls of pollen mother cells close and the cytoplasmic channels are cut off. The pollen mother cells undergo meiosis as isolated cells. Callose walls are formed between dyads and between microspores of the tetrad (Fig. 2.5C).

Cytomixis

In *Oenothera gigas* and *O. biennis* Gates (1911) observed frequent migration of nuclear material from one pollen mother cell to another and termed the phenomenon as cytomixis. Since then it has been reported in several other plants. A multi-nucleate, polyploid and aneuplod pollen mother cells are the convergence of cytomixis. The extrusion of chromatin and cytomixis in pollen mother cells affect the viability of gametes. It is associated with high pollen sterility. It is now clear that cytomixis is a pathological phenomenon.

Cytokinesis

Cytokinesis in pollen mother cells is of two types.

1. **Successive:** The pollen mother cell undergoes meiosis I followed by wall formation. Cell plate is formed in the center and extends laterally. The wall formation thus is centrifugal. Callose wall is then formed on both sides of the cell plate. The two dyad cells undergo meiosis II and wall is formed as in division I resulting in a tetrad (Fig. 2.6). Successive cytokinesis is common in monocots.
2. **Simultaneous:** The first meiotic division is not accompanied by wall formation. The two-nucleate condition is immediately followed by meiosis II. The callose wall formation occurs by centripetally growing furrows which meet in the center and separates the four microspores. Simultaneous cytokinesis is common in Dicots.

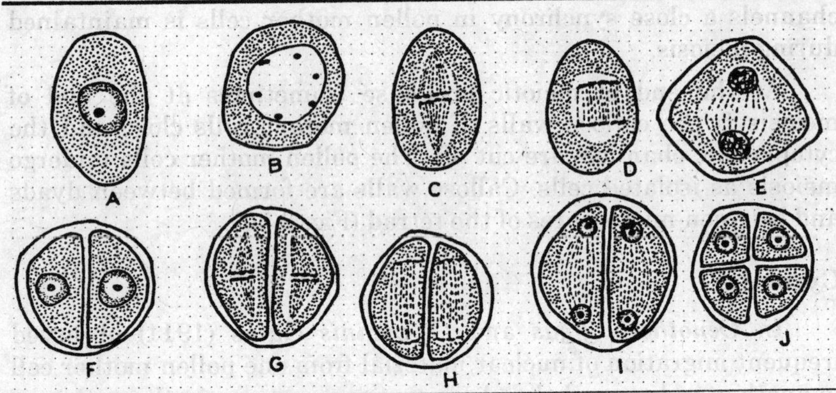

Fig. 2.6. Microsporogenesis in *Brachiaria eruciformis* (after Febulaus and Pullaiah, 1998).

Microspore tetrads

The arrangement of microspores in a tetrad exhibits five patterns. Usually, they are tetrahedral or isobilateral (Fig. 2.7 A, B). In addition, they may be decussate, T-shaped and linear (Fig. 2.7C-I). All five types are reported in *Aristolochia elegans* and *Sparganium erectum*.

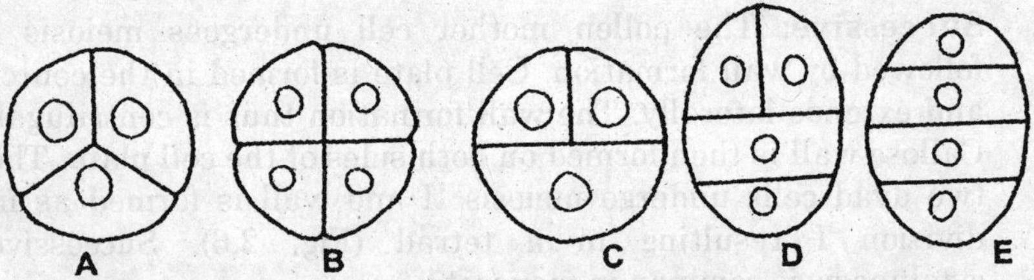

Fig. 2.7. Type of microspore tetrads. **A.** Tetrahedral; **B.** Isobilateral; **C.** Decussate; **D.** T-shaped; **E.** Linear

The microspores in a tetrad are surrounded by callose wall. Due to the activity of callase enzyme released by the anther tapetum the callose wall around the individiual spores is degraded and the microspores are released in the anther locule. In Podostemaceae pollen grains are held in pairs. However in families like Winteraceae (Fig. 3.1), Ericaceae, Monimiaceae, Philydraceae and Juncaceae the pollen grains remain in permanent tetrads. In Asclepiadaceae and some Orchidaceae, all the microspores in a sporangium remain together to form a pollinium.

Polyspory

Occurrence of more than four microspores is called polyspory. Such polyads have been reported in *Eupatorium odoratum* (Maheswari Devi and Pullaiah, 1975) (Fig. 2.8) and *Hyphaene indica* (Mahabale and Chennaveeraiah, 1957) and *Thunbergia mysorensis* (Khaleel and Boraiah, 1973). Polyspory is considered as an abnormality.

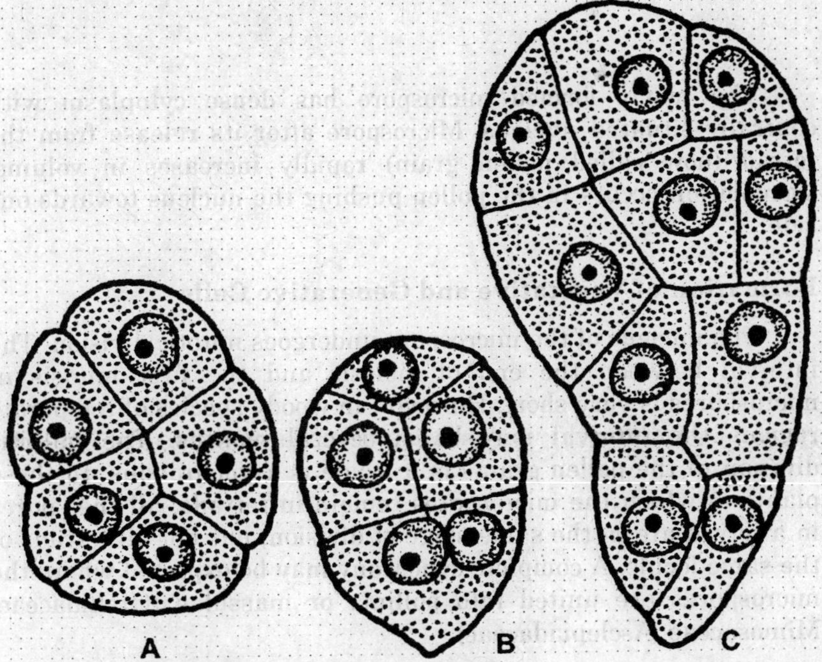

Fig. 2.8. Polyspory in *Eupatorium odoratum*. **A-B.** Pentads, **C.** Decad (after Maheswari Devi and Pullaiah, 1977).

CHAPTER - III

MALE GAMETOPHYTE

The newly formed microspore has dense cytoplasm with centrally situated nucleus. Microspore after its release from the tetrad (now called pollen grain) rapidly increases in volume. Vacuolation occurs in the pollen pushing the nucleus towards one side.

Formation of Vegetative and Generative Cells

The nucleus of the microspore undergoes mitotic division. The interval between the end of meiosis and the first mitosis in microspores is very short. However, in woody species of temperate regions the interval may be of several months. The mitoitic division in the pollen grains in a locule is asynchronous. In those plants in which the microspores remain in tetrads, all the spores in a tetrad are in the same stage of division, but not all tetrads of the same locule. A complete synchrony may be expected where the microspores are united into pollinia or massulae (Orchidaceae, Mimosaceae, Asclepiadaceae).

The nucleus of the pollen grain moves towards the side where it undergoes division (Fig. 3.2A-C). This position is genetically controlled. Thus, the generative cell, may be cut off on the outer side (*e.g.*, *Spiranthes*) (Fig. 3.1A) or towards the inner side (*e.g.*, *Drimys,* Fig. 3.1B, *Tradescantia*). The inner or outer position of the generative cell can be identified only when the microspores remain in tetrads during pollen mitosis. The spindle of pollen mitosis is asymmetric, the more pointed pole being directed towards the middle of the microspore and the blunt pole towards the wall of the microspore.

A curved cell plate separates the two nuclei, and extends up to the intine (Fig. 3.2D). The curved wall grows between the intine and plasmalemma of the generative cell. The mitotic divisions are always unequal resulting in a large vegetative cell and a small generative cell (Fig. 3.2E,F). The generative cell gets separated from the wall of the pollen grain, and comes to lie in the cytoplasm of the vegetative cell.

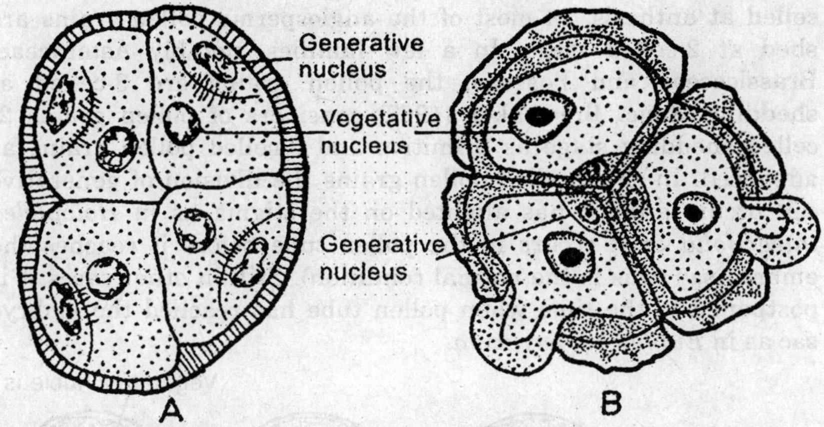

Fig. 3.1. Pollen mitosis tetrads. A. *Spiranthes australis*, B. *Drimys winteri*.

Vegetative cell

After pollen mitosis the vegetative cell continues to grow and accumulates reserve food in the form of starch and fat. It becomes rich in protein and its nucleus may undergo endoduplication and become highly polyploid. The vacuole disappears. The nuclear membrane becomes highly convoluted. At mature stage the nucleus often lacks a nucleolus.

Generative cell

The cytoplasm of the generative cell contains usual cell organelles like mitochondria, ribosomes, endoplasmic reticulum, microtubules and dictyosomes. Plastids are absent in most of the species (*e.g.*, *Beta, Haemanthus*) while in some they are present (*Oenothera*). The generative cell is spherical when it is detached from the pollen wall. It elongates to such an extent, that it appears vermiform at maturity. The elongated shape probably facilitates its movement into the pollen tube.

Divisions of the generative cell

The generative cell undergoes mitotic division forming two sperms (Fig. 3.2 G,H). This may occur when the pollen grains are in the anther or the division may occur after the pollen grains are released from the anther. Thus pollen grains may be 3-celled or 2-celled at anthesis. In most of the angiosperms pollen grains are shed at 2-celled stage. In a few families, notably Asteraceae, Brassicaceae and Poaceae, the pollen grains are 3-celled at shedding stage. Brewbaker (1967) considers of pollen grains 2-celled shedding stage as primitive and 3-celled pollen grains as advanced. In the 2-celled pollen grains the division of generative cell occurs after it has alighted on the stigma as in *Holoptelea integrifolia* or it occurs in the pollen tube before it reaches the embryo sac (this is the normal condition). Still in other cases it is postponed to the time when pollen tube has reached the embryo sac as in *Euphorbia terracina.*

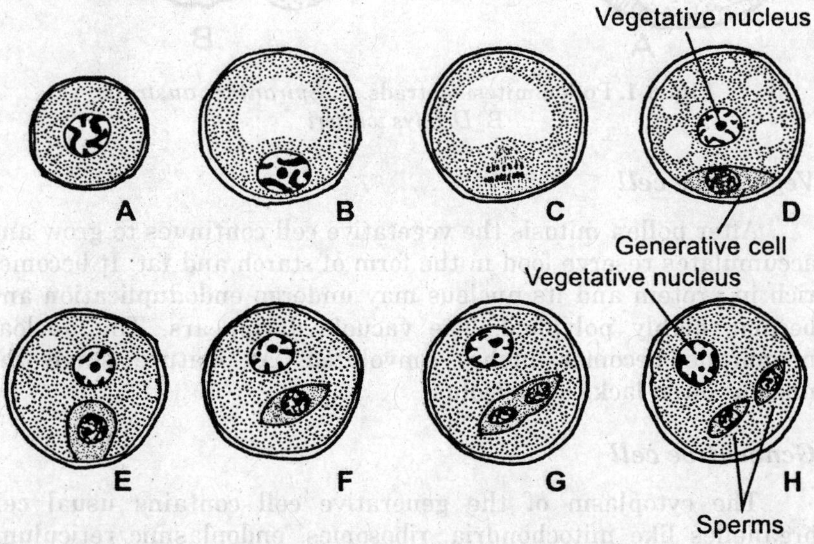

Fig. 3.2. Stages in the development of male gametophyte (after Maheshwari, 1950).

Generative cell undergoes a more or less prolonged mitosis. During the period, the generative cell undergoes elongation to acquire its characteristic vermiform appearance. In species with binucleate pollen grains, the division of the generative cell is postponed until pollen germination and it passes through a stage of temporary dormancy.

The sperm cells are oval to fusiform in shape with varying size, ranging from 14 mm in length in *Zea mays* to 40 mm in length in *Helianthus annuus*. Each sperm cell consists of a large nucleus surrounded by a thin sheath of cytoplasm externally limited by a cell membrane. Cell wall is absent. The two male gametes frequently remain connected to each other and become located near the vegetative nucleus, forming the male germ unit (MGU).

According to Brewbaker (1957) homomorphic plants with 2-celled pollen grains show gametophytic incompatibility, while sporophytic incompatibility is linked with 3-celled pollen grains. Pollen tubes of 2-celled pollen grains are inhibited in the style, while in 3-celled grains on the stigma. The form of sperm cells varies from oval to spindle shaped.

Male Germ Unit/ Male Dimorphism

The concept of Male Germ Unit (MGU) was first introduced by Dumas *et al.* (1984). The generalized feature of MGU is "connected sperms in close association with the vegetative nucleus". Russell and Cass (1983) and Dumas *et al.* (1984) made significant observation in *Plumbago zeylanica* and *Brassica oleracea,* respectively, and reported that the sperms are markedly dimorphic. One sperm is larger and contain almost all the mitochondria, while the small sperm contains nearly all the plastids. In some species the sperms are isomorphic (barley).

A common cross wall or evagination of plasma membrane links the two sperms and the larger sperm is also connected with vegetative nucleus by cytoplasmic extension. Thus, a 3-celled male germ unit (MGU) is formed (Fig. 3.3). Once the MGU is organized it travels through the pollen tube as one entity. On reaching the ovule the vegetative nucleus and sperms get separated. The destiny of each sperm is also predetermined. The larger sperm fuses with the egg, while the smaller one participates in triple fusion (Russell, 1985).

The formation of MGU is of wide occurrence in angiosperms. However, the time of appearance of this unit may vary with the species. MGU functions as a vehicle for transmission of sperms and it also reduces the incidence of heterofertilization. Other examples, which show MGU are *Acacia retinoides, Nicotiana tabacum*, *Sambucus nigra*, etc.

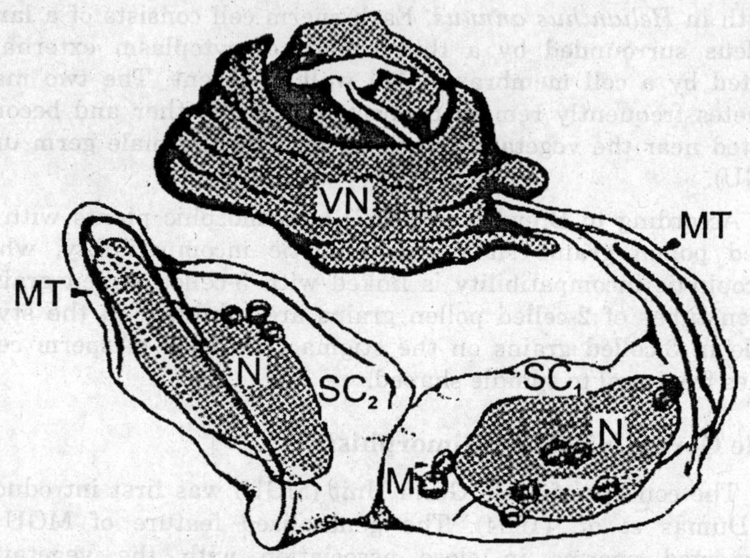

Fig. 3.3. Computer-assisted three-dimentional schematic diagram of the male germ unit from a pollen grain of *Brassica campestris*, showing placement of vegetative nucleus (VN) and sperm cells (SC_1, SC_2) with nuclei (N), mitochondria (M) and arrays of microtubules (MT) (after McConchie *et al.*, 1985)

Isolation of Sperm Cells

Isolation of large numbers of live sperm cells of flowering plants has opened up a new field of research in plant biology. It provides the opportunity to investigate the structure, physiology and biochemistry of these cells.

The procedure for sperm isolation depends on the condition of pollen at dehiscence. In plants with 3-celled pollen, sperm isolation involves osmotic bursting of freshly collected pollen. However, in plants with 2-celled pollen, it is possible only after pollen germination.

Intact sperm cells *Plumbago zeylanica* were isolated by Russel (1985) from mature pollen grains near anthesis using osmotic shock with sucrose (20%). Sperm cell yields of up to 75 percent can be attained by differential centrifugation using a clinical centrifuge with concentrations of up to 8.8 × 10' cells/milliliter. Such 'gametoplasts' remain intact for up to 24 hours according to Evan's blue exclusion and can be used for characterization or physiological manipulation.

The method for isolation of sperm cells from 3-celled pollen of maize is as follows. Pollen are collected by shaking tassels over a tray, sieved through 80μm nylon mesh and hydrated in moisture-saturated atmosphere for about 20 min, at room temperature. One gram of hydrated pollen is incubated in 10 ml in Brewbaker and Kwack's medium containing sucrose (15%) and pH 5 (bursting medium), at room temperature. Within 20 minutes, the pollen grains burst, releasing their contents, including the sperms, through the germ pores. Gentle shaking (130 rpm) during incubation in the bursting solution favours the release of the sperms. The bursting medium is then filtered through 50μm nylon mesh to remove the exine fragments. Further cleaning, to remove starch grains and other cytoplasmic contents of the vegetative cells, is done by discontinuous percoll gradient centrifugation. The percoll gradient is prepared by layering 30%, 15% and 10% percoll solution (prepared in the bursting medium) in 15 ml centrifuge tubes, and 5-10 ml of the filtrate is loaded on the top of 10 percent percoll. After centrifugation at 3000g, for 60 min, at 4°C, sperm cells are collected at the interface between 15 percent and 30 percent percoll. The sperm cells are pipetted out, washed with the bursting medium and centrifuged at 3000g for 7 min. The supernatant is discarded and the sperms, settled at the base are collected in 2 ml of the bursting solution used as desired.

Pollen Embryo Sacs

Nemec (1898) reported that in the petaloid anthers of *Hyacinthus orientalis* the pollen grains sometimes form large eight nucleate structures showing resemblance to embryo sacs. De mol (1923) also observed this so-called 'Nemec phenomenon' in the same species. Such pollen embryo sacs have also been reported by Stow (1930, 1934).

The microspore here increases in size to form a sac like structure. The nucleus undergoes three mitotic divisions to form 8 nuclei. Of these, 3 lie at the end where exine is still intact, 3 at the opposite end and 2 in the middle. Stow observed that when the pollen embryo sacs were placed on an agar medium, together with some normal pollen grains of another variety, the pollen tubes from the latter coiled around the pollen embryo sacs.

Pollen Wall

Structure

The pollen wall comprises of two layers - the acetolysis resistant exine and an acid degradable intine. Chemically the exine consists of two layers - the inner endexine and outer ektexine. The two layers may not always correspond to the two morphologically recognized layers, namely sexine and nexine. The ektexine is composed of foot layer, baculum and tectum.

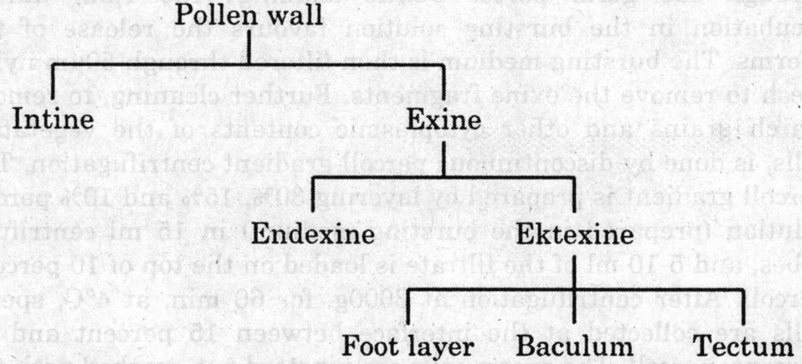

The exine is composed of sporopollenin, a highly resistant material believed to be produced by oxidative polymerization of carotenoid pigments and carotenoid esters. The exine is sculptured with varying degrees of elaboration. The sculptured part of the exine is made up of radially oriented baculae (columella) which may remain open (pilate grain) or covered by a roof or tectum (tectate grain). The baculae (in pilate grain) may be enlarged above and stand free, or may be fused together to give a raised wall often disposed in a reticulate pattern.

The intine is pectocellulosic. The cellulose component is microfibrillar, with the microfibrils oriented in a plane parallel to the surface. Beads, ribbons or plates of enzymatic proteins are present in the intine especially near the vicinity of the germ pores.

Development

Soon after the completion of meiosis, the blue print for the exine, the primexine, is formed, while the tetrads are still enclosed in the callose wall. The primexine is distinguishable from the callose by a difference in electron opacity. It has a matrix. The cellulose microfibrils of this layer are deposited in between the convoluted plasmalemma of the spore and the callose wall (Fig. 3.4 A,B).

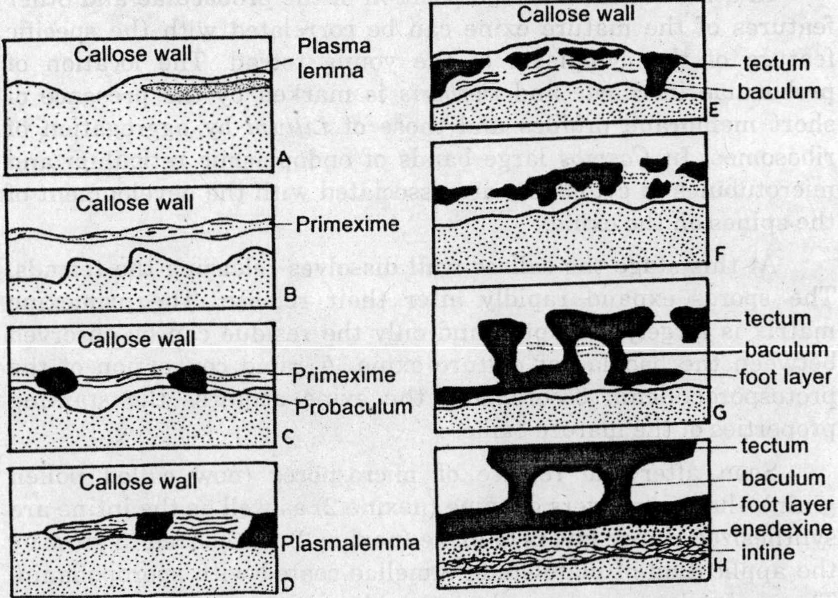

Fig. 3.4. Development of pollen wall in *Helleborus foetidus.* **A.** Plasmalemma is directly surround by callose wall; **B.** Cellulose primexine has appeared between plasmalemma and callose wall; **C-D.** Probacula have penetrated the primexine, E. Bacula and tectum are formed; **F-G.** The callose wall has disappeared and ektexine is well developed; **H.** The wall is fully developed (after Echlin, 1968).

The germinal apertures (pores/colpae) are demarcated during the formation of the primexine by the presence of endoplasmic reticulum oriented parallel to the plasma membrane. These regions remain free of primexine deposition. When the cellulosic primexine has reached a certain thickness additional gaps appear at the surface of the plasmalemma (Fig. 3.4C). These columns are called probacula. Now the cytoplasm of the spore starts synthesizing the precursors of sporopollenin which are polymerized and deposited on the surface of the lamellae (Fig. 3.4D-G). The probacula become connected at their bases to form a foot layer (Fig. 3.4 G,H). Depending on the taxon, the bacula remain free above or form a roof over the primexine matrix. The roof over the bacula is called tectum (Fig. 3.4H).

In a number of taxa the position of the probaculae and other features of the mature exine can be correlated with the specific feature of the cytoplasm of the young tetrad. The location of probaculae in *Silene* and *Berberis* is marked by the presence of short membrane profiles and those of *Lilium* by aggregation of ribosomes. In *Cosmos* large bands of endoplasmic reticulum and microtubules in cytoplasm are associated with the development of the spines on the exine.

At this stage the callose wall dissolves releasing the tetrads. The spores expand rapidly after their release. The primexine matrix is largely disrupted and only the residue can be observed between the baculae of mature exine. A rapid conversion of the protosporopollenin occurs, and the exine acquires the staining properties of the mature exine.

Soon after the release of microspores (now called pollen grain), the inner layers of exine (nexine 2) as well as the intine are synthesized. Inner layers of exine (nexine 2) is reported to grow by the application of platelets of lamellae coated with sporopollenin. These platelets are formed at the plasmalemma, become coated with sporopllenin, and pass across the developing intine. Growth of the intine is initiated after the release of microspores from the callose wall and is associated with dictyosome activity. Certain proteinaceous plates or ribbons are incorporated in the intine in the vicinity of germ pores. These proteins show enzymatic activity.

While the spore wall formation is going on inside the tetrad wall, some curious rounded bodies of lipid nature appear in the cytoplasm of the tapetal cells. These bodies, called proubisch bodies, migrate to the cell surface where they get coated with sporopollenin and are released into the anther locule. The ubisch bodies, which are coated with sporopollenin, are involved in external thickening of the exine.

In majority of the angiosperms the bulk of exine develops only after release of tetrads. In such instances the tapetum supplies the precursors for exine. However, in *Tradescantia bracteata* the entire exine is deposited before the break down of tetrads. In this species, as there is no further deposition of exine after the release of tetrads, the tapetum does not contribute directly to the integral structure of the exine.

Pollenkitt

The pollenkitt is an oily layer found on the outside of the mature pollen grains of many insect-pollinated species. The pollenkitt is made up of lipoidal material, flavonoids, carotenoids and degenerative products of tapetal proteins. Carotenoids are responsible for the characteristic yellow or orange colour of the pollen grains. The enzymes involved in the biosynthesis of flavonoids are localized predominantly in the tapetal cells. Lipoidal material of pollenkitt is synthesized in plastids, which disintegrate prior to tapetal degeneration. Pollenkitt is quite conspicuous in insect pollinated pollen.

The function of pollenkitt is not clear. However, it may be contributing in the following ways.

1. It is supposed to act as an insect attractant and help in pollen dispersal.
2. It protects the pollen against the damaging effects of UV rays.
3. Deposition of flavonoids appears to be essential for normal functioning of pollen. Pollen grains of maize, which lack flavonoids are incapable of achieving fertilization.
4. Lipoidal components of pollenkitt probably have an important role, in controlling water loss from germ pores and microspores after pollen dispersal.
5. Pollenkitt may be functioning as the pollen-borne substances involved in sporophytic incompatibility.

Tryphine is often distinguished as different from pollenkitt. Tryphine is a complex mixture of hydrophilic substances derived from the breakdown of tapetal cells, whereas pollenkitt is principally made up of hydrophobic lipids containing species - specific carotenoids.

Gene Expression During Male Gametophyte Development

A substantial part of genome is expressed in the male phase and the gametophytic-sporophytic genetic overlap has been estimated to exceed 60 percent (Sari-Gorla *et al.*, 1986; Willing and Mascarenhas, 1984). However, all the studies have considered only mature pollen as male gametophyte and there is very little data concerning gene expression during microspore development.

In this report seven genes coding for enzymes were analyzed throughout pollen development, from 5 days after meiosis (DAM) to full maturity (anthesis). Six of these enzymes are present in mature pollen. Four (ADH-1, GOT-1. GOT-2, CAT-1) are multimeric and therefore suitable for the detection of the type of expression, haploid or diploid, of their encoding genes (Sari-Gorla *et al.*, 1986). Two (GOT-3 and ^{2}-GLU-2) are invariant and possibly monomeric. For these only the onset of enzymatic activity during development was determined. One (^{2}-GLU-1), absent in mature pollen, was included to verify its sporophyte-specificity.

The analysis indicates that ADH-1 and GOT-2 show haplo-diploid expression from the earliest stages examined. CAT-1 shows haplo-diploid expression only starting 7th day after meiosis. No activity at all could be detected earlier. An additional nodal band showing catalase activity appears in mature microspores only. The electrophoretic mobility and the absence of heterodimers with CAT-1 suggest this isozyme may be CAT-4. For GOT-1, 3 bands typical of sporophytic genetic control appear in the early stages (up to 9th day after meiosis), while the 2-band pattern expected for haploid expression was found from 9th day after meiosis. On GOT-2 and ^{2}-GLU-2 were clearly detectable from 5th day after meiosis. ^{2}-GLU-1 was absent throughout pollen development, thus confirming its sporophyte-specificity.

On the whole the data indicate that:

- The male gametophyte is genetically largely independent from the sporophyte even in the first stage of development.
- In some cases (GOT-1) a sporophytic control over gametophytic functions may persist at least for the early steps of microspore development.
- Some genes appear to be developmentally regulated (CAT-1, CAT-4) in the "pollen formation phase".

MALE STERILITY

It is the absence or lack of fertile pollen in male plants. Such male sterile plants play an important role in plant breeding programs, because male sterile plants can be used as female parent without emasculation and facilitates easy cross pollination without bagging.

The male sterility in angiosperms may arise due to several reasons:

a) Anther suppression, abortion, phyllody, pistillody – this has premeiotic determination.

b) Aberrant meiosis or gametophytic development.

   i) Determined by gametophyte

   ii) Determined by sporophyte

c) Failure of anther dehiscence

d) Premature dissolution of callose wall

e) Fungal or viral infection

Gametophyte-determined pollen sterility

Gametophyte-determined pollen sterility is usually due to meiotic abnormalities, such as meiosis in polyploids, which results in an unequal distribution of chromosomes in the daughter cells. It results in the production of genomically imbalanced spores, which are non-viable.

Sporophyte-determined pollen sterility

Sporophyte-determined pollen sterility is due to

a) Genetic factors

b) Cytoplasmic factors

c) Environmental factors.

A) *Genetic Male sterility:* Genetic pollen sterility is spread among angiosperms. It is due recessive genes. This can be best explained with the example as follows:

When a male sterile (mm) plant is crossed to a male fertile (MM) plant, the obtained F1 progeny will be male fertile, which are heterozygous (Mm), when the F1 plants are selfed, the F2 progeny segregated in the ratio similar to Mendel's monohybrid cross.

From the above example, it is clear that, there should be two recessive genes for the expression of male sterile character.

B) *Cytoplasmic Male Sterility (CMS):* In several crops like maize, cytoplasmic control of male sterility is known. In such cases if the female parent is male sterile and F1 progeny would always be male sterile because the cytoplasm is mainly derived from egg obtained from the male sterile plant. The cross can best be explained as follows:

Maize shows genetic as well as cytoplasmic pollen sterility.

C) *Environmental Pollen sterility:* Environmental factors such as photoperiodic treatment (*Glycine max, Silene pendula, Zea mays*) and temperature (*Oryza sativa*) are known to induce male sterility. Certain chemicals, such as auxins, maleic hydrazide and dalapon selectively induce male sterility.

In most cases, the pollen sterility has been attributed to the malfunctioning of tapetum.

i. Inhibition of normal RNA synthesis and increase in DNA content.

ii. Hypertrophy of the tapetal cells. They enlarge acquiring supranormal cytoplasmic RNA and become multinucleate.

iii. The cells invade anther locule, and spores or meiocytes are crushed or even reabsorbed.

iv. Premature degradation of the tapetum depriving the developing spore of its nutrition.

Production of Male Sterile Lines

Flavonoids are essential for normal pollen development and flavonoid deficiency prevents pollen maturation. Chalcone synthase (CHS) is a key enzyme of flavonoid biosynthesis. In *Petunia*, antisense construct of gene encoding CHS has been transferred and transgenic plants with negligible CHS activity, white flowers and non-functional pollen developed.

It is possible to induce male sterility in a wide range of angiosperms by inserting bacterial gene BARNASE linked with a tapetum-specific promoter PTA29 from tobacco. The chimeric gene causes premature breakdown of tapetum resulting in the arrest of microspore development.

The male fertility of barnase male steriles is restored by another gene BARSTAR of the bacterium *Bacillus amyloliquefaciens*.

Pollen grains in Cyperaceae

In all the Angiosperms pollen mother cell undergoes meiosis resulting in a pollen tetrad and all the four microspores are functional. Cyperaceae presents a peculiar type of pollen development. The meiotic division of the pollen mother cells results in four haploid nuclei (Fig. 3.5 A-D). Of the four nuclei, only one is functional. The functional nucleus remains in the centre, while the three nonfunctional nuclei are cut off on one side of the cell (Fig. 3.5E). The three nuclei may undergo a further division, but they soon degenerate. However, the functional microspore remains enclosed in the wall of the tetrad, and is often described as pseudomonad (Fig. 3.5 F). The functional nucleus divides to form a vegetative and a generative cell (Fig. 3.5 G-H).

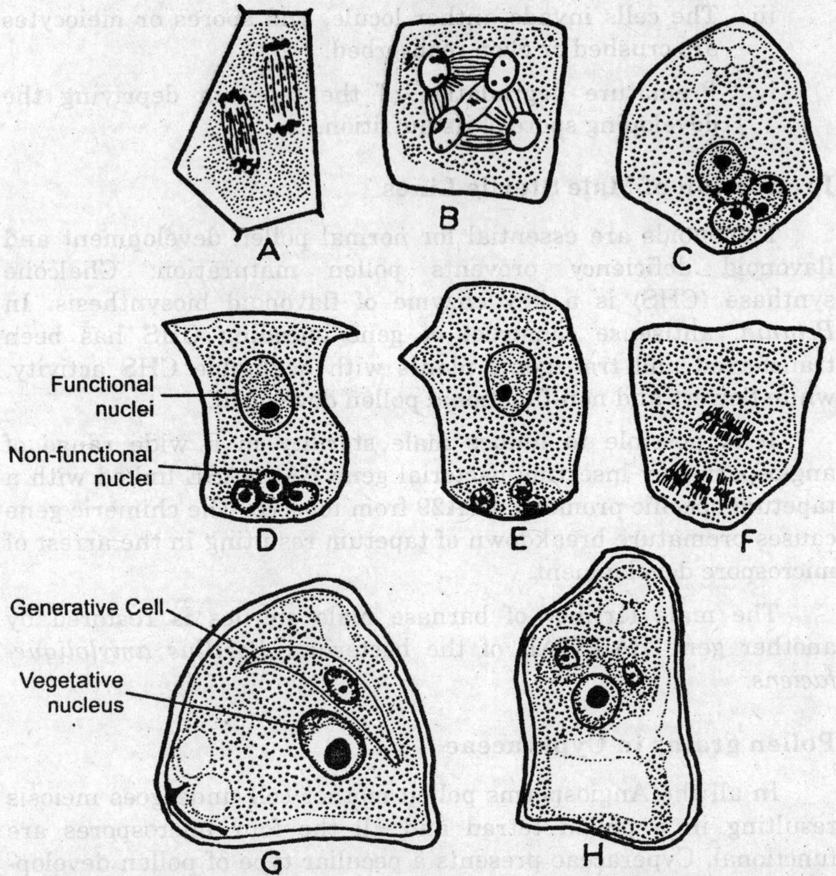

Fig. 3.5. Development of pollen in *Carex*-Cyperaceae. **A-B.** Meiosis II in microspore mother cell; **C.** Microspore mother cell with four identical nuclei; **D.** One functional and three non-functional nuclei; **E.** Delimitation of non-functional nuclei; **F.** All the four nuclei in metaphase; **G.** Two-celled pollen grain; **H.** Three - celled condition (Pseudomonad) (after Shah, 1962).

CHAPTER – IV

PALYNOLOGY

The study of external morphological features of pollen and spores is known as Palynology. The term palynology, first used by Hyde and Williams (1845), is derived from the Greek word '*palynein*' which means to spread and distribute. Although the functional importance of pollen has been realised by the ancient people as early as 717 B.C., the potentialities of pollen and spores as a morphological entity in plants have become increasingly understood since the invention of microscope by Hooke (1665). The first comprehensive book on Palynology was written by Wodehouse (1935) and subsequently Erdtman (1943) contributed a treatise on pollen analysis. The first manual, *Pollen Morphology and Plant Taxonomy*, appeared in 1952 which deals with angiosperms (Erdtman, 1952). This book provided a summary of pollen characteristics of all angiosperm families. Erdtman (1969) also brought out *Handbook of Palynology*. In India Nair has brought out several books on Palynology which include *Advances in Palynology* (1964), *Essentials of Palynology* (1966) and *Pollen Morphology of Angiosperms: A Historical and Phylogenetic Study* (1970).

Pollen grains are initially formed in tetrads. A schematic illustration of pollen grain illustrated in Fig. 4.1 is called Palynogram. Pollen grains have two poles at opposite ends of the polar axis. The proximal pole (P.P.) is at the centre of the proximal face (toward the centre of tetrad), whereas the distal pole is at the centre of the distal face (away from the centre of the tetrad) (Fig. 4.1). The pollen grains are called isopolar if their two faces are similar and heteropolar if different. In heteropolar pollen grains, one face has an aperture, while the other has none. The equator runs round the surface of pollen grains midway between the poles.

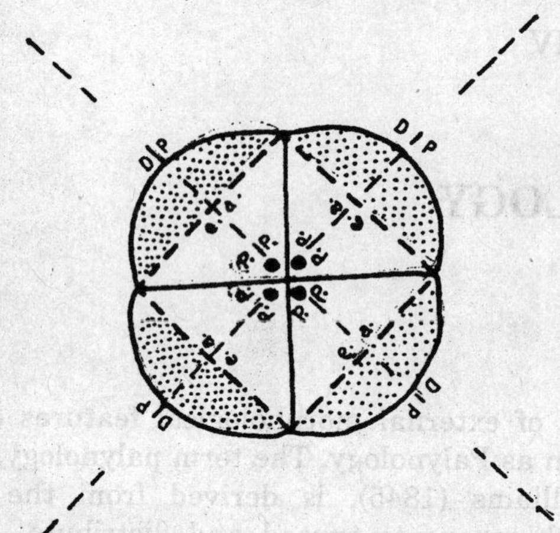

Fig. 4.1. P.P. : Proimal Pole, D.P. : Distal Pole, e. a. : Eqvuatorial axis

Aperture

Aperture is any weak area on the pollen surface associated with its germination. Most angiospermous pollen grains fall into only two general types, uniaperturate and triaperturate from which some less common types have been derived. Uniaperturate pollen grains have a single germinal furrow or pore situated at the proximal or distal position. Proximal form of aperture is found in primitive Angiosperms while majority of angiosperms have distal aperture. Uniaperturate pollen grains are characteristic of Monocotyledons and some Magnoliidae.

NPC System

NPC refers to Number (N), Position (P) and Character (C) of aperture in pollen grains (Fig. 4.2). In this system the term 'treme' has been used in place of 'aperture' for the classification of pollen grains. Pollen grains without aperture are called atreme and represented as No. Based on the number of apertures pollen grains are classified as monotreme (N_1), ditreme (N_2), tritreme (N_3), tetratreme (N_4) (Fig. 4.2), pentatreme (N_5) and hexatreme (N_6). Pollen grains with more than six apertures are referred to as polytreme (N_7). When the apertures irrespective of their number are irregularly spaced the pollen grains are described as anomotreme (N_8).

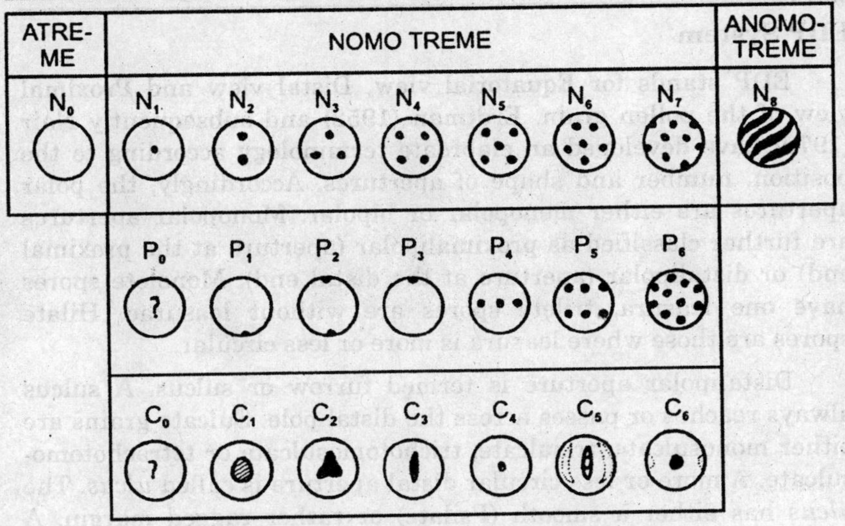

Fig. 4.2. NPC system (after Erdtman, 1969)

Based on the position of apertures angiospermous pollen grains are classified into six types (Fig. 4.2). Po refers to those grains whose position of aperture is not known. When the aperture is on the proximal face the pollen are designated as catatreme (P_1) if it is on both distal face and proximal face it is referred to as anacatatreme (P_2). If the aperture is on the distal face the pollen grains are called anatreme (P_3). When the apertures are located on the equator, the pollen grains are referred to as zonotreme (P_4) and dizonotreme (P_5) if they are in two rows on the equator. Pantotreme (P_6) pollen grains are those where apertures are more or less uniformly distributed all over the pollen surface.

Based on the character of aperture pollen grains are classified into six types (Fig. 4.2). C_1 pollen have an aperture like thin area, or leptoma. Pollen with one leptoma are monolept. The leptoma may be on the distal face (analept) or on the proximal face (catalept). Trichotomocolpate (C_2) pollen grains are those with 3 slit colpus. Pollen grains with elongate apertures are called colpate (C_3), while those with circular apertures are called porate (C_4). In some plants both types of apertures occur as one superimposed above the other and the grains having these are termed colporate (C_5). Pollen grains with compound pores are referred to pororate (C_6) pollen.

EDP System

EDP stands for Equatorial view, Distal view and Proximal view of the pollen grain. Erdtman (1952) and subsequently Nair (1970) have developed an elabroate terminology according to the position, number and shape of apertures. Accordingly, the polar apertures are either monopolar or bipolar. Monopolar apertures are further classified as proximalipolar (aperture at the proximal end) or distalipolar (aperture at the distal end). Monolete spores have one leasura, trilete spores are without leasurae. Hilate spores are those where leasura is more or less circular.

Distalipolar aperture is termed furrow or sulcus. A sulcus always reaches or passes across the distal pole. Sulcate grains are either monosulcate, trisulcate, trichotomosulcate or tetrachotomo-sulcate. A more or less circular distal aperture is called *ulcus*. The *ulcus* has either a smooth (Psilate) or rather ragged margin. A bisulcate grain has two furrows parallel to each other with the distal pole in the centre between the furrows. In trisulcate grains the furrows form a triangle around the pole.

Equatorial elongate apertures with one of their axes crossing the equator at right angles are termed *Colpi*, while circular apertures are known as pores. Pollen grains with such apertures are termed colpate or porate, respectively. In some plants both types of apertures occur as one superimposed above the other and the grains having these are termed colporate. Global apertures are known as rugae if they are elongate and foraminae if they are circular. Grains having these are then termed rugate or forate, respectively (Fig. 4.3).

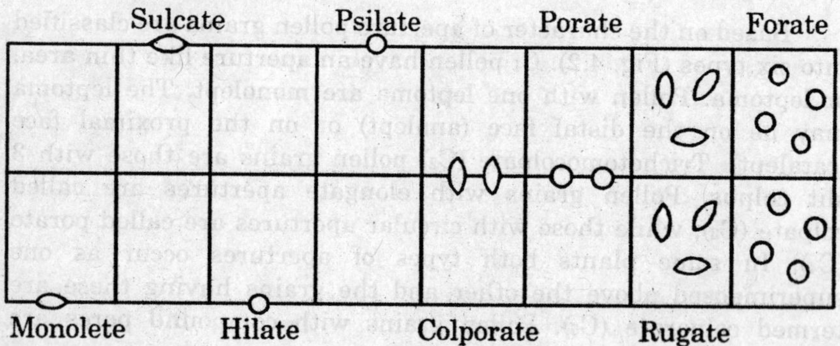

Fig. 4.3. EDP system (after Erdtman, 1952).

Chemical Composition of Pollen

The outer walls of pollen are composed of a pectinaceous substance called "Pollenin", the chemical formula being $C_{90}H_{129}OH_5$. The protoplasm of the grains contain proteins, amino acids, carbohydrates, lipids, vitamins of particularly the 'B group', hormones and enzymes apart from traces of inorganic substances, such as K, Mg, Ca, Co, Fe, Si, P, S and Cl. The chemical composition of pollen is sometimes suggestive of plant relationships. The relatively high content of pantothenic acid and inositol in the pollen of *Zea mays* and riboflavin in the pollen of *Alnus glutinosa* and *Alnus incana* are said to be true characteristic of these pollen types.

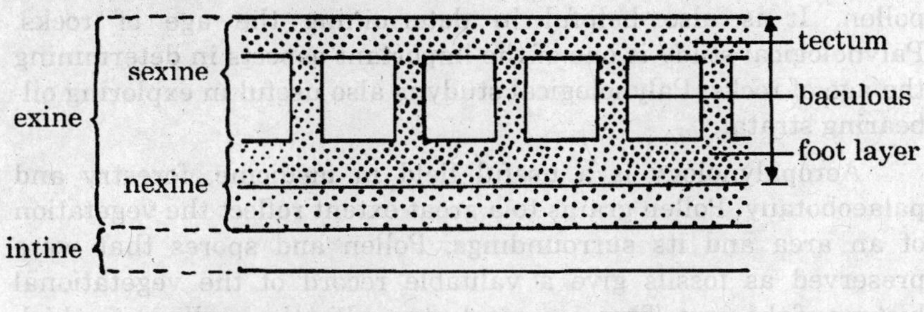

Fig. 4.4. Diagram showing wall stratification in a tectate-imperforate, non-acetolysed pollen grain, as seen in cross section (after Walker, 1976).

Shape and Size

The shape of pollen grains can vary from very flat (peroblate), where equatorial diameter is more than twice the length of the polar axis, to very elongate (perprolate) where the diameter is less than half the length of the polar axis. Although the variations in the shape of pollen are quite characteristic for taxonomic and phylogenetic considerations, these are less important than apertures. The size of the pollen grains varies from nearly 10 µm in *Myosotis* to as much as 200 µm in some Cucurbitaceae and Nyctaginaceae.

SCOPE OF PALYNOLOGY

Research work on palynology can be either basic or applied. The basic aspects belong to the pollen morphology and its relation to taxonomy, while applied aspects include geopalynology (fossil pollen grains), aeropalynology (pollen found in atmospheric air), iatro-palynology (medical aspects of palynology like hayfever, criminology etc.) and melittopalynology or melissopalynology (study of pollen in honey).

Applied aspects of Palynology

Stratigraphical problems like correlation of coal seams and oil fields can be understood by studying the distribution of spores and pollen. It is also helpful in determining the age of rocks. Palynological study is one of the important aspects in determining the age of rocks. Palynological study is also useful in exploring oil-bearing strata.

Aeropalynology is a useful field in medicine, forestry and palaeobotany. Pollen grains to a great extent reflect the vegetation of an area and its surroundings. Pollen and spores that were preserved as fossils give a valuable record of the vegetational history of the past. They are not destroyed in the medium in which they are preserved because the exine of the grains is formed from one of the most extraordinarily resistant material known to the organic world namely sporopollenin. The resistance capacity of the protective walls can be evaluated from the fact that pollen and spores were recovered even from the earliest geological formations. The data provided by pollen grains contained in fossiliferous rocks have been successfully made use of a tool in interpreting the past vegetation, particularly of the quaternary period in different parts of the world.

The composition of pollen in the air or 'pollen spectrum' at a particular locality provides an arbitrary picture of the local flora.

Forensic Palynology

Forensic palynology relates to the application of the study of pollen and spores to legal matters, often used to establish links between objects, people and places based on the analysis and identification of pollen.

Palynological samples can be recovered from a wide range of sources, including people, such as on their clothes, in their hair or even in their nasal passage, vehicle tyres, air filters in cars, on objects and in mud. Due to the dispersal mechanisms of some plants, pollen can be readily picked up and transferred. A person can easily inadvertently pick up pollen from a crime scene, whether it be in mud on their shoes or on their clothes from directly brushing against a plant in the area.

With this in mind, a primary use of palynology in a forensic investigation is to establish a link between two places, objects or people. For instance, it may be possible to link a suspect to an object, a vehicle to a crime scene, or even link two separate incident scenes. If a suspect was present at a particular crime scene at which pollen can be found (for instance a field or garden), they may have picked up pollen on their clothing or in their hair. Since pollen is so resilient, it can often stick to other objects even after that object has been washed. If the pollen recovered from the suspect matches pollen collected from the crime scene, this could suggest that the suspect was in fact present at that scene. However, it must be considered that although the presence of pollen may establish a link, the lack of pollen does not necessarily prove that there is not a link.

Similarly, palynology may be able to determine the location of a crime scene if it is not known. For instance, a body that is believed to have been moved may carry pollen grains that can be analysed and traced to a likely location. This may particularly be suspected if the body carries large amounts of a particular pollen that is not found at the location in which the body was found.

The study of pollen can also be used to determine the travel history of an item. In some cases it may be necessary to ascertain where an item has originated from, especially illicit drugs, money, antiques and even food. By analysing pollen recovered from suspect items, it may be possible to trace that item to a particular country if the pollen grains identified are sufficiently distinctive. Although this application of palynology may not necessarily be able to establish an exact location, it may be least be possible to rule out certain geographical locations and point the investigation in the right direction.

It may even be possible to estimate the time of year at which a crime took place. In the investigation of a somewhat older crime scene, pollen collected may actually be released at a different time of year, indicating the crime occurred during this period.

Of course despite the links palynology may establish, further evidence may be needed to support any conclusions reached. Pollen recovered from a suspect that happens to match that of a crime scene may simply suggest that the individual had visited that area at some point recently, not necessarily prove that they have committed a crime.

Although forensic palynology has been utilised for decades, it is unfortunately not widely accepted as a reliable forensic technique, instead frequently seen as a last result failing more 'standard' investigative techniques. In addition to this, there are very few people properly trained to analyse palynological samples, thus samples are often collected and handled by untrained staff, inevitably leading to issues of sample preservation and contamination.

Case Study: The first documented use of the analysis of pollen and spores to a forensic investigation was in Austria in 1959. Whilst on a trip down the Danube River, an Austrian man disappeared. His body could not be found. A friend and business partner of the victim soon fell under suspicion, and was arrested and charged with murder. Unfortunately without a body, and of course the suspect proclaiming his innocence, there was not much of a case against this man. During a search of the man's cabin, a pair of muddy boots were recovered, providing the authorities with a new avenue of investigation. Palynologist Wilhelm Klaus of the University of Vienna was called upon to provide his expert opinion regarding a mud samples recovered from these shoes. Klaus was able to identify a number of modern pollens in the mud, including spruce, willow and alder pollen, along with a significantly older fossil pollen. Only a small area north of Vienna was consistent with this combination of types of pollen. The defendant was confronted with this new piece of information, at which point he finally cracked and confessed to the murder. He led police to the clandestine grave which was, interestingly, in the region selected by Wilhelm Klaus.

Melittopalynology (Melissopahlnology)

Plants offer pollen grains and also nectar to their visitors. The honey bees collect pollen and nectar for making 'bee bread' which is an important item of food of both the adult bees and the brood. The study relating to the various aspects of the palynology of honeys and related substances is known as Melittopalynology or Melissopalynology.

Nair (1963) made an extensive study of Indian honeys collected from various places. Such an analytical study has provided useful data about pollen and nectar yielding plants. A trained melittopalynologist can easily distinguish between the honey from two different localities. He can also tell the time when a particular honey was prepared.

Honey is said to be 'unifloral' if it is dominated (50% or more) by pollen of only one plant and 'multifloral' if it contains various types of pollen in considerable percentages. Nair and Chaturvedi (1974) examined microscopically 76 samples of Indian honey. They showed the presence of pollen belonging to *Eugenia, Sapindus, Putranjiva, Citrus, Brassica* (all nectar yielding plants), *Holoptelia, Alnus, Borassus* and other palms (all pollen yielding plants). Such an analytical study provides useful data about the pollen and nectar yielding plants of a locality, which should be helpful in establishing apiary garden.

Pollen and Spore Allergy

Some fungal spores and pollen grains initiate allergic response to susceptible individuals. Allergic people react to potential allergens being hypersensitive to them causing several types of eye, skin and respiratory disorders. There are two types of major, common respiratory allergic diseases, namely asthma and hay fever. The allergen is not the pollen grain itself, but factors located on or within it. These factors induce allergic diseases in the form of hypersensitive reaction.

Allergens in pollen grain contain proteins or glycoproteins. Respiratory allergy is an acquired hypersensitive reaction to specific components of the air spora. Two types of allergy are recognized. (1) Immediate allergy (after exposure) and the other, (2) Non Immediate (prolonged exposure to spores mostly in occupational environments). The allergens of pollen grains are located in extracellular wall sites *i.e.*, in intine and exine. This can be established using fluorescent antibody methods.

A growing number of populations appear to be hypersensitive to pollen, spores, moulds and other microorganisms. By long term monitoring, the occurrence of these allergens, their dispersal, interaction with environmental factors, annual and seasonal variations etc. can be made known to the sufferers and their doctors. With this knowledge, coupled with pollen and spore forecasts an allergic person will be in a better position to avoid unnecessary exposure, take the right medicine in time and minimize some of the daily troubles.

The plants known for high pollen productivity are more significant for pollen allergy problems. A survey conducted at Bangalore has revealed that *Casuarina equisetifolia* which has been extensively planted under afforestation programmes being useful as a wind breaker, fire wood and building material. However, it is high pollen producer and allergic manifestations due to this pollen are high. Similarly *Amaranthus spinosus* known for high pollen production and dissemination by wind brings about fairly high positive skin reactions. *Ricinus communis* and *Eucalyptus* also show positive skin reactions. Sensitivity was also observed against some high pollen producing taxa, such as *Prosopis juliflora, Argemone mexicana* and *Azadirachta indica.*

Hey fever is another case of allergy. This is a special case of human hyper sensitiveness towards an alien kind of protein. Atmospheric pollen and spores are main cause of hey fever. Pollen grains vary in their toxicity or power to sensitize. The allergenic activity of gymnospermous pollen is negligible. But conifers like *Pinus* pollen are mildly toxic. Each person is sensitive to a particular kind of pollen, some to grass pollen, others to *Ambrosia* pollen and so on. Suffering from hey fever is directly related to the amount of sensitive pollen inhaled by the individuals. Moderately sensitive hey fever patients get symptoms for 40 - 50 pollen grains and highly sensitive ones for 3-4. In USA, one person out of 20-30 is allergic to pollen of some kind or the other.

Hey Fever Tests

Patients are tested as to which kind of pollen one is allergic for. The tests are as follows: (1) Scratch Test: The skin of the fore arm is first washed with alcohol. After it dries off, very superficial scratches are made at that place. A drop of NaCl 0.9% solution is then placed and pollen grains are stirred into the scratches with a glass rod and the reaction studied. (2) Intracutaneous Test: In this method 0.005 cc of pollen extract is injected in the superficial layer of the patients and the reaction studied. In USA, the main hey fever plants are *Ambrosia, Salix, Juglans, Casuarina, Quercus, Ulmus, Morus* etc. In Norway, the disease coincides with the flowering of a number of Amentiferae and Gramineae. In England and S. Africa high percentage of grass pollen was noticed in the air during the outbreak of allergic diseases. In India, pollen of *Holoptelea, Parthenium* and fungus spores such as those of *Alternaria* and *Helminthosporium* are allergic.

Pollen Free Areas

Extensive pollen analytical survey of the geographic incidence of pollen in the air have made it possible to advise hey fever sufferers on pollen free vacation resorts. With regard to the ragweed (*Ambrosia*) which is the main cause of hey fever in the USA and Canada, it has been found that the Western part of the area is pollen free or nearly so. Also suitable clearance of weeds is made to control the incidence of hey fever.

Medicinal Properties of Pollen and Spores

This has been realized since early days. *Lycopodium* powder consisting of their spores was known commercially as "Vegetable sulphur". This has been used medicinally for stomach disorders by the Arabian physicians. The spores of *L. clavatum* are diuretic, demulcent, antispasmodic and emmenagogue. Similar properties are known for the spores of *Adiantum*. *Adiantum* spores are renowned as a specific remedy for coughs. The pollen of *Cycas circinalis* are narcotic and those of *Typha* have astringent properties. Pollen grains have also been observed to possess bacteriostatic activities on the intestinal flora of mice. There are several reports regarding the useful properties of certain pollen tablets, tonics, creams etc. Pollen tablets are used in the treatment of prostatitis (inflammation of the prostate gland).

Pollen samples from the stomach contents or excrements have provided a clue to several deaths in Brazil. The deaths were ascribed to use of poisonous honey. The pollen of poisonous plant *Sarjania lethalis* (Sapindaceae) were found in the stomach contents of one of the victims who died after taking poisonous honey.

PALYNOLOGY AND TAXONOMY

Pollen morphology offer reliable data and evidence of taxonomic relationships. Palynological evidence may be used to place taxa of uncertain affinities, to suggest rearrangements, withdrawals and separations as well as corraborating other lines of evidence. Sometimes, the study of fine structural details in fossil pollen opens up possibility of new sources of phylogenetic evidence.

The major characters of pollen grains that are used in solving taxonomic and phylogenetic problems are (1) Apertural patterns, (2) Exine stratification and ornamentation, (3) Pollen association and (4) Pollen nuclear number.

The taxonomic and evolutionary importance of pollen morphology may be at specific, generic or higher levels. In many cases the type of pollen of a taxon is characteristic and constant; such a taxon is called *Stenopalynous* or unipalynous (Nair, 1970) and may be exclusive of that group, *e.g.*, thick walled pollen grains of the Gyrostemonacee of Australia. In other cases, the types of pollen may vary considerably in size, aperture, stratification of exine etc. Such taxa are termed *eurypalynous* or *multipalynous*. Stenopalynous taxa are generally considered to be very natural. Asclepiadaceae, Asteraceae, Poaceae, Brassicaceae and Lamiaceae are some of the stenopalynous families.

Eurypalynous condition, seen in Malvaceae, is primitive, while stenopalynous condition, seen in Poaceae, is an advanced character.

Apertural Patterns

By using the NPC system, formulae can be derived to individual taxa and families. Families with more closely similar formulae must be considered more closely related than those with less similar formulae.

Engler and Diels (1936) included Canellaceae in the order Parietales. When NPC formulae of all the families included in the Parietales are observed, they appear to be more or less uniform having $N_3P_4C_5$ (345). But Canellaceae appears to be an exception - with the formula 133. Subsequently, Melchoir (1964) divided Partietales into two orders the Violales in which the NPC is essentially 345 throughout and others in Guttiferales.

The genus *Theobroma* provides an example of a genus with two distinct types of pollen. One is small, flattened and is found in the majority of the species including *T. cacao.* The other type is larger, more elongate. This is one of the reasons for the recent removal of the species with larger pollen grains to a distinct genus *Herrania.*

Caltha palustris var. *palustris* and *C. palustris* var. *alba* can readily be distinguished by their aperture.

C. palustris var. *palustris*	*C. palustris* var. *alba*
Tricolpate	Pantoporate
Position of apertures equatorial	Position of apertures global
Size of apertures - large	Size of apertures - small.

In the genus *Sesamum*, the number of colpi is species specific. Eleven in *S. indicum*, nine in *S. prostratum* and eight in *S. laciniatum*. Similarly *Butomopsis umbellatus* can be differentiated from *B. lanceolata* by the nature of apertures, the former having monocolpate and the later pantoporate grains.

Salix and *Populus* of Salicaceae can readily be distinguished from one another on the basis of aperture. The former has 3-colpate pollen and the latter inaperturate one.

Similarly, *Phytolacca* and *Rivinia* of Phytolaccaceae can be distinguished from one another on pollen apertural patterns. *Phytolacca* has 3-zonocolpate pollen while *Rivinia* has 5-colpate pollen.

The genus *Nelumbo* has till recently been included in Nymphaeaceae. *Nelumbo* has 3-colpate pollen, while Nymphaeaceae have uniformly monosulcate pollen. Hence *Nelumbo* has been segregated into a separate family Nelumbonaceae.

Bentham and Hooker (1862-83) and other earlier taxonomists have included *Alisma*, *Butomus* and *Butomopsis* in the family

Alismataceae. But Hutchinson (1969, 1973), Cronquist (1988) and Takhtajan (1980) have segregated *Alisma* into Alismataceae and *Butomus* and *Butomposis* into Butomaceae. This is supported on Palynological grounds. *Alisma* have pantoporate pollen, while *Butomus* and *Butomopsis* have monocolpate one.

Apertural characters are also useful in understanding the phylogeny. The monosulcate (uniaperturate) pollen found in some Ranales indicate primitiveness, Triaperturate and polyaperturate pollen are found in advanced families like Asteraceae.

Exine Sculpturing

The exposed surface-details of the pollen wall constitute the sculpturing. Some of the more important types are: Psilate (smooth), foveolate (pitted), fossulate (grooved), scabrate (very fine projections), verrucate (warty), baculate (rod-like elements with swollen tips), gemmate (sessile pillar), echinate (spiny), rugate (elongate elements irregularly distributed tangentially over the surface), striate (elongate, more or less parallelly distributed tangentially over the surface), punctate (minute perforations) and reticulate (elements forming open network) (Fig. 4.5).

Exine patterns have occasionally been found useful in differentiating species of the genus. In the genus *Bauhinia, B. racemosa* has reticulate, *B. purpurea* has reticulate-tuberculate, *B. malabarica* has spinulate *B. acuminata* has pilate, *B. krugii* striate and *B. retusa* has verrucate pollen grains.

The genus *Cicer*, usually placed in the tribe Vicieae of the Fabaceae, has many characters which suggested that this position is anomalous. Clarke and Kupicha (1976) studied the pollen grains of *Cicer*, of the Vicieae and the allied tribes, Trifolieae and Ononideae. They found that *Cicer* has pollen that is more similar to that of the Ononideae than that of Vicieae. They suggested the transfer of the genus *Cicer* from the Vicieae to the Ononideae.

In general, Angiosperm pollen grains have a massive exine and a thin intine. However certain taxa (*e.g.*, Amaryllidaceae, Cannaceae, Iridaceae, Musaceae, Zingiberaceae, Costaceae, Marantaceae, Strelitziaceae and Heliconiaceae among Monocotyledons) are unique in having a highly reduced exine and elabroate intine. In *Heliconia*, the exine is relegated to a few spinules and a thin layer over intine, which is almost a hundred times as thick.

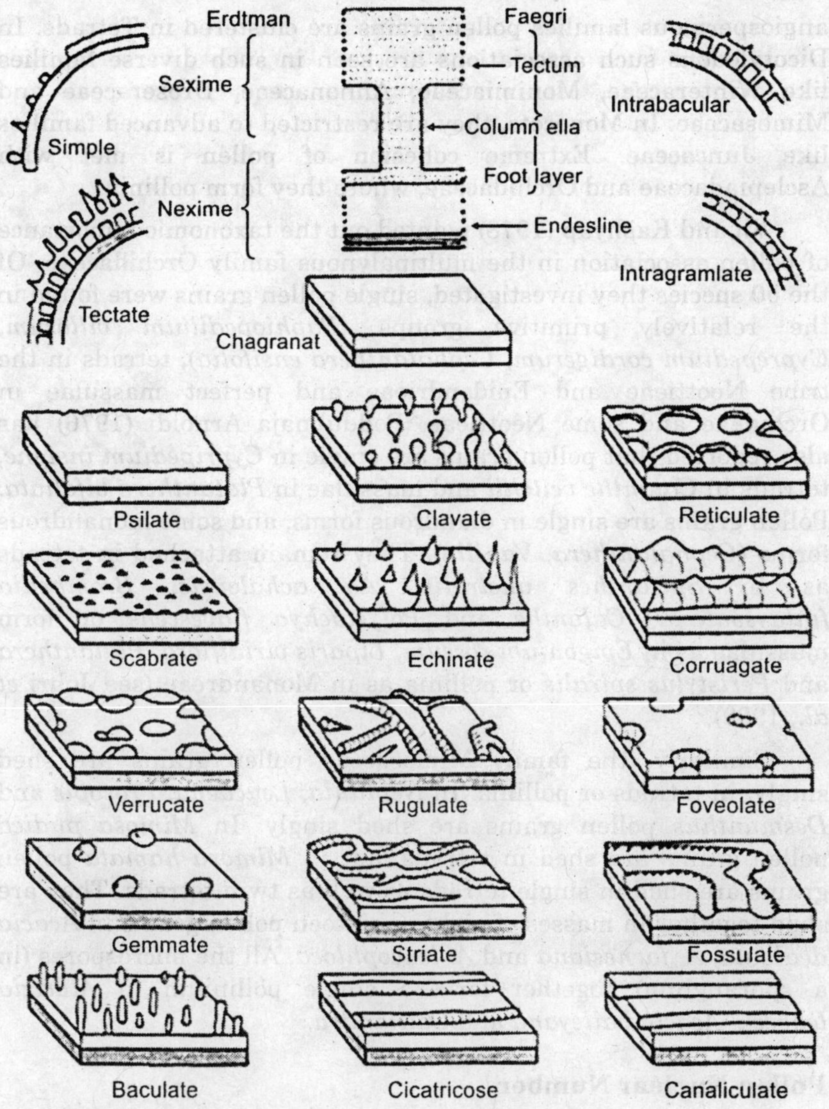

Figure 4.5. Exine Patterns in Angiosperms

Pollen Association

In majority of the Angiosperms pollen grains occur singly. However, in Podostemaceae pollen grains occur in pairs. In some angiospermous families pollen grains are clustered in Tetrads. In Dicotyledons such associations are seen in such diverse families like Winteraceae, Monimiaceae, Annonaceae, Droseraceae and Mimosaceae. In Monocots, they are restricted to advanced families like Juncaceae. Extreme cohesion of pollen is met with Asclepiadaceae and Orchidaceae, where they form pollinia.

Vij and Kashyap (1978) pointed out the taxonomic importance of pollen association in the multipalynous family Orchidaceae. Of the 50 species they investigated, single pollen grains were found in the relatively primitive groups (*Paphiopedilium villosum, Cyprepedium cordigerum, Cephalanthera ensifolia*), tetrads in the tribe Neottieae and Epidendrieae and perfect massulae in Orchideae and some Neottieae. Poddubnaja Arnoldi (1976) has also reported that pollen grains are single in *Cypripedium insigne*, tetrads in *Calanthe veitchii* and massulae in *Platanthera bifoliata.* Pollen grains are single in diandrous forms, and some monandrous forms (*Cephalanthera, Vanilla*). They remain attached in tetrads as in *Spiranthes australis, Aa achalensis, Bromhedia finlaysoniana, Calanthe* and *Polystachya flavescens* or form massulae as in *Epigonium roseum, Liparis viridiflora, Platanthera* and *Peristylus spiralis* or pollinia as in Monandreae (see Johri *et al.*, 1992).

Similarly, the family Mimosaceae pollen grains are shed singly, in tetrads or pollinia. In *Neptunia, Leucaena, Prosopis* and *Desmanthus* pollen grains are shed singly. In *Mimosa pudica* pollen grains are shed in tetrads, but in *Mimosa hamata* pollen grains are shed in single tetrads as well as twin tetrads. They are shed together in masses of eight or sixteen pollen grains in *Acacia dealbata, A. farnesiana* and *A. leucophloea.* All the microspores (in a sporangium) together form a single pollinium in *Albizzia lebbeck, Acacia baileyana* and *Calliandra.*

Pollen Nuclear Number

In Angiosperms pollen grains are shed either at 2-nucleate stage or 3-nucleate stage. This character is valuable in taxonomy and phylogeny of angiosperms.

Brewbaker (1967) made an extensive survey of about 2000 species belonging to the 265 families. According to Brewbaker two-nucleate pollen grains are more primitive in angiosperms as a whole and within individual taxa, three-nucleate pollen grains have arisen from the two-nucleate ones.

In the Monocotyledons, the three-nucleate pollen grains occur in highly reduced wind pollinated groups, aquatics and achlorophyllous saprophytes. In the Dicotyledons, however, there is no definte pattern. Certain orders, *e.g.*, Caryophyllales, Plumbaginales, Polygonales and Asterales, have consistently tri-nucleate pollen grains (Dahlgren, 1975). Magnoliales, Laurales and Violales have consistently two-nucleate pollen grains. Most other orders are quite heterogeneous in having both two and three-nucleate ones. Both these types may occur in one and the same genus, as has been observed in *Burmannia, Lobelia* and *Ipomoea* (Brewbaker, 1967).

Recently, an important connection discovered between bi- and tri-nucleate pollen grains, and types of self-incompatibility systems (see Brewbaker, 1957) opens up further possibilities. Particularly in the dicotyledonous taxa, the groups with binucleate pollen grains tend to have the gametophytic type of self-incompatibility and those with trinucleate pollen grains the sporophytic type. In Lamiaceae, Kooiman (1972) has also demonstrated the correlation between binucleate and tricolpate pollen grains and between trinucleate and hexacolpate pollen grains.

References

Erdtman, G. 1969. *Handbook of Palynology*. Hafner Publication Company, New York.

https://biocyclopedia.com/index/palynology.php

Nair, P.K.K. (ed.) 1964. *Advances in Palynology*. National Bot. Gardens, Lucknow.

Nair, P.K.K. 1966. *Essentials of Palynology*, Asia Publishing House, Bombay.

Nair, P.K.K. 1970. *Pollen morphology of Angiosperms. A Historical and Phylogenetic Study*. Scholar Publishing House, Lucknow.

Wodehouse, R.P. 1935. *Pollen grains: Their structure, identification and significance in Science and Medicine*. McGraw-Hill, New York.

CHAPTER – V

OVULE AND MEGASPOROGENESIS

The megasporangium along with its protective coats, namely integuments, is known as the ovule. The ovule is the forerunner of the seed. Therefore, the study of the megasporangium or the ovule is most important for an understanding of the structure and function of the seed.

The first classical microscopists, Grew (1671) and Malpighi (1675) studied the ontogeny of the young ovules and the developing seeds. They have described the most conspicuous parts of the ovules only, because the resolving power of their microscopes was not sufficient to distinguish the various parts. Turpin (1806), after an analysis of the seeds of 1200 species, concluded that the micropyle is of universal occurrence and forms an essential part of all monocotyledons and dicotyledons. To the micropyle, he attributed a function associated with fertilization. Brown (1826) gave the first correct definition of an angiospermous ovule which was also confirmed by Brongniart (1827) and Mirbel (1829, 1830). Mirbel introduced the term *nucellus* for the central core of the ovule and exo- and endostome for the openings in the outer and inner integuments, respectively. He also made an extensive survey of the ovules and recognised four different types of ovules - orthotropous, anatropous, campylotropous and amphitropous. Schleiden (1837, 1839) recognised the hemianatropous type of ovule. Van Tieghem (1898) introduced the terms 'tenuinucellate' (with a thin nucellus) and 'Crassinucellate' (with a bulky and more persistent nucellus).

Generally, the ovule consists of a nucellus, enveloped by one or two integuments and is found attached to the placenta, on the inner wall of the ovary, by a stalk known as the funicle. The part of the funicle that runs parallel with the nucellus is called the 'raphe'. An ovule ready for fertilization consists of nucellus enveloped by one or two integuments, leaving a small opening at the apical end. This opening, called *micropyle,* is the main passage for the entry of the pollen tube into the ovule. The *chalaza* is the region where the integuments are inserted. In fact, funicle, raphe, chalaza and nucellus form a continuous tissue, the central axis of the ovule and no sharp demarcations can be drawn between them.

Types of Ovules

Mainly, based on the position of the micropyle with respect to the funiculus, the mature ovules are classified into 6 main types.

1. *Orthotropous ovule or Atropous ovule:* In this, the micropyle and funicle lie in a straight line (Fig. 5.1A). This type of ovule is seen in about 20 families (Davis, 1966) and prominent among them are: Polygonaceae and Piperaceae.

2. *Anatropous ovule:* The ovule becomes completely inverted and as a result of which, the micropyle and the funicle lie very close to each other (Fig. 5.2B). This type of ovule is most common among the angiosperms and is found in 204 families (Davis, 1966), *e.g.*, Asteraceae, Lamiaceae and other Sympetalae.

3. *Campylotropous ovule:* The ovule is curved (Fig. 5.1C) in this type and is found in 5 families, *e.g.*, Chenopodiaceae, Capparaceae etc.

4. *Hemianatropous or Hemitropous ovule:* In this, the funicle lies at right angles to the nucellus and the integuments (Fig. 5.2D). It is present in 13 families including Malpighiaceae and Primulaceae.

5. *Amphitropous ovule:* The curvature of the ovule is more pronounced here and the curvature also extends to the embryo sac which becomes horse-shoe shaped (Fig. 5.1E). *e.g.*, Amaranthaceae, Alismataceae, Butomaceae etc.

6. *Circinotropous ovule:* It is a very peculiar type of ovule in which the funicle is exceptionally long and forms a complete circle around the ovule but is free from it except for a small area of attachment at the end of funicle (Fig. 5.2). It occurs mostly in the members of Cactaceae and also in a few taxa of Plumbaginaceae.

Table 5.1. Data on the occurrence of various features of ovules in Angiosperms (after Davis, 1966)

Feature	No. of families showing	Examples
TYPES OF OVULE		
Anatropous	204	Sympetalae
Orthotropous	20	Polygonaceae, Piperaceae, Cistaceae Urticaceae
Hemianatropous	13	Malpighiaceae Primulaceae
Campylotropous	5	Capparaceae, Chenopodiaceae
Amphitropous	4	Crossosomataceae, Leitneriaceae
Circinotropous	1	Cactaceae
OVULE (ENVELOP)		
Bitegmic	208	Amaranthaceae, Fabaceae, Cucurbitaceae
Unitegmic	90	Asteraceae, Lamiaceae, Acanthaceae
MICROPYLE FORMED BY		
Inner integument	88	Centrospermae Cucurbitaceae Poaceae
Outer integument	4	Podostemaceae, Euphorbiaceae
Both integuments	74	Pontederiaceae
NUCELLUS		
Crassinucellate	179	Polygonaceae, Aizoaceae, Cucurbitaceae, Amaranthaceae
Tenuinucellate	105	Asteraceae Lamiaceae
Pseudocrassinucellate	11	Anemone

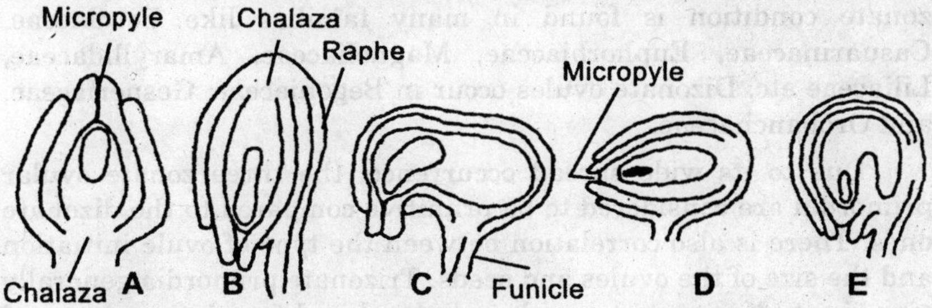

Fig. 5.1. Types of ovules. **A.** Orthotropous; **B.** Anatropous, **C.** Campylotropous; **D.** Hemianatropous; **E.** Amphitropous (after Maheshwari, 1950).

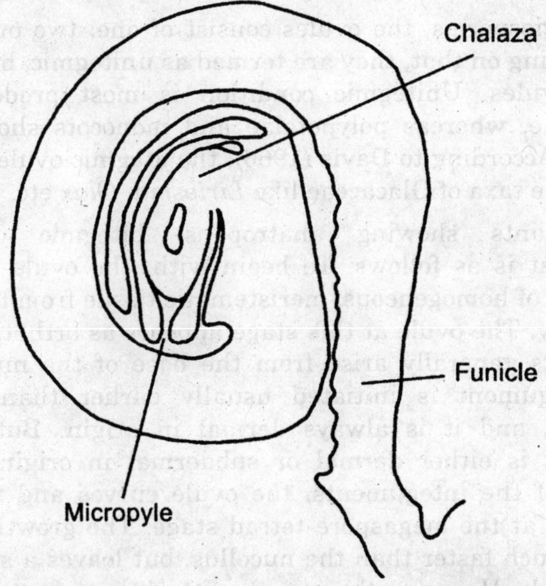

Fig. 5.2. Circinotropous ovule in *Opuntia* (after Maheshwari, 1950).

Ovule Initiation

Ovule primordia are initiated by periclinal divisions of the second or third layer of the placenta and the histogenic organisation is similar to that of shoot apex (according to the concept of tunica and corpus). Basing on its origin either from the second or third zone of the placental meristem, the primordia are classified as dizonate and trizonate ovular primordia. The three zonate condition is found in many families like Betulaceae, Casuarinaceae, Euphorbiaceae, Magnoliaceae, Amaryllidaceae, Liliaceae etc. Dizonate ovules occur in Begoniaceae, Gesneriaceae, and Orobanchaceae.

Due to its wide spread occurrence, the three zonate ovular primordia are considered to be primitive compared to the dizonate ones. There is also correlation between the type of ovule initiation and the size of the ovules and seeds. Trizonate primordia generally give rise to larger ovules with well developed funicles, raphes and nucelli than two-zonate primordia.

Integuments

In angiosperms, the ovules consist of one, two or no integuments. Basing on that, they are termed as unitegmic, bitegmic and ategmic ovules. Unitegmic condition is most predominant in gamopetalae, whereas polypetalae and monocots show bitegmic condition. According to Davis (1966), the ategmic ovules are found only in some taxa of Olacaceae like *Liriosma*, *Olax* etc.

In plants showing anatropous, bitegmic ovules the development is as follows: To begin with, the ovule arises as a small mass of homogeneous, meristematic tissue from the placenta of the ovary. The ovule at this stage appears as orthotropous. The integuments generally arise from the base of the nucellus. The inner integument is initiated usually earlier than the outer integument and it is always dermal in origin. But the outer integument is either dermal or subdermal in origin. After the initiation of the integuments, the ovule curves and reaches the final shape at the megaspore tetrad stage. The growth of integuments is much faster than the nucellus, but leaves a small gap in the micropyle. However, the growth of the integument is very slow in some Proteaceae members like *Macadamia* and it never covers the nucellus completely. Usually, the outer integument is massive and thicker than the inner integument in bitegmic ovules. Sometimes, in certain families like Cucurbitaceae and Aizoaceae the inner integument grows beyond the outer integument.

Bitegmic ovules are considered to be primitive than the unitegmic ovules. According to Maheshwari (1950), the unitegmic condition may arise by the suppression and elimination of one integument or the fusion of two separate primordia. The unitegmic condition must have been achieved in different ontogenic pathways during evolution and hence the unitegmy is not homologous throughout the angiosperms.

Fusion of various parts of the ovule is seen in some cases. Sometimes, the two integuments of the bitegmic ovule are fused or the inner integument may fuse with the funiculus up to various lengths. Most often, in the anatropous ovules, the funiculus and the outer integument are closely fused together, so that the distinction between them seems impossible.

During seed development various structural modifications occur in the integuments. The changes may occur on their cell contents or wall structure. Occasionally, some or all the integumentary layers may be destroyed during seed development. Sometimes, both the integuments, but often one or only a part of it participates in the formation of a seed coat.

In the members of Cactaceae, a conspicuous air space is present between the outer and inner integuments in the chalazal region of the ovule. It forms an important diagnostic feature of the family. A similar space is also found at the chalazal region of the ovule of *Trianthema argentina, Bassia* and *Tetragenia tetragonioides.*

Stomata are seen on the outer integument of *Michelia, Magnolia, Cleome* etc., but their role is not understood. In *Gossypium* also stomata appear two days before flowering and they are believed to be concerned with respiration. Rudimentary stomata are reported in *Isomeris* on the funicular side of the ovule. In *Hymenocallis,* extensive chlorophyllous tissue, capable of photosynthetic activity, is seen in addition to stomata. Similarly, chlorophyllous integuments are also seen in *Gladiolus, Lilium, Moringa* etc. In some taxa of Caesalpiniaecae, the micropylar end of the ovule becomes red or deep pink at the time of fertilization due to the presence of anthocyanin. In several plants belonging to the families. Cistaceae and Ericaceae, symbiotic or parasitic fungi are seen inhabiting the ovules and seeds. The fungus is found either in the testa or other tissues of the seed and is also transmitted to the offspring so that the association becomes hereditary. In *Ardisia,* symbiotic bacteria are seen in the seeds and their presence seems to be essential for the successful germination of the seeds.

Aril

These are the structures associated with the integuments. Several authors have used these terms vaguely and sometimes indiscriminately. Maheshwari (1950) defined the aril as a new structure that arises from the base of the ovule and forms a sort of third integument. It is a secondary collar like out growth which develops from the funiculus and surrounds the ovule almost completely in fertilization stages.

The presence of an aril is considered as a primitive feature. Basing on this Corner (1949, 1953, 1954) proposed the 'Durian Theory' which states that all the arillate families are primitive and others are advanced. But Pijl (1952) and Parkin (1953) strongly criticised this theory and according to them, the presence of an aril has no phylogenetic significance.

Raju (1956) observed aril in the members of Passifloraceae where it is made up of thin-walled cells containing oil, starch and yellow-red chloroplasts which give it a brownish-pink coloration. In *Crossosoma californicum,* the aril develops numerous finger like projections and gives a fimbriate appearance. The aril is highly vascularised in *Myristica fragrans.*

Arillode

Pijl (1957) recorded presence of Arillode in a number of Sapindaceous taxa. An outgrowth arising from the tip of the outer integument grows downward and covers the seed either completely or in part. This can also be regarded as an excessively developed caruncle. Kadry (1946) described the development of an aril in *Cardiospermum halicacabum*, but Pijl (1957) considered it as an arillode.

Sarcotesta

The outer integument, sometimes, becomes juicy or fleshy and it is known as Sarcostesta. It is often pigmented and rich in fats in the members of the Magnoliaceae (Pijl, 1955). Examples of seeds with a sarcotesta are pomegranate (*Punica granatum*) seeds. The sarcotesta of pomegranate seeds consists of epidermal cells derived from the integument.

These structures namely, aril, arillode and sarcotesta are attractive to animals like birds and, mammals and help in the dispersal of the seeds.

Endothelium

In several taxa of the Gamopetalae, where mostly unitegmic and tenuinucellate ovules are seen, cells of the inner epidermis of the integument elongate radially, acquire dense cytoplasm, show prominent nuclei and differentiate into endothelium, otherwise known as integumentary tapetum (Fig. 5.3). In most of the cases endothelium is differentiated at the time of megaspore tetrad formation (Fig. 5.3A) or just after the tetrad is formed. As the ovule grows the nucellar epidermis degenerates and the embryo sac comes in direct contact with the endothelium. Kapil and Tiwari (1978) observed the endothelium in 65 families of dicots and in 47 families it is an important identification character.

The endothelium is generally single-layered and the cells are uninucleate. Multiseriate and multinucleate endothelium has been reported in many members of Asteraceae. In *Tithonia rotundifolia* (Pullaiah, 1978) endothelium becomes multilayed at the chalazal region and the cells become occasionally multinucleate (Fig. 5.3B,C). In *Zinnia angustifolia* the endothelium becomes biseriate throughout the length of the embryo sac (Fig. 5.3D). In *Lagascea mollis* the endothelium becomes multiseriate in the middle region of the embryo sac (Fig. 5.3E). In sunflower, Savchenko (1960) reported 10-12 layered endothelium which is in actively dividing condition. Since frequent degeneration of the embryo sacs is seen in such ovules where multilayered endothelium is present, he concluded that the nutrient substances entering into the ovule were not utilized by the degenerating embryo sacs, but accumulated in the endothetial cells and stimulated them to grow. In the Solanaceae (*Datura*), the endothelial cells digest and absorb the integument present outside it, until finally the endothelium itself is absorbed by the developing endosperm.

In *Pedicularis*, the endothelial cells become polyploid (32n). Multinucleate condition of the endothelium is also seen in *Balanites*. The endothelium normally surrounds the entire embryo sac or sometimes, it may be restricted either to the micropylar or chalazal end of the embryo sac. It usually degenerates after fertilization, but in some families it persists up to the seed stage.

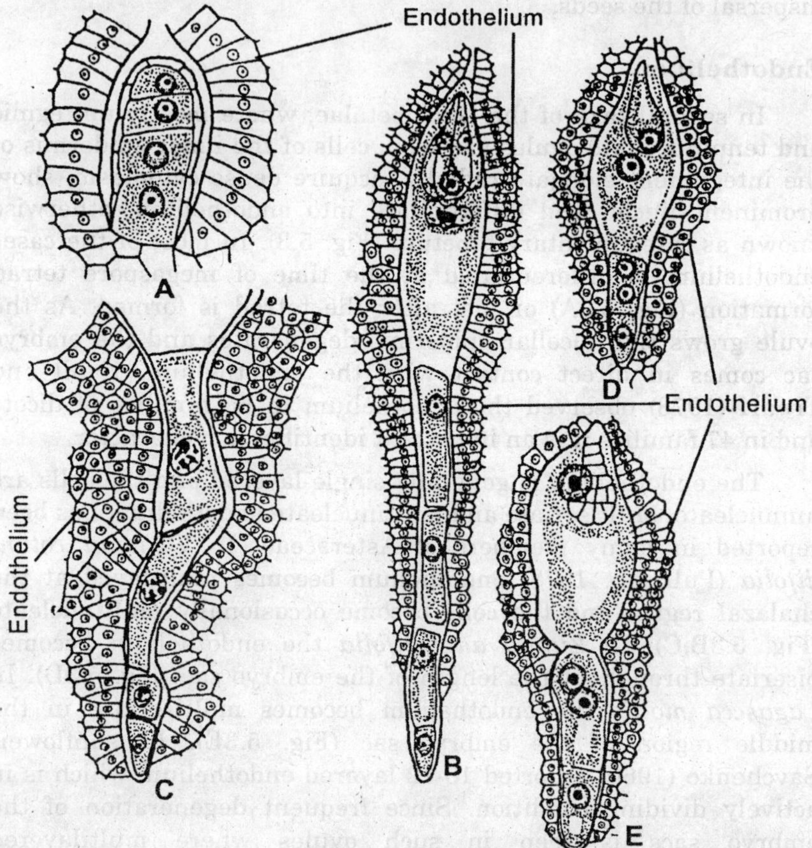

Fig. 5.3. Endothelium. **A.** Linear megaspore tetrad and integumentary tapetum; **B.** Mature embryo sac with multinucleate endothelial cells; **C.** Chalazal portion of embryo sac invested by multilayered, endothelium, antipodal cells are polyploid; **A-C.** *Tithonia rotundifolia*; **D.** Two layered endothelium in *Zinnia angustifolia*; **E.** Two-layered integumentary tapetum in the middle region in *Lagascea mollis* (after Pullaiah, 1978, 1981).

Ultrastructural studies of the endothelium have indicated that it is similar to the secretory or meristematic cells. They are rich in RNA, proteins, ascorbic acid and enzymes. The cells are interconnected and also have plasmodesmatal connections with the other cells of the integument. Callose is found on the endothelial walls of *Petunia* and this may be helpful in the isolation of the embryo sac from the rest of the sporophyte tissue during fertilization. The excessive growth and proliferation of the

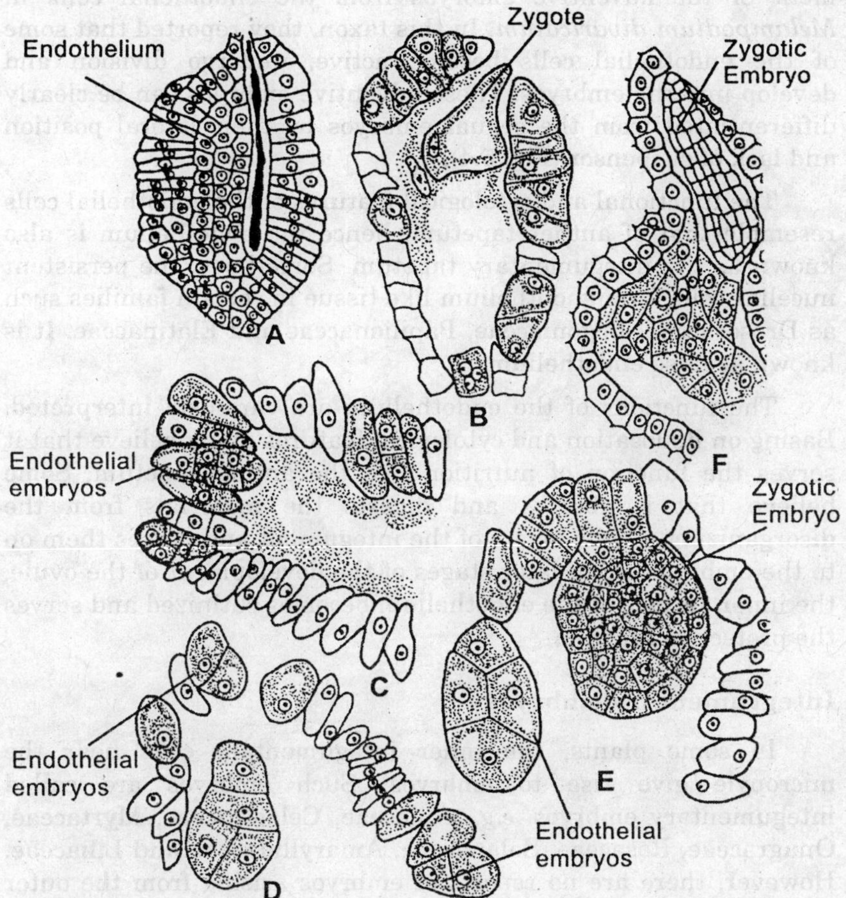

Fig. 5.4. Endothelial embryony in *Melampodium divaricatum.* **A.** Proliferating integrumentary tapetum and degenerating embryo sac; **B.** Embryo sac with degenerated egg cell and secondary nucleus, endothelium shows adventive proembryos at different stages of development; **C-E.** Stages in the development of adventive embryos from endothelial cells; **F.** Upper portion of embryo sac with adventive embryo (after Maheswari Devi and Pullaiah, 1976).

endothelium may sometimes results in the absorbtion of the embryo in the incompatible crosses of *Datura, Solanam, Lycopersicon* and *Nicotiana*.

Maheswari Devi and Pullaiah (1976) reported the development of the adventive embryos from the endothelial cells in *Melampodium divaricatum*. In this taxon, they reported that some of the endothelial cells become active, undergo division and develop into the embryos. These adventive embryos can be clearly differentiated from the sexual embryos by their lateral position and lack of suspensor (Fig. 5.4).

The functional and cytological features of the endothelial cells resemble that of anther tapetum, hence the endothelium is also known as the integumentary tapetum. Sometimes, the persistent nucellar cells form endothelium like tissue in certain families such as Droseraceae, Begoniaceae, Papilionaceae and Elatinaceae. It is known as false endothelium.

The functions of the endothelium are variously interpreted. Basing on its location and cytological features many believe that it serves the function of nutrition like the anther tapetum. Some believe that it absorbs and digests the materials from the disorganizing middle layers of the integument and passes them on to the embryo sac. In later stages of the development of the ovule, the inner surface of the endothelium becomes cutinized and serves the protective function.

Integrumentary embryos

In some plants, the inner integumentary cells near the micropyle give rise to embryos. Such embryos are called integumentary embryos. *e.g.*, Apiaceae, Celastraceae, Myrtaceae, Onagraceae, Rosaceae, Solanaceae, Amaryllidaceae and Liliaceae. However, there are no reports of embryos arising from the outer integument so far.

Operculum

Hegelmaier (1868) recorded the presence of a stopper like structure on the micropylar end of the seed of *Lemna* and named it as the operculum. Caldwell (1899) believed that it originated partly from the inner end partly from the outer integument. But according to S.C. Maheshwari (1954, 1956), it is formed only from the persistent apical cells of the inner integument (Fig. 5.5A). The cells of the operculum are thick walled and show an orange coloured pigment in them. In *Lemna* and *Spirodela,* the operculum is enclosed by the outer integument. Operculum also is seen in families like Araceae and Droseraceae.

Caruncle or Strophiole

The caruncle is formed by the enlargement and proliferation of the cells of the outer integument (Fig. 5.5B). It's cells are initially thin walled, but later becomes thick walled. Caruncle is characteristic and diagnostic feature of the Euphorbiaceae. In *Acalypha,* the caruncle is found only from the seeds developed from the lateral flowers of the spike, but not from the terminal flowers. Caruncle helps in the dispersal of seed.

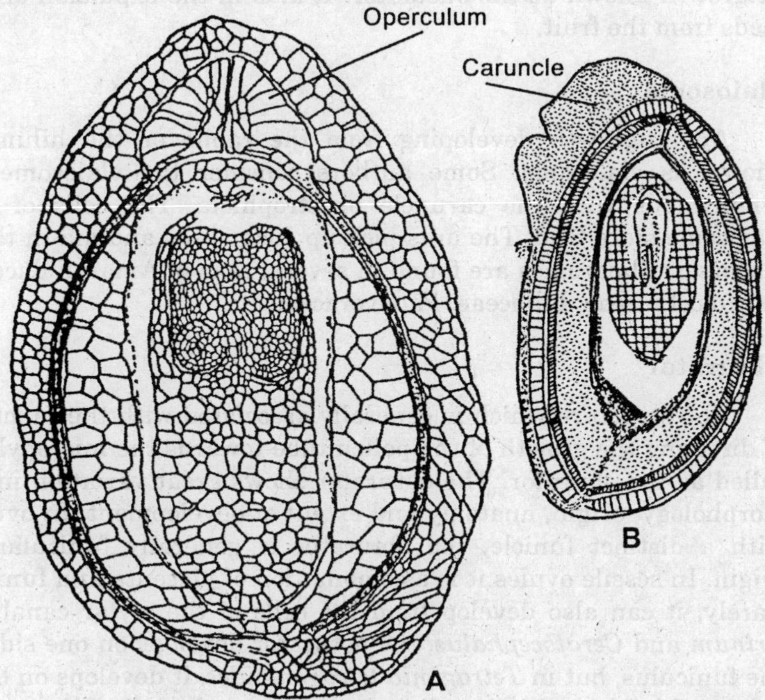

Fig. 5.5. A. Operculum in *Lemna* (after S.C. Maheshwari, 1956); **B.** Caruncle in *Acalypha* (after Johri & Kapil, 1953).

Coma

In Asclepiadaceae members, a tuft of hairs or coma is seen originating from the integumentary tissue near the funicle. The detailed development of coma is studied in *Asclepias syriaca, Caralluma attenuata, Oxystelma esculentum* (Maheswari Devi and Lakshminarayama, 1978). These hairs develop from the epidermal cells of the outer integument. The hairs may grow up to the length of 4-5 cms in *Caralluma*. Coma is also seen *Tamarix*. Seed hairs of unusual length occur in *Gossypium*. In some varieties of cotton, the staple length is influenced by the pollen.

Jaculator

In majority of the members of the Acanthaceae, some of the funicular cells elongate and differentiate into stiff, hook-like outgrowth known as the Jaculator. It aids in the expulsion of the seeds from the fruit.

Elaiosome

An outgrowth developing from the raphe or the hilum is known as elaiosome. Some authors consider the elaiosome as synonymous with the caruncle or strophiole. The cells of the elaiosome store food. The ants pick up such seeds and aid in their dispersal. Elaiosomes are found in several taxa of Amaryllidaceae, Aristolochiaceae, Liliaceae, Papaveraceae etc.

Obturator

The structure which is generally associated with the function of directing the growth of the pollen tube towards the micropyle is called as the oburator. The obturator shows great variation in its morphology, origin, anatomy and extent of development. In ovules with a distinct funicle, the obturator is generally funicular in origin. In sessile ovules it is a combination of placenta and funicle. Rarely, it can also develop from the cells of the stylar canal. In *Crinum* and *Ceratocephalus*, the obturator develops on one side of the funiculus, but in *Tetragonia tetragonioides*, it develops on both sides and is better developed on the side away from the micropyle.

In Cuscutaceae and Euphorbiaceae, the obturator develops from the placenta (Fig. 5.6). In genera like *Chrozophora*, it is composed of small, thin walled, compactly arranged cells, while in *Euphorbia* and *Acalypha*, the cells are loose, elongated and hair like. In *Aegle*, the obturator is partly funicular and partly placental in origin.

The obturator is stylar in origin in Thymeleaceae. The cells of the stylar canal become papillate, filled with starch and grow up to the micropyle. Very rarely, it develops from the wall of the ovary as in *Rivinia*. In *Myriocarpa*, the obturator develops from the integument. In *Rhus* and *Elatostema*, it develops from the inner epidermis of the ovary wall.

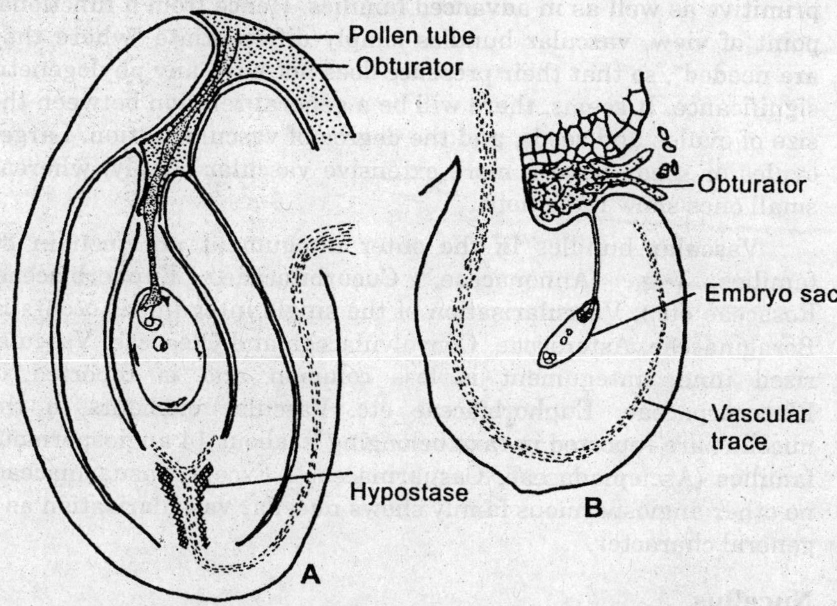

Fig. 5.6. Placental obturator. **A.** *Acalypha indica* (after Johri & Kapil, 1953); **B.** *Cuscuta reflexa* (after Johri & Tiagi, 1952).

The cells of the obturator are densely cytoplasmic and are rich in endoplasmic reticulum, dictyosomes, vesicles and other organellels. These cells produce a surface exudate and provide nutrition, mechanical and chemical guidance to the pollen tube. The obturator is devoid of any vascular tissue and the cells shrink and disappear after fertilization.

Vascular Supply of the Ovule

Most of the ovules show a very simple type of vascularization. The funicular vascular supply generally terminates in the chalaza, but sometimes enters into the nucellus or integuments.

Vascular bundles are reported in both the integuments in *Cremophyllum* or only in the inner integument in *Ricinus*. In large number of families, the outer or single integument becomes vascularised. The presence of integumentary vascular bundles was considered to be a primitive feature. But they were seen in primitive as well as in advanced families. Hence from a functional point of view, vascular bundles simply differentiate "where they are needed", so that their presence does not have any phylogenetic significance. It seems, there will be a general relation between the size of ovules and seeds, and the degree of vascularisation. Larger ovules or seeds have a more extensive vascular supply, whereas small ones show reductions.

Vascular bundles in the outer integument are seen in 80 families (*e.g.*, Annonaceae, Cucurbitaceae, Euphorbiaceae, Rosaceae etc.). Vascularisation of the single integument occurs in Boraginaceae, Asteraceae, Convolvulaceae and Oleaceae. Vascularized inner integument is less common and is reported in Elaeocarpaceae, Euphorbiaceae etc. Vascular elements in the nucellus are reported in taxa belonging to about 14 angiospermous families (Asclepiadaceae, Casuarinaceae). Except Casuarinaceae, no other angiospermous family shows nucellar vascularisation as a general character.

Nucellus

The nucellus develops from the apex of the ovular primordium and it varies in quality, in its extent and dimension. The ovule has only one nucellus, but rarely as in *Aegle marmelos*, twin nucelli occur enclosed by the common fold of integuments. This type of abnormality has no significance except in the formation of False polyembryony. In the majority of the sympetalae members, the nucellus is represented by a single layer of cells, namely nucellar epidermis and it encloses the megaspore mother cell. Such ovules where the sporogenous cell (or MMC) is hypodermal in origin and the nucellar epidermis around it remains single layered is termed as tenuinucellate.

Sometimes, the archesporial cell undergoes a transverse division and cuts off an outer parietal cell and an inner sporogenous cell. The former by further divisions gives rise to the parietal tissue by means of which the sporogenous cell becomes deep seated in the massive nucellus. Such ovules where the sporogenous cell becomes deeply embedded are known as Crassinucellate ovules. In a few cases, as in *Anemone*, the sporogenous cell becomes embedded in the nucellar tissue, not by the formation of parietal tissue, but the divisions of the nucellar epidermis; such ovules are termed as pseudo crassinucellate ovules (Davis, 1969). Of all these, the tenuinucellate ovule is considered to be more advanced than the others.

Davis (1966) gave the distribution of these three conditions in various families of the angiosperms. Crassinucellate ovules are seen in 179 families, tenuinucellate ovules in 105 families and pseudocrassinuclellate ovules in 11 families. In other families, the type of ovule, varies with the taxa and the nucellus forms only a specific or generic character.

The ultrastructure of the nucellus is studied in *Ornithogalum* in which the cells are lignified, columnar, richly cytoplasmic and consists of large amounts of RNA, smooth endoplasmic reticulum and microtubules with swollen cisternae.

A prominent nucellar cap is seen in the members of the Lemnaceae due to divisions in the nucellus. In *Trapa, Euphorbia, Chrozophora, Polygonum* and certain taxa of the Cucurbitaceae, a prominent nucellar beak is formed and projects into and sometimes beyond the micropyle.

In the majority of angiosperms, the nucellus is consumed during the development of the embryo sac and the endosperm. But in some plants like *Piper nigrum*, members of Amaranthaceae, Aizoaceae, Chenopodiaceae, it remains persistent and takes up the function of storage, in such case it is referred as the perisperm. In some angiosperms, the nuclei of the nucellus wander into the embryo sac and incorporated into it *e.g.*, *Cocos nucifera* and *Tamarix*.

Pseudo-embryo sac

In members of the family Podostemaceae like *Dicraea*, the

nuclellar cells lying below the megaspore mother cells break down and form a large cavity called the pseudo-embryo sac. The pseudo-embryo sac is formed after fertilisation in taxa like *Terniola* and *Indotristicha*. The pseudo-embryo sac contains both cytoplasm and nuclei, and nourishes the growing embryo in the absence of the usual endosperm in this family (Fig. 5.7).

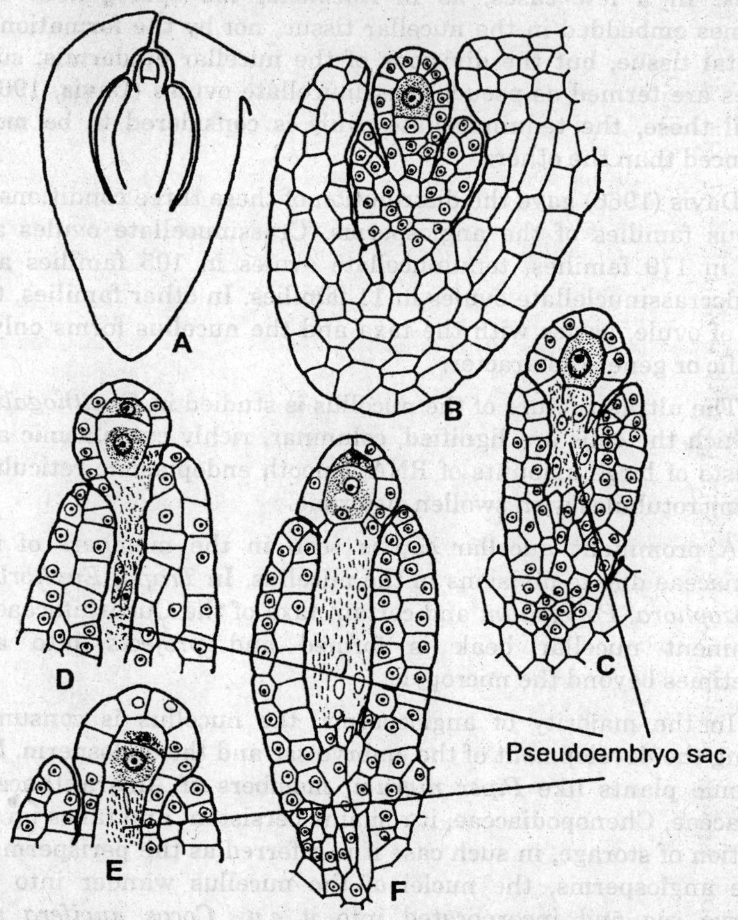

Fig. 5.7. Ovule and pseudoembryo sac in *Podostemon ceratophyllum* (after Hammond 1937). **A.** L.S. Ovule, diagrammatic; **B.** L.S. Young ovule showing archesporial cell; **C.** Older state showing formation of Pseudoembryo sac by disintegration of nucellar cells lying just below the megaspore mother cell; **D.** Formation of dyad cells; **E.** Degeneration of upper dyad cell; **F.** Two nucleate embryo sac.

Hypostase and Epistase

Just at the level of origin of the integuments and directly below the embryo sac a group of nucellar cells have partially lignified or suberised walls. Van Tieghem (1901) was the first to identify this tissue and named it as hypostase. This hypostase is situated above the chalazal vasculature. The hypostase shows different structural features in different taxa. The cells show dense cytoplasm and sometimes accumulate tannin like substances. They are often thick walled and impregnated with cutin, suberin, lignin or callose. Histochemical studies have indicated the presence of starch, proteins and lipids in the cells of hypostase. It is found in many families, *e.g.*, Amaryllidaceae, Cyperaceae, Liliaceae, Anacardiaceae, Euporbiaceae, Sapindaceae etc.

The hypostase is organised after fertilization in the members of Elaegnaceae, on the other hand it occurs at the archesporial cell stage itself in the Loranthaceae. Sometimes as in *Aristolochia*, it persists up to the mature seed stage. Several functions have been attributed to the hypostase: 1. It may act as a barrier tissue and prevents the encroachment of the embryo sac (Van Tieghem, 1901). 2. It connects the chalazal vascular supply with the embryo sac and helps in the transport of nutritional materials (Venkata Rao, 1953). 3. It may produce certain hormones or enzymes which play an important role in the water balance of the dormant seeds. 4. It may take up a protective function in the mature seeds. Thus the hypostase shows a variable structure and is multifunctional. The 'Podium' (Dahlgren, 1940) and 'Postament' (Westermaier, 1890) are structures more or less similar to the hypostase and are derived from the chalazal nucellar tissue.

At the micropylar end of the embryo sac, sometimes, the nucellar remnants persist and function as the epistase. This is derived from the nucellar epidermis or its derivatives which become cutinized and persists during the development of the embryo and plugs the micropyle. It occurs only a few families like Zingiberaceae, Nymphaeaceae and in some taxa of the order Myrtales.

MEGASPOROGENESIS

Archesporium is hypodermal in origin. A cell of the nucellus situated below the epidermis, becomes more conspicuous by its larger size, dense cytoplasm and prominent nucleus and functions as the archesporium. The archesporial cell may divide to form primary parietal cell and a primary sporogenous cell (crassinucellate ovules) or it may directly function as the megaspore mother cell (tenuinucellate ovule). The primary parietal cell may remain undivided or undergoes further divisions and forms a variable number of wall layers. The primary sporogenous cell generally functions as the megaspore mother cell without any further divisions.

The archesporium, most commonly, is single-celled. In certain families like Onagraceae, Paeoniaceae and Crossosomataceae, the archesporium is multicelled, but only one of them is functional. Multiple archesporial cells of *Sedum* function as multiple megaspore mother cells and after meiosis results in the formation of multiple megaspore tetrads. Similar condition is also seen in members of Oleaceae (Maheswari Devi and Lakshminarayana, 1978).

The megaspore mother cell undergoes usual meiotic divisions and forms the tetrad. The first division is always transverse and results in the formation of a dyad. Generally, the second division is also transverse and results in the formation of a linear tetrad of four megaspores. Sometimes, vertical division in the micropylar dyad and transverse division in the chalazal dyad occurs leading to the formation of a T-shaped tetrad as in *Orchis, Rheum, Rumex* etc. Isobilateral and Tetrahedral megaspore tetrads are very rare. Sometimes, four different kinds of tetrads namely - Linear, T-shaped, inverted T-shaped and isobilateral types are reported in the genus *Musa*. Frequently a row of only three cells (triad) is seen instead of four, due to omission of the second meiotic division in one of the dyad cells. Such triads are reported from Amaranthaceae, Cactaceae, Orchidaceae etc.

Functioning Megaspore

Generally, the chalazal megaspore of the tetrad functions and develops into an embryo sac (Fig. 6.3). But in Onagraceae and few other taxa, the embryo sac is formed by the micropylar megaspore. In *Aristotelia* (Elaeocarpaceae) the third megaspore from the micropylar end is functional. Rarely in *Gloriosa, Poa* and *Casuarina*, all or any of the four megaspores is functional and begin to enlarge and divide.

Megaspore Haustoria

An uncommon and notable phenomenon is the formation of megaspore haustoria by the megaspores in some taxa of Crassulaceae, Rubiaceae, Rosaceae etc. In *Rosularia, Sedum* (Fig. 5.8A-C), and *Potentilla*, the megaspores give out lateral tubes which grow upwards and function as haustoria. In *Putoria* and *Galium*, they enter into the tissues of the integument and draw nourishment to the megaspores (Fig. 5.8D). In *Rosularia*, two to three megaspore tetrads are formed, and as every megaspore give rise to a haustorium, the upper part of the nucellus shows a tangle of haustorial processes competing with one another (Fig. 5.8 E). One of them penetrates through the nucellar epidermis and forms a vesicle near the base of the micropyle. The megaspore from which the tube arises alone forms the embryo sac.

The knowledge regarding the physiological aspects of the megasporogenesis is very little when compared to the entire process of microsporogenesis. This is due to the limited number of Megaspore mother cells and their deep seated nature.

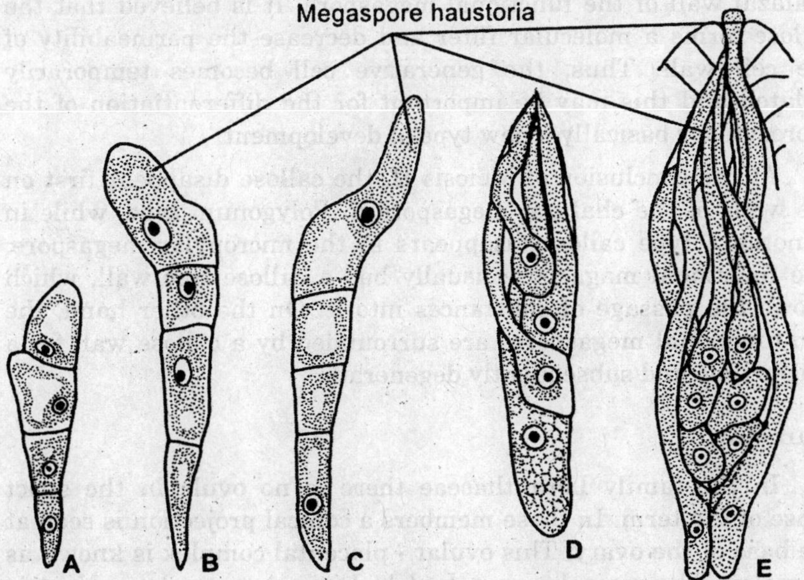

Fig. 5.8. Megaspore haustoria. **A-C.** *Galium lucidum* (after Fagerlind, 1937); **D.** *Sedum sempervivoides;* **E.** *Rosularia pallida* (after Mauritzon, 1933).

Callose wall formation during Megasporogenesis

The knowledge that is gained through the use of the recent advanced techniques, such as fluorescence microscopy and electron microscopy has helped for a more detailed study of megasporogenesis. Rodkiewicz (1967) for the first time demonstrated the formation of callose during megasporogenesis in angiosperms. Of the 43 species examined by Rodkiewicz (1970), 40 species (39 with monosporic embryo sac and 1 with bisporic embryo sac) showed the appearance of callose. The remaining 3 species, which bear the tetrasporic embryo sac, did not show this feature. The callose formation starts during the prophase of the meiosis. It is first deposited in the chalazal side in the Polygonum type of embryo sac. The entire meiocyte is surrounded by a callose containing wall by early metaphase. The cross walls formed after the meiotic divisions are also rich in callose. Actually the callose is abundant in the cross walls than the side walls. Finally, the callose formation starts decreasing from the chalazal end of the functional megaspore. In orchids, the callose persists for a long time in the chalazal wall of the functional megaspore. It is believed that the callose forms a molecular filter and decrease the permeability of the cell wall. Thus, the generative cell becomes temporarily isolated and this may be important for the differentiation of the sporocyte for basically a new type of development.

At the conclusion of Meiosis II, the callose disappear first on the walls of the chalazal megaspore in Polygonum type, while in Oenothera type callose disappears in the micropylar megaspore. The functional magaspore usually has a callose free wall, which allows the passage of substances into it. On the other hand, the non-functional megaspores are surrounded by a callose wall for a longer time and subsequently degenerate.

Mamelon

In the family Loranthaceae there is no ovule in the strict sense of the term. In these members a conical projection is seen at the base of the ovary. This ovular - placental complex is known as mamelon. The mamelon may be lobed in genera such as *Nuytsia* and *Lysiana* and unlobed in *Helicanthes.* In the lobed mamelon the archesporium differentiates hypodermally in each lobe, whereas in the unlobed placenta the entire subepidermal tissue functions as the achesporium. The genera *Dendrophthoe* and *Helixanthera* show even more obscure condition, in that there is no trace of any projection at the base of the ovary. The archesporium differentiates from a compact tissue in the hypodermis of the ovarian cavity. This archesporium directly functions as the megaspore mother cells and latter develops into the female gametophyte. The reduction of mamelon is diagramatically represented in Fig. 5.9.

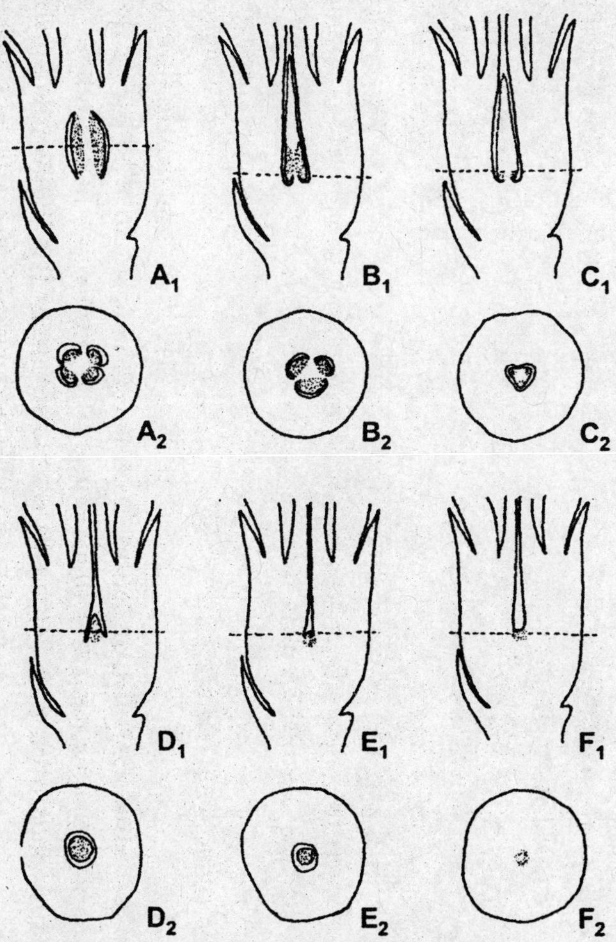

Fig. 5.9. L.S. and T.S. of ovary showing the reduction of the mamelon in the Loranthaceae. **A.** *Lysiana*; **B.** *Nuytsia*; **C.** *Macrosolen*; **D.** *Helicanthes*; **E.** *Amyema*; **F.** *Helixanthera* (modified after Maheshwari *et al.*, 1957).

Reference

Kapil, R.N. and S.C. Tiwai 1978. The integumentary tapetum. *Bot. Rev.* 44: 457-490.

CHAPTER – VI

THE FEMALE GAMETOPHYTE

The ovule is the source of the female gametophyte and the progenitor of the seed. Differentiation of the megasporocyte, production of a functional megaspore (megasporogenesis), formation of the embryo sac (megagametogenesis) and embryogenesis, all occur within the ovule. The functional megaspore, after some mitotic divisions give rise to a small sac like, coenocytic structure which later organizes into the female gametophyte. In general, the embryo sac or megagametophyte is a 7-celled and 8-nucleate structure and consists of an egg apparatus, a central cell and antipodal cells (Fig. 6.1). The female gametophyte is a highly specialized structure and shows polarity from micropyle to chalaza.

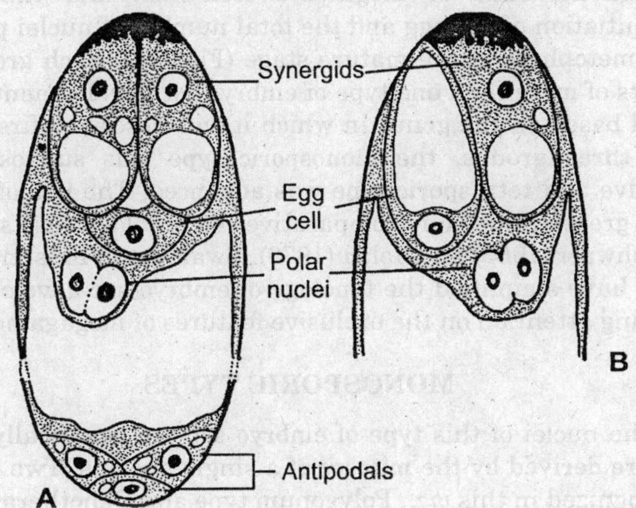

Fig. 6.1. Organised embryo sac - diagrammatic representation (after Willemse and Van Went, 1984).

During the development of the female gametophyte, the nutrients required for its growth and development are supplied by various ovular and other maternal tissues. The growth of the female gametophyte ceases when it is fully differentiated. Further development of the embryo sac can only be seen after the arrival of the pollen tube on the stigma. The process of megasporogenesis and the formation of the megagametophyte is genetically determined. To some extent it can also be influenced by environment factors. The induction and probably the duration of the process show the genetic base. Nutrition and temperature are influences from outside which act on the sequence of embryo sac development.

Types of Embryo Sacs

Depending on the number of megaspores or megaspore nuclei taking part in the development of the embryo sac, three main groups of embryo sac development can be discernible in the angiosperms - monosporic, bisporic and tetrasporic type. In the first, only one of the four megaspores take part in the development of the gametophyte, in the second two megaspore nuclei take part in its formation and in the third all four of them. Further subdivision is based on the number of nuclear divisions intervening between the time of megaspore formation and the time of differentiation of the egg and the total number of nuclei present in the gametophyte at the mature stage (Fig. 6.2). Each group again consists of more than one type of embryo sac development and was named based on the genus in which it was described first. Among these three groups, the monosporic type was supposed to be primitive and tetrasporic type was advanced. The use of typology is of great value for comparative embryology as shown by Maheshwari (1963) and Johri (1970). Swamy and Krishna Murthy (1975) have simplified the typology of embryo sac development by focussing attention on the exclusive features of megagametophyte.

MONOSPORIC TYPES

The nuclei of this type of embryo sac are genetically alike as they are derived by the mitoses of a single nucleus. Two types can be recognized in this *viz.*, Polygonum type and Oenothera type.

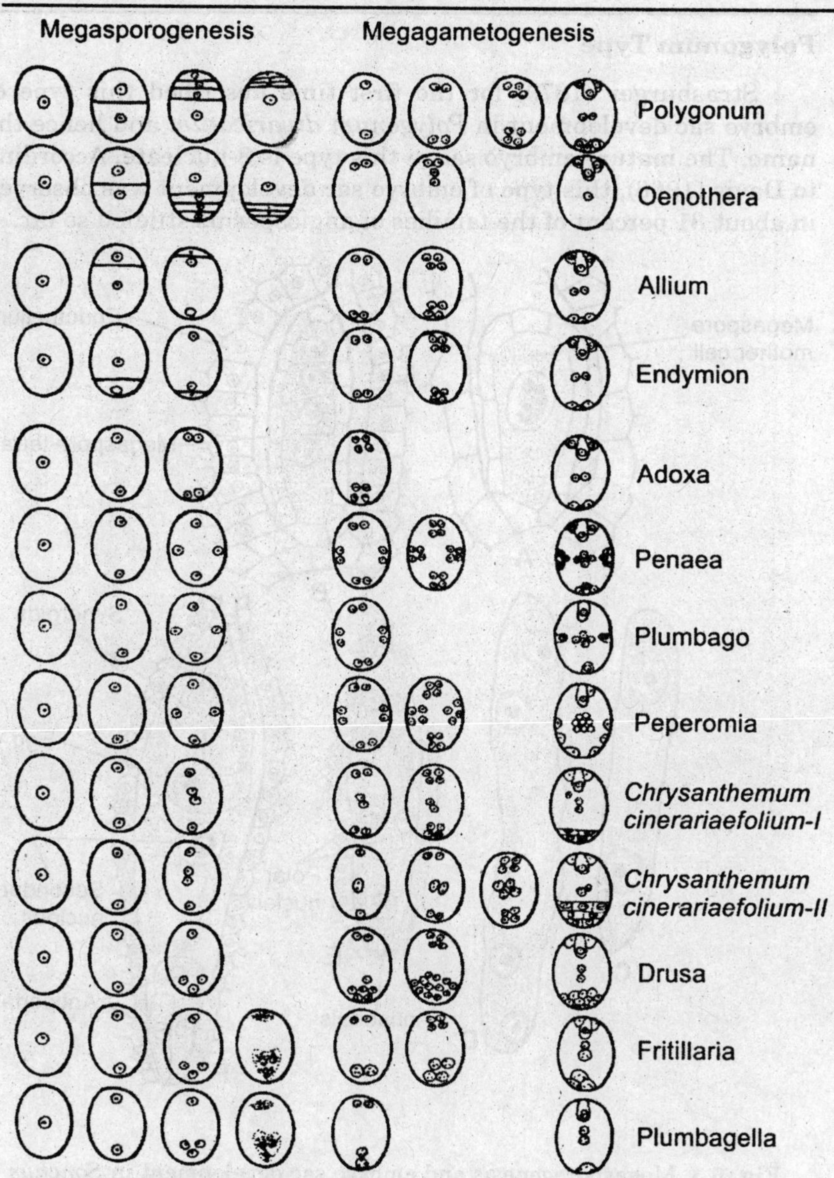

Fig. 6.2. Schematic representation to show origin and development of different types of embryo sac. [modified after Johri (1963) and Erdelska (1981)].

Polygonum Type

Strasburger (1879) for the first time described this type of embryo sac development in *Polygonum divaricatum* and hence the name. The mature embryo sac in this type is 8-nucleate. According to Davis (1966), this type of embryo sac development was observed in about 81 percent of the families of angiosperms studied so far.

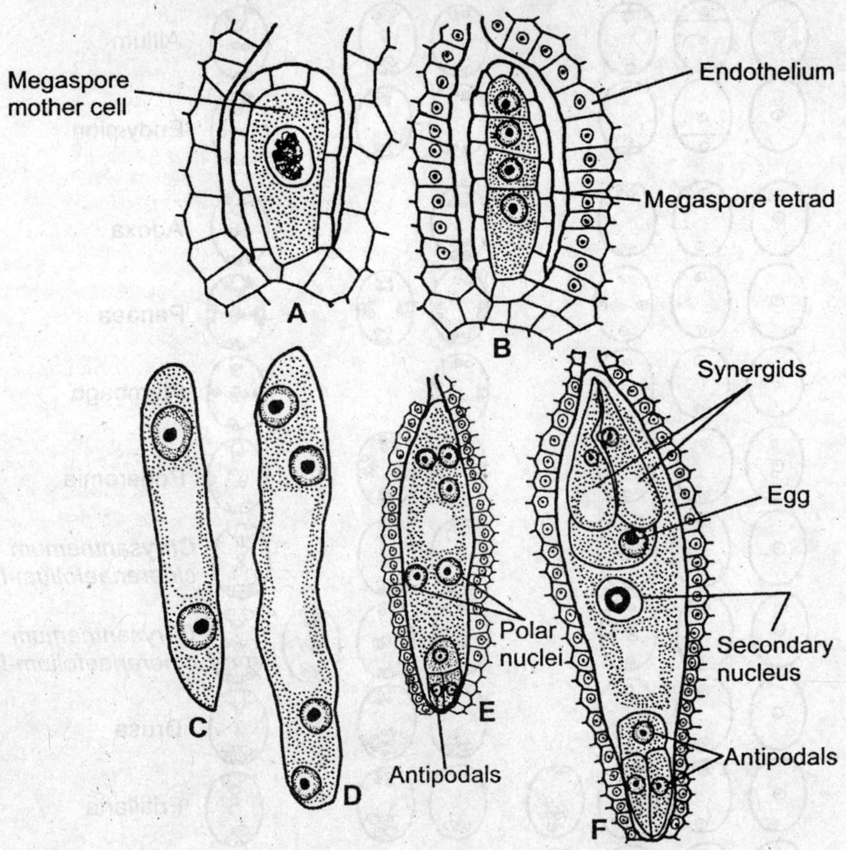

Fig. 6.3. Megasporogenesis and embryo sac development in *Sonchus oleraceus*. **A.** Megaspore mother cell; **B.** Linear megaspore tetrad; **C.** 2-nucleate embryo sac; **D.** 4-nucleate embryo sac; **E.** 8-nucleate embryo sac; **F.** Organised embryo sac (after Pullaiah 1982).

The development of embryo sac begins with the elongation of the functional megaspore which is usually the chalazal one of the tetrad. The elongation is largely in the micropylar-chalazal axis. Initially, the megaspore cytoplasm is non-vacuolate, but later smaller vacuoles appear which may fuse together to form a large vacuole. The spindle of the first nuclear division in the megaspore is oriented along the long axis of the cell. Wall formation does not follow the nuclear division. A large vacuole appears between the two daughter nuclei as a result of which the two daughter nuclei (the primary micropylar and primary chalazal nuclei) are pushed towards the opposite poles of the cell (Fig. 6.3C). Most of the cytoplasm is aggregated around them and the rest forms a thin peripheral layer. The second division produces one pair of nuclei at each pole (Fig. 6.3D) and the third division results in two groups of 4 nuclei at the opposite poles of the elongated embryo sac (Fig. 6.3E). Of the 8 nuclei arising in this manner, the micropylar quartet differentiates into 3-celled egg apparatus and the upper polar nucleus and the chalazal quartet into a group of three antipodals and the lower polar nucleus. The two polar nuclei fuse to give rise to a secondary nucleus (Fig. 6.3 E,F).

Occasionally, embryo sacs are found with less than the normal quota of 8 nuclei. This is usually because of early degeneration of antipodals or non-functioning of some of the nuclei of 2-4 nucleate stages.

The reverse condition, *i.e.*, the occurrence of more than 8 nuclei in the embryo sac is less frequent and may arise either due to fusion of two or more embryo sacs or migration of nucellar cells into the embryo sac or occurrence of secondary divisions of some of the first formed 8 nuclei.

Oenothera Type

The mature embryo sac in this type is only 4-nucleate. Geerts (1908) reported this, for the first time in *Oenothera lamarckiana*. This type of embryo sac is usually formed by the micropylar megaspore of the tetrad which undergoes only two nuclear divisions, instead of the usual three occurring in the Polygonum type. The organization of the mature embryo sac is, an egg apparatus and an uninucleate central cell. Since there is no third division and all the nuclei are situated in the micropylar part of the developing embryo sac, there is neither a lower polar nucleus nor any antipodal cells. This type of embryo sac development is found to be a characteristic and constant feature of the entire family Onagraceae. Outside the family, the Oenothera type occurs only in *Schisandra chinensis. S. chinensis* is slightly different in that the functional megaspore is the chalazal one and not the micropylar as in the Onagraceae members (Yoshida, 1962; Swamy, 1962).

Another interesting feature in the Oenothera type of embryo sac is the concurrent growth of more than one megaspore of the tetrad. However, eventually the micropylar megaspore functions, but sometimes it may be chalazal and occasionally both grow simultaneously forming 'twin' embryo sacs.

Rarely, more than 4 nuclei may be seen in an embryo sac. This condition results, either from the incorporation of an adjacent megaspore and its contents or divisions in the original 4 nuclei of the embryo sac. Embryo sacs with fewer than 4 nuclei are rare. In *Hartmannia tetraptera* (Johansen, 1929) and *Jussiaea repens* (Khan, 1942), two 3-nucleate embryo sacs having a single synergid, an egg cell and a polar nucleus have been reported. Their origin can be explained by the lack of division in the primary synergid nucleus of the 2-nucleate stage.

BISPORIC EMBRYO SACS

In this type of embryo sac development, the first meiotic division is accompanied by wall formation, so that a dyad is formed. Only one of the dyads undergoes the second meiotic division and the other degenerates. In the functional dyad, cell wall formation does not occur after meiosis-II and both the resulting megaspore nuclei contribute to the formation of the embryo sac. Each megaspore nucleus undergoes two mitotic divisions forming 8 nucleate embryo sac and the final organization is similar to that of the Polygonum type.

Normally, segregation occurs during meiosis and the four resulting nuclei are genetically different. Since a bisporic embryo sac is derived from two megaspore nuclei whose genetic constitution is different, four nuclei are of one type and the other four of different type. Bisporic embryo sacs are of two main types: Allium type and Endymion type.

(a) Allium type

This type of embryo sac development was described in *Allium fistulosum* by Strasburger (1879) and was later confirmed in several species of the genus *Allium*. The megaspore mother cell undergoes meiotic division I resulting in a dyad. Of these two dyad cells the chalazal becomes functional, while the micropylar degenerates. The nucleus of the functional dyad cell undergoes meiotic division II forming two nuclei. These two nuclei move to the opposite poles on the long axis and are separated by a vacuole. These two nuclei divide to form 4 and then 8 nuclei. They organize into 3-celled egg apparatus, three antipodal cells, while the remaining two nuclei function as polar nuclei. This type of embryo sac is seen in families Butomaceae, Alismataceae and Viscaceae. It is also seen in members of Liliaceae, Amaryllidaceae and Orchidaceae.

(b) Endymion type

This type was first described in *Endymion hispanicus* (= *Scilla hispanica*). In this type megaspore mother cell undergoes meiosis I forming a dyad. The micropylar dyad cell is functional, while the chalazal dyad cell degenerates. Its nucleus divides to form 2, 4 and finally 8-nucleate embryo sac comprising egg, two synergids, two polars and three antipodals.

Variations

A different type of development was observed in *Ammobium alatum* of family Compositae by Davis (1962). Here the first meiotic division is not followed by wall formation and the dyad cell is represented by a single binucleate cell. Both the nuclei undergo the second meiotic division simultaneously and the daughter nuclei take up a 2 + 2 position at telophase and their persistent spindles form a joint structure stretching from one pair of nuclei to the other. Cell plate formation is initiated across the equator of the united spindles which then extends to the lateral walls, so that each cell of the dyad is binucleate. Further development follows Allium type described above.

In *Tridax trilobata* (Hjelmquist, 1951), embryo sac development may follow either monosporic or bisporic type. In this species the megaspore mother cell undergoes the first meiotic division resulting in two nuclei which is not followed by cell division. The two nuclei then undergo second meiotic division and this is followed by the formation of two cell plates in such a fashion that the central cell contain two nuclei and the other two cells contain a single nucleus each. If the embryo sac develops from the central cell it is of the bisporic type. While, if the embryo sac develops from the basal megaspore then the development of embryo sac is of the Polygonum type. Both these conditions occur in *Tridax trilobata* (Hjelmquist, 1951).

The chief variation in development is a tendency towards reduction in the number of nuclei at the chalzal end. This has been clearly demonstrated in the Alismataceae, Butomaceae, Podostemaceae and some members of the Orchidaceae. In *Machaerocarpus californicus* (Maheshwari and Singh, 1943) the development normally proceeds up to 4 nucleate stage. After this only the 2 micropylar nuclei divide again resulting in a 6 nucleate stage comprising an egg apparatus, two polar nuclei and a single antipodal nucleus. In some plants, the reduction has gone still further and only 5 nuclei are formed, four at the upper end and the undivided primary chalazal nucleus at the lower. The mature embryo sac comprises an egg apparatus, an upper polar nucleus and a single antipodal nucleus; lower polar nucleus is absent.

A reduced Allium type of development (tetranucleate embryo sac) is seen in *Podostemon ceratophyllum* (Hammond, 1937). Mukkado (1962, 1963) reported the reduced Allium type in *Indotristicha* and the *Terniola*. A significant feature of these two taxa is the persistence of the chalazal nucleus which is generally ephemeral.

A special situation occurs in *Schisandra chinensis*. In some of the ovules the development of the female gametophyte is bisporic, but the final organisation of the embryo sac is seen similar to Oenothera type (Swamy, 1962).

TETRASPORIC EMBRYO SACS

In this group both the meiotic divisions are not accompanied by cell plate formation hence all the four haploid nuclei at the end of meiosis are enclosed in a common cytoplasm forming a coenomegaspore. All the four nuclei of the coenomegaspore take part in the formation of the embryo sac hence the tetrasporic embryo sac is more heterogeneous than a bisporic embryo sac.

Much variability was observed in the nuclear behaviour of the tetrasporic embryo sacs. Basing on the number of post-meiotic mitoses, number of groups of nuclei in the mature gametopyte and presence or absence of nuclear fusions during the development, Maheshwari (1950) recognized seven principal types of tetrasporic embryo sacs namely Adoxa, Peperomia, Penaea, Drusa, Fritillaria, Plumbagella and Plumbago.

(i) Adoxa type

It was described for the first time by Jonsson (1879-80) in *Adoxa moschatellina* and later confirmed by Lagerberg (1909) and Fagerlind (1938a). There are only five genera in which this type of embryo sac development is more or less regular feature: *Adoxa, Sambucus* and some species of *Erythronium, Tulipa* and *Ulmus.*

In this type of embryo sac development, the nuclei of the coenomegaspore take up 2+2 arrangement. All the nuclei of the coenomegaspore undergo just one more division to form an 8-nucleate embryo sac. The mature female gametophyte has normal egg apparatus, three antipodals and two polar nuclei.

An interesting variation was reported in some species of *Tulipa.* In *T. sylvestris* (Bambacioni-Mezzetti, 1931) vacuolation frequently commences at the megaspore mother cell stage and all the 4 megaspore nuclei gather at the micropylar end of the cell, where they divide to give rise to a group of 6 cells (one of which is egg) and 2 free nuclei. *T. tetraphylla* (Romanov, 1938) show almost a similar type of development. After the meiotic divisions are over, 3 nuclei go to the micropylar pole and one to the chalazal due to the formation of vacuole. All of them divide again so that there are 6 daughter nuclei in the upper part of the sac and 2 in the lower. Cell plates laid down later resulting in the formation of 5 cells at the micropylar end (one of these is egg) and one cell at the chalazal end leaving 2 free nuclei (polars) in the centre. Since this peculiar mode of development occurs only in the Eristemones section of the genus *Tulipa,* it is known as the 'Eristemones form' of the Adoxa type.

(ii) Peperomia type

Campbell (1899a,b) and Johnson (1900) reported that, in *Peperomia pellucida,* there are two mitotic divisions in the 4 nuclei of the coenomegaspore after taking up tetrapolar arrangement and results in a 16 nucleate megagametophyte. According to Johnson, in the final organization at the micropylar end, form the egg and a synergid, 8 fuse to form the secondary nucleus and the remaining 6 are cut off at the periphery of the embryo sac. Campbell also observed a similar type of embryo sac development and according to him the egg apparatus is invariably 3-celled, approximately 8 nuclei form the secondary nucleus and the remaining 4-6 nuclei are cut off as antipodal cells. Fagerlind (1939) made a detailed study of *P. pellucida.* He found that the outline of the coenomegaspore may be more or less spherical or pear-shaped and in the latter case there is slight protruberance at the micropylar end of the coenomegaspore. The mature embryo sac, derived from the spherical coenomegaspore shows 2-celled egg apparatus, 8 polar nuclei and six peripheral cells and the embryo sac derived from the pear shaped coenomegaspore shows a 3-celled egg apparatus. Fagerlind's (1939) observations thus help us to understand the slight divergence between the accounts of Johnson and Campbell. The former saw only one synergid, while the latter believes that there were more than one. It appears that both conditions *i.e.,* 2 and 3-celled egg apparatus are possible and depends on the form which the embryo sac takes during its growth and development. In the pear-shaped there are invariably two synergies and only one in spherical coenomegaspore.

Johnson (1914) reported a different type of embryo sac development in *Peperomia hispidula.* At the 8-nucleate stage, 2 nuclei are seen at the micropylar end and 6 nuclei at the chalazal end. At the 16 nucleate stage, 4 lie at the micropylar end and 12 at the chalazal end. Two nuclei of the micropylar group now form the egg and single synergid as in other species. But the remaining 2 nuclei of the micropylar group and all the remaining 12 nuclei of the chalazal group meet at the centre and fuse to form a single large secondary nucleus and there are no antipodals. As this is different from the *P. pellucida,* it is recognized as subtype namely *P. hispidula* type. Thathachar (1952) also reported a similar development in *Acalypha lanceolata.*

In *P. maculosa* (Martinoli, 1948) the embryo sac may resemble that of either *P. hispidula* or *P. pellucida.* Martinoli attributed this variation to the irregularities in the polarisation of the nuclei. Murthy (1959a), who studied several species of *Peperomia,* stated that in the mature embryo sac the egg apparatus may be 2 or 3-celled and there may be 6-9 polar nuclei and 4-6 peripheral (antipodal) cells. Maheshwari (1963) opined that in view of such variability, it was not necessary to maintain the two sub-types — *P. hispidula* and *P. pellucida* types.

(iii) Penaea type

Stephens (1909) described this type of embryo sac develop-ment in 3 genera of Penaeaceae namely *Penaea, Brachysiphon* and *Sarcocolla.* Here the nuclei of the coenomegaspore take up tetrapolar arrangement and divide twice to form a 16 nucleate embryo sac. The 16 nuclei are arranged crosswise, 3 nuclei of each quartet later cut off as cells, while the fourth remains free and moves to the centre. There are thus four 'triads' and 4 polar nuclei. Generally, the central cell of the micropylar triad alone functions as the egg. The central cell of the other triads, also simulate the egg. This type of embryo sac development is found characteristic of the family Penaeaceae, some members of the Malpighiaceae and Euphorbiaceae.

Maheshwari and Johri (1941) studied the embryo sac development in *Acalypha indica* which is almost similar to Penaea type, but differs slightly in the organisation of the mature embryo sac. Here each peripheral group consists of two cells and 2 nuclei of each quartet remain free and migrate to the centre and fuse to form the secondary nucleus. Thus, there are four groups of two cells each at the periphery and 8 free nuclei in the centre in this type of variation.

(iv) Plumbago type

Haupt (1934) reported this type of embryo sac development in *Plumbago capensis.* The 4 nuclei of the coenomegapore are arranged in a cross wise fashion. They undergo another division resulting in 8 free nuclei arranged in 4 pairs. A nucleus of the micropylar pair is now cut off to form the lenticular egg cell. Of the remaining 7 nuclei, 4 (probably one member from each of the 4 pairs) undergo a slight increase in size and gradually approach one another and function as polar nuclei. The remaining 3 nuclei degenerate or occasionally 1, 2 or all 3 of them are cut off at the periphery to form cells which persist and assume an egg like appearance. Synergids are entirely absent.

The Plumbago type also occurs in *P. coccina* and *P. scandens* (Boyes and Battaglia, 1951a). This type was also found in two other genera of the Plumbaginaceae namely *Ceratostigma* and *Vogelia,* and is not known outside this family.

(v) Drusa type

Hakansson (1923) in *Drusa oppositifolia* (Umbelliferae) recorded a 16-nucleate embryo sac of different nature. After the two meiotic divisions were over, the 4 nuclei of the coenomega-spore take up 1+3 arrangement and this is followed by 2+6 arran-gement and then 4+12 nucleate stages. The 4 micropylar nuclei give rise to an egg apparatus and the upper polar nucleus and the 12 chalazal nuclei to a lower polar nucleus and 11 antipodals. Later this type of development was also noticed in *Mallotus japonicus* (Ventura, 1934), *Crucianella latifolia* (Fagerlind, 1937), *Smilacina stellata* (Eunus, 1950) etc. Sometimes, due to the failure of division of some chalazal nuclei at the 8 nucleate stage, there are sometimes fewer than 11 antipodals.

The embryo sac of *Mainthemum canadense* (Swamy, 1949) also shows Drusa type, but in about 13 per cent of the ovules, during the division following the 8 nucleate (2+6) stage, the chalazal spindles fuse in pairs resulting in 6 diploid nuclei instead of 12 haploid nuclei.

In *Tanacetum vulgare,* the 4 megaspore nuclei of the coenomegaspore are arranged in a linear row and another division results in 8 nuclei. Sometimes 1 or 2 chalazal nuclei of the coenomega-spore fail to divide, as a result the embryo sac contains 7 or 6 nuclei. Depending on the number of nuclei taking part in the next division the mature gametophyte is 14 or 12 nucleate with the 3-celled egg apparatus, two polar nuclei and varying number of antipodals (Fagerlind, 1941).

(vi) Fritillaria type

This type was first described in *Fritillaria persica* and *Lilium*. The coenomegaspore in this type of embryo sac development takes up 1+3 arrangement. The 3 chalazal nuclei are present very close to each other. During the next division, the micropylar nucleus divides normally, but the three chalazal spindles fuse to form a single common spindle, so that, at the close of the division, there are 2 haploid nuclei at the micropylar end and 2 triploid nuclei at the chalazal side. One more division occurs, resulting in 8 nuclei, of which the 4 chalazal are triploid and the 4 micropylar are haploid. The mature embryo sac finally consists of an egg apparatus of which all the cells are haploid, the three triploid antipodals and a tetraploid secondary nucleus formed by the fusion of two polar nuclei, one haploid and the other triploid.

The fusion of 3 chalazal nuclei of the coenomegaspore during the division was first observed by Bambacioni (1928) hence, the fusion process is known as Bambacioni's effect. Cooper (1935) found the Bambacioni effect and the formation of Fritillaria type of embryo sac development in the entire tribe Lilioideae and several other genera belonging to diverse families : *Piper, Tamarix, Cornus, Rudbeckia, Gaillardia* (Fig. 6.4), *Tulipa, Clintonia* etc.

The fusion of the 3 chalazal megaspore nuclei may take place when they are either in the prophase stage or in early metaphase. In the former case, the secondary 4-nucleate stage is preceded by a secondary 2-nucleate one and the sequence then is as follows: Megaspore mother cell, primary 2-nucleate stage, primary 4-nucleate stage, secondary 2-nucleate, secondary 4-nucleate and lastly the 8-nucleate stage.

Normally all the 4 megaspore nuclei are of the same size, but in some plants the micropylar nucleus is larger than the rest in which case the nuclei of the secondary 2- and 4- nucleate stages show no appreciable difference in size in spite of their triploid nature. Rarely the chalazal nuclei are smaller than the micropylar although they are triploid.

Sometimes the basal nuclei of the secondary 4 nucleate stage fails to divide as a result only 6-7 nucleate gametophytes are found in some species of *Statice* (Fagerlind, 1939) and *Gagea* (Romanov, 1936). Depending on this, the mature gametophyte has either 1 or 2 antipodals.

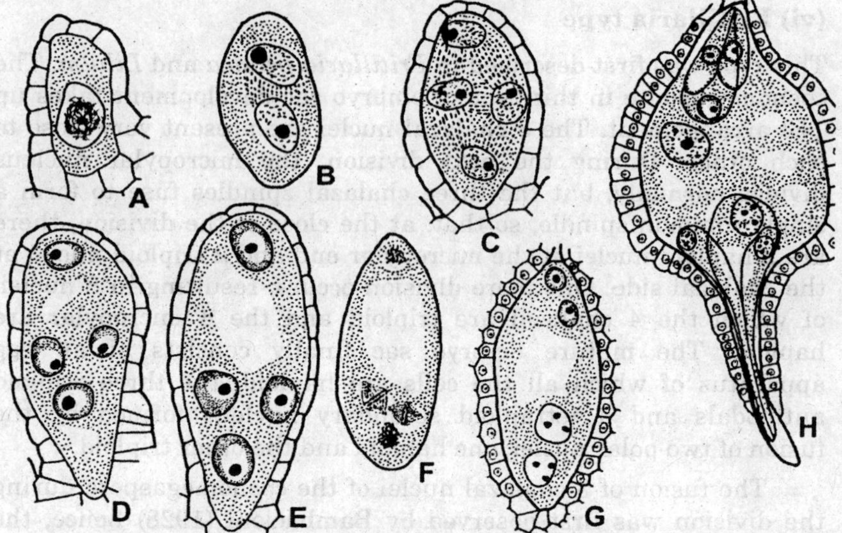

Fig. 6.4. Fritillaria type of embryo sac development in *Gaillardia lorenziana* (after Pullaiah, 1976). **A.** L.S. nucellus showing mega spore mother cell in meiosis; **B.** Two-nucleate stage; **C.** Primary four-nucleate stage; **D-E.** Megaspore nuclei showing 1+3 arrangment; **F.** Same showing spindles; **G.** Secondary four-nucleate Stage; **H.** Mature embryo sac.

(vii) Plumbagella type

This is a reduced Fritillaria type of embryo sac and is reported only in *Plumbagella micrantha* (Fagerlind, 1938; Boyes, 1939). The 4 megaspore nuclei take up 1+3 arrangement. A large vacuole separates the 3 chalazal nuclei from the micropylar. The 3 chalazal nuclei eventually fuse to form a single triploid nucleus and results in secondary 2-nucleate stage. This is followed by a secondary 4-nucleate stage in which the 2 micropylar nuclei are haploid and the 2-chalazal are triploid. There are no further divisions. The nucleus nearest the micropylar end organizes into egg, the triploid nucleus nearest the chalazal end organizes into a single antipodal cell and the remaining nuclei one haploid and the other triploid fuse to form a tetraploid secondary neuleus. This mode of development is similar to the Fritillaria type and the only difference is that in *Plumbagella,* the development is arrested at secondary 4-nucleate stage and the fourth division is omitted.

ABERRANT AND UNCLASSIFIED TYPES

In addition to the seven fairly distinct and well established types of tetrasporic embryo sacs, there are few which appear to be more or less isolated. Of this the embryo sac development in *Chrysanthemum cinerariaefolium* was special and interesting.

Chrysanthemum cinerariaefolium Type

The embryo sac development is somewhat peculiar in this taxon and is reported by Martinoli (1939). The embryo sac is tetrasporic and the megaspore nuclei take up 1+2+1 arrangement so that there is 1 nucleus at each pole and 2 nuclei in the middle. The two central nuclei become separated from the terminal nuclei by vacuoles and may either fuse to form a single diploid nucleus or may simply remain close to each other without undergoing any fusion. Based on the behaviour of the two central nuclei there are two variations within this.

(i) First variation: The two central nuclei fuse to form a diploid nucleus. After fusion all the three nuclei divide twice to form three groups of 4 nuclei each. From the micropylar quartet are produced the egg apparatus and the upper polar nucleus, all haploid. The chalazal quartet gives rise 4 antipodals, all of which are haploid. The central quartet is composed of diploid nuclei. One of these functions as the lower polar nucleus and the remaining three are organized as additional antipodals. Occasionally fewer than 12 nuclei are found either because of failure of some divisions at the chalzal end or because the central diploid nucleus undergoes only one division instead of two.

(ii) Second variation: In the second of the two alternatives *i.e.*, the two central megaspore nuclei do not fuse, but only lie in contact with each other and function directly as the polar nuclei. The micropylar as well as the chalazal megaspore nuclei divide twice forming 4 nuclei at each pole. Thus the embryo sac is 10-nucleate. All the 4 nuclei from the chalazal quartet organize as antipodals. Of the micropylar quartet, three nuclei organize into egg apparatus, while the fourth one, along with the two central nuclei function as polar nuclei. However, there is no regularity in the behaviour of the chalazal nuclei. Sometimes there is only one division or none and basing on that the mature female gametophyte is 9 or 7 nucleate.

Occurrence of more than one type of embryo sac development in the same species or even in the same individual is common in some cases. Kapil and Prakash (1966) in *Delosperma cooperi* (Aizoaceae) reported 14 per cent ovules showing Polygonum type of embryo sac, 18 per cent Endymion type and 68 per cent various tetrasporic types. Similarly, Abe (1971) recorded Allium (31%), Adoxa (53%) and Polygonum (16%) type of embryo sacs in the orchid, *Oreorchis patens*. But the interesting thing was that the mature embryo sac contained only 6 nuclei in all of them. In the genera like *Chrysanthemum, Erigeron* and *Euphorbia*, the development of megagametophyte may be mono-, bi- or tetrasporic.

The environmental conditions to some extent can influence the type of embryo sac development in some cases. Hjelmquist and Grazi (1965) have shown that temperature influences the type of embryo sac. For example, *Ulmus glabra* show Adoxa and Drusa types of embryo sacs in addition to some rare types. According to them low temperatures (15°-19°C) considerably enhanced the frequency of 8-nucleate Adoxa type of embryo sacs. They explained that the temperature affects nuclear division more than the cell growth and vacuolation. At low temperature, the nuclear division is delayed in relation to cell growth so that the central vacuole appears after meiosis-I separating the two megaspore nuclei to the ends of the embryo sac. By a single mitosis after meisosis-II, an 8-nucleate Adoxa type of embryo sac is formed. At higher temperatures the nuclei divide rapidly in relation to cell growth so that vacuolation appears only after the second meiotic division separating a single nucleus at the micropylar end from 3 at the chalazal end. By two more divisions a 16 nucleate embryo sac of the Drusa type is formed. In *Capsicum* (Dharamdhaj and Prakash, 1978), it was reported that the embryo sacs formed in winter were bisporic and those formed in summer were of the monosporic types. Thus the environmental factors like the temperature can influence the type of embryo sac development.

ULTRASTRUCTURE OF THE EMBRYO SAC

Strasburger (1877) was the first person to describe the female gametophyte in angiosperms. Since then large number of species were studied by Light microscopic methods and several interesting reviews on this subject were found in the literature (Maheshwari,

1950, 1963; Johri, 1963, 1970; Rutishauser, 1969; Cocucci, 1973; Kapil and Tiwari, 1978; Kapil and Bhatnagar, 1981 etc.). But the studies on the ultrastructure of the female gametophyte are relatively recent and resulted due to the advancement in the techniques of microscopy (Jensen, 1963, 1964, 1965; Diboll and Larson, 1966; Van Went and Linskens, 1969, Van Went 1970; Schulz and Jensen, 1968a, b, 1973; Cocucci and Jenson, 1969, Newcomb, 1972, 1973).

In majority of the angiosperms (Polygonum, Allium, Endymion, Adoxa and Fritillaria types) the mature embryo sac consists of 7 cells; the egg cell, two synergids, the central cell and the three antipodals. In others, the embryo sac consists of fewer or more cells. The variation in the number of cells might have resulted from the absence of one or both synergids, reduced or increased number of antipodal cells etc. Generally all the nuclei of the embryo sac except that of the central cell are haploid. The central cell consists of two halploid nuclei called the polar nuclei which later form the diploid secondary nucleus. However, the ploidy level of the central cell and antipodals varies considerably in some taxa.

The egg apparatus is present attached to the extreme micropylar part of the embryo sac and consists of the egg cell and the synergids. The cells of the egg apparatus are arranged in triangular fashion and share some common features. Generally, the synergids are in direct contact with the micropyle and the egg lies some microns below the apices of the synergids at one side of the embryo sac. The central cell occupies the largest portion of the embryo sac and extends from the micropylar pole to the chalazal pole. The antipodals are the smallest cells of the embryo sac and mostly they are ephemeral in dicots. All the cells of the embryo sac show a specific morphology and a characteristic structural organization.

The structure of the embryo sac wall is studied in several taxa and it is interesting. In cotton it is PAS positive and rich in pectic substances. It is multilayered and pectocellulosic in maize (Chebotaru, 1975). Diboll and Larson (1966) opined that the embryo sac wall is two layered, the innermost layer represents the wall of the functional megaspore and the outer formed from the crushed nucellar cells. The wall is three-layered in *Stipa* according to Maze and Lin (1975); the innermost is gametophytic, the middle and outer layers represent the degenerated megaspore and nucellar cells, respectively. All the cells of the embryo sac are interconnected by plasmodesmata. But the embryo sac is not connected to the outer layers through plasmodesmata. The embryo sac wall is cutinized in some cases.

1. Synergids

The synergids are elongated cells present at the micropylar end of the embryo sac. The two synergids, if present, are attached to one another and also cover the egg. Sometimes a hook like structure is found on the micropylar end of the synergid. In majority of the taxa studied the synergids are surrounded by partial walls. However, considerable variation in thickness was reported. The micropylar one third of the cell is surrounded by a distinct wall which gradually thins towards the chalazal side and the chalazal one-third of the synergid lacks a wall of its own. At this point the synergid and the central cell are separated by a double membrane, one of the synergid and the other of the central cell. There are some exceptions to this rule. In *Epidendrum* (Cocucci and Jensen, 1969) the synergid cell wall is complete. Sometimes, cutin deposition occurs in the micropylar portion of synergids in *Jasione*. Some authors (Linskens, 1968; Van Went, 1970; Jensen, 1972) believe that the absence of a cell wall at the chalazal region of the synergid plays an important role in the fusion of male and female gametes.

Haberlandt (1906) for the first time described a prominent structure called filiform apparatus in the micropylar end of the synergid (Figs. 6.1A, 8.5, 8.6) which was later confirmed by several embryologists. The morphology of the filiform apparatus is highly variable. It is spherical in *Petunia* (Van Went, 1970a) and wedge shaped in *Helianthus* (Newcomb and Steeves, 1971), while in most of the species it forms a mass of finger like projections of the wall into the cytoplasm. In *Petunia* and *Torenia* the filiform apparatus shows a homogeneous structure and smooth surface. However, in most of the species studied with Electron microscope, the filiform apparatus shows very complex organization and heterogenous structure. It consists of a mass of wall projections extending deep into the cytoplasm. Due to this organization there is tremen-dous increase in the absorptive surface. The wall projections show two structural phases. They consist of a central electron-dense core surrounded by an electron translucent layer. According to Schulz and Jensen (1968), the dense core has tightly packed microfibrils surrounded by a non-fibrillar sheath. The filiform apparatus of *Capsella* gave a strong positive reaction for insoluble polysaccharides and negative reaction for nucleic acids and proteins. Chao (1971) showed that the main component of the filiform apparatus is hemicellulosic in *Paspalum*.

The cytoplasm of the synergid is highly polarised. Most of the cytoplasm and nucleus are present on the micropylar part and one large or few smaller vacuoles are present in the chalazal part (Fig. 8.4B). The vacuoles are rich in calcium salts in *Gossypium* (Jensen, 1965) and give a negative reaction for carbohydrates, nucleic acids and proteins. The synergid cytoplasm shows complex organization and is very rich in cell organelles. Large number of mitochondria, dictyosomes, extensive ER and abundant ribosomes are found in the synergid cytoplasm and especially their concentration is greatest near the Filiform Apparatus. *Gossypium* and *Nicotiana* have numerous plastids which contain starch. In *Paspalum,* starch is present in the immature synergid and disappear when it becomes mature. Lipids, RNA and proteins are also abundant in the synergid cytoplasm.

Synergids are ephemeral structures and one of the synergids degenerates prior to the arrival of the pollen tube. The degeneration results in a decreased volume, disappearance of the vacuole and an increased stainability and density of the cytoplasm. There is some information concerning the possible relation between the pollination and synergid degeneration. Jensen and Schulz (1976) and Jensen *et al.* (1977), in cotton, observed degeneration of synergids even in cultured ovules which were not pollinated. This indicates that the process of degeneration is autonomous, although it probably can be influenced by pollination. The second synergid retains its original constitution. In *Petunia, Helianthus, Capsella* etc. both the synergids remain healthy until the entry of the pollen tube into the embryo sac. However, one of the synergids show degeneration after the pollen tube has penetrated and ejected its contents. Occasionally one of the synergids may persist for considerable period after fertilization. Its nucleus may enlarge and show polyploidy. In *Allium angulosum* and *A. pulchellum* the synergid nucleus becomes octoploid.

The synergids are considered as highly active cells. The presence of numerous organelles like mitochondria, dictyosomes, abundant ribosomes and extensive Rough endoplasmic reticulum indicate an intense metabolism. Histochemical studies show that the synergid cytoplasm is rich in protein and RNA. Malik and Vermani (1975) observed intense cytochrome oxidase and succinic dehydrogenase activity together with acid phosphatase. ATPase, alkaline phosphatase, phosphorylase and lipase activity was also found in *Zephyranthes* and *Lagenaria.*

(a) Synergid haustoria

Generally, the synergids are confined to the embryo sac. However, in Compositae, they break through the embryo sac and project into the micropyle. One of the synergids become swollen and haustorial in nature in *Cotula australis.* The synergids in *Ursinia chrysanthemoides, U. anthemoides* and *U. cakelifolia* (Ahlstrand, 1978) are elongated, when the egg cell has been fertilized, one of the synergids seem to degenerate, while the other one extends half-way up the micropyle and later penetrates the surrounding tissues on the side facing funiculus. The basal part of the synergid is then bladder like. The synergid obviously has a haustorial function. In *Elephantopus scaber* (Pullaiah, 1979) both the synergids become elongated and protrude into the micropyle and function as haustoria (Fig. 6.6C). These haustoria persist up to globular stage of the embryo. Synergid haustoria also occur in *Sedum sempervioides.* Most extensive synergid haustoria are found in *Quinchamalium chilense* of Santalaceae. The synergids are conspicuously large in this species and their tips elongate and grow into the stylar tissue (Fig. 6.5). They reach up to one-third the length of the style and attain the maximum length of 1200 µm. Philipson (1981) reported extensive synergid haustoria in *Cortaderia* where they arise as finger like projections from the Filiform Apparatus and penetrate the micropyle.

(b) Functions

The structure and concentration of cell organelles in the synergid shows that they are highly active metabolically. It is

believed that the synergids perform three important functions:

1. **The synergids are believed to play an important role in directing the pollen tube growth by secreting some chemotropically active substances.**

2. **The synergids are believed to play an important role in the entrance and discharge of the pollen tube into the embryo sac and the subsequent transfer of the sperm cells to the egg and central cells.**

3. **According to Jensen (1965), the synergids play an important role in the nutrition of the embryo sac. They absorb nutrients from the metabolite-laden nucleus through filiform apparatus. The energy necessary for the absorption is supplied by the mitochondria. The presence of many plastids containing starch in the synergid near the filiform apparatus also supports its absorptive function. The extensive endoplasmic reticulum of the synergid is thought to function as an internal transport system, through which the nutrients are transported towards the egg and central cells. This transfer of material to the latter cells may be facilitated by the absence of cellulosic wall at the chalazal end of the synergids. The presence of the cuticle over the micropylar wall of the embryo sac in *Jasione* supports this view. The currently prevailing view is that the entry of metabolites into the embryo sac occurs mainly through the chalaza. The features mentioned as indicative of absorption can also be interpreted as an indication of a secretory function. In this Thomas (1970) who found a close resemblance between synergids and stylar canal cells show a similar polarity both in structure and function as do the synergids. Enzyme production and secretion have also been postulated by Chao (1971) for the synergids of *Paspalum*. The secreted enzymes are presumed to be responsible for the degradation of integumentary cells leading to the release of PAS positive substances into the micropyle.**

2. Egg

Egg cell occupies a central position in the egg apparatus and is surrounded by the synergids and the central cell. The egg cell also has a partial cell wall. The egg cell wall is thickest at the micropylar region and gradually thins towards the chalazal region. The chalazal half is limited by plasma membrane only. In *Caspsella* (Schulz and Jensen, 1988b), the chalazal region of the egg is surrounded by a thin, irregular, PAS positive cell wall. However, wall appears to be present all over the egg cell in *Epidendrum scutella.* The lateral walls of the egg cell join the central cell at the micropylar end and so actually the end wall of the egg cell is the wall of the central cell. The egg is connected to two synergids and the central cell by plasmodesmata.

The egg cell is also characterized by distinct polarity. Most of the cytoplasm and nucleus are found aggregated in the chalazal end of the cell, while the micropylar end is occupied by a large vacuole (Fig. 6.1B). This type of distribution of the cytoplasm and vacuole in the egg cell is just the opposite condition seen in synergids. In *Epidendrum* and *Zea,* the nucleus of the egg cell takes up a central position and many small vacuoles are present along the periphery of the cell. According to Jensen (1965b), the vacuole of the *Gossypium* egg cell does not show any ash and in this respect it is different from that of the synergid vacuole.

The ultrastructure of the egg cell cytoplasm varies considerably among the various species studied. In *Gossypium* the egg cell contains large amount of endoplasmic reticulum, numerous ribosomes, numerous mitochondria with few cristae and only few dictyosomes and plastids (Jensen, 1965b). In *Petunia,* endoplasmic reticulum is scarce and polysomes, mitochondria, plastids and dictyosomes are few (Van Went, 1907b). However, the egg cell of *Zea mays* (Diboll and Larson, 1966) and *Helianthus* (Newcomb, 1973) contains many mitochondria and plastids, but only few dictyosomes and a small amount of endoplasmic reticulum. A common feature of the egg cell cytoplasm is the presence of only limited number of relatively inactive dictyosomes. Egg cells are rich in protein. Many authors believe that the egg cell is in a relatively inactive stage on the basis of the cytoplasmic constitution.

The egg cell of *Plumbago* (Cass, 1972; Cass and Kara, 1974) has a filiform apparatus like structure in the micropylar region. As the synergids are absent in *Plumbago,* it is presumed that the egg cell also take up the role of synergid. The *Plumbago* egg cell is considered as a highly active cell metabolically and combines both gametic and synergid functions.

3. The Central Cell

The central cell occupies the largest portion of the embryo sac and extends between the micropylar and chalazal poles. The lateral wall of the central cell forms the main part of the embryo sac wall and is in contact with the surrounding tissue. After the last nuclear division, the embryo sac considerably becomes enlarged due to the inflation of the large vacuole of the central cell. The vacuole of the central cell is a reservoir of sugars, amino acids and inorganic salts (List and Steward, 1965; Ryczkowski, 1964).

The central cell is highly vacuolate and its cytoplasm is confined to a thin layer along the embryo sac wall and accumulates near the egg apparatus and the antipodal cells. The micropylar accumulation contains the two polar nuclei. The time of fusion of the polar nuclei varies from species to species and it may occur before or during double fertilization to form secondary nucleus.

Like the cells of egg apparatus, the central cell is surrounded only by a partial cell wall. The cell wall is absent at the micropylar pole near the base of the egg apparatus. The wall is generally thick and regular against the nucellus or integuments. The central cell is connected with all the other cells of the embryo sac through plasmodesmata but no such connections exist between the central cell and the adjacent nucellar cells. Cell wall projections are present in the micropylar and chalazal region of the central cell and aid in drawing nutrition from the surrounding nucellus or integuments.

The cytoplasm of the central cell is complex and is rich in organelles. There are numerous mitochondria with well developed, cristae extensive rough endoplasmic reticulum, large number of dictyosomes which appear active in the production of vesicles; numerous ribosomes both as monosomes or polysomes. There are plastids containing starch and sometimes protein and phytoferretin. Besides starch, usually lipids are present as reserve substances. In *Capsella,* the central cell possess numerous sphaerosomes associated with glyoxysomes that probably convert fat into sugar. The central cell contains sufficient food reserves that are available for use during fertilization and early stages of endosperm development.

In *Aquilegia vulgaris* skirt-like evaginations arise from the central cell. These haustorial extensions enter the ovular tissue and aid in the absorption of nutrients. Sometimes, these extensions also secrete enzymes that help in the digestion of the nucellar tissue (Rifot, 1973).

4. Antipodals

Among all the cells of the embryo sac, the antipodals show much variation. In many dicots, they degenerate before or during the maturation of the embryo sac. In Sapotaceae and Thismiaceae the antipodal nuclei degenerate even without organizing into cells. In some cases the antipodals persist even during post fertilization development. In grasses, they proliferte and form a multicellular tissue. In *Capsella* (Schulz and Jensen, 1971) the antipodals are small inactive cells and are surrounded by cell walls of uniform thickness. The antipodals are interconnected and also are connected to the nucellus and central cell. This indicates a possible pathway of metabolites from the nucellus through the antipodals. The antipodal cell walls bordering the nucellus have finger like wall projections, similar to those observed in the central cell and the filiform apparatus. The cytoplasm contains abundant organelles such as mitochondria, dictyosomes, plastids, ribosomes and extensive endoplasmic reticulum.

Histochemical tests on antipodals indicate that they are highly active metabolically. Malik and Vermani (1975) established enzyme activity such as peroxidase, succinic dehydrogenase, cytochrome oxidase and lipase in the antipodal cells of *Zephyranthes* and *Lagenaria.* These cells also contain high amounts of protein, polysaccharides, lipids and RNA. Sometimes, the antipodals also contain starch, which is used during fertilization and subsequent development of the endosperm and embryo.

Lot of variation was observed in the cellular organization of the antipodal tissue. Generally they are three in number, but in majority of Compositae only two were seen, of which, one is uninucleate and the other is binucleate (Pullaiah, 1984; Kapil and Bhatnagar, 1976). In *Zea mays,* the number of antipodals increases up to 20 and each cell contains 1-4 nuclei. In *Sasa paniculata* highest number of antipodal cells *i.e.,* about 300 is reported. In *Tagetes, Stackhousia,* the antipodals are multinucleate. In *Zea mays* (Diboll and Larson, 1966) and *Linum* (Vazart, 1969), the antipodal cells are partly separated by wall. These incomplete walls later dissolve and a syncytium is formed. Antipodal nuclei may also become polyploid due to endopolyploidy, or polyteny. In *Scilla,* the level of endopolyploidy reaches as much as 1,024n (Firsch and Nagl, 1979).

Generally three main functions are attributed to antipodals:

1. They may be involved in the transfer of nutrients and serve as a pathway for metabolites from the nucellus to the central cell. Further evidence for this may be based on the presence and location of plasmodesmata and the presence of cell wall projections in the antipodals cells and metabolically active cytoplasm.

2. They may serve as storage cells for the developing endosperm and embryo. The presence of abundant starch, lipid and proteins in the antipodals gave further evidence of this function. Degeneration and breakdown of antipodals can serve as an additional food supply.

3. The antipodals may also have a secretory function. This is indicated by the presence of extensive rough endoplasmic reticulum and abundant ribosomes. The antipodals possibly produce and secrete growth-controlling substances which regulate the development of the adjacent endosperm.

(a) Antipodal haustoria

Antipodals are mostly confined within the embryo sac. But in some members of Asteraceae, antipodals show out growths forming Antipodal haustoria (Fig. 6.5). Antipodal haustoria have been reported in *Grindelia squarrosa* (Norris, 1892), *Aster novae-*

anglieae, Senecio vulgaris (Small, 1919), *Artemisia tridentata* (Diettert, 1938), *Bidens pilosa, Coreopsis tinctoria, Cosmos sulphureus* (Maheswari Devi, 1963), *Coreopsis lanceolaria* and *Elephantopus scaber* (Pullaiah, 1979, 1983).

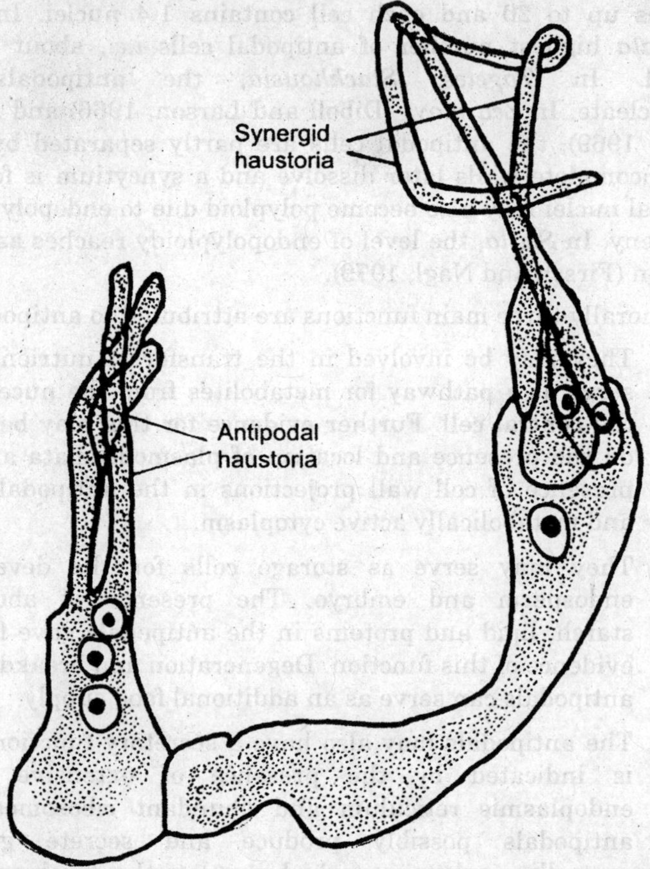

Fig. 6.5. Mature embryo sac of *Quinchamalium chilense* showing synergid and antipodal haustoria (after Johri and Agarwal, 1965)

In *Elephantopus scaber* (Pullaiah, 1979) the three antipodal cells are arranged in a linear row (Fig. 6.6A). Later on they become multinucleate. Simultaneously, with the growth of embryo sac the antipodal cells also increase in size and invade the chalazal tissue of the integument (Fig. 6.6B, D-F). The micropylar and chalazal antipodal cells, later on also the middle one, produce

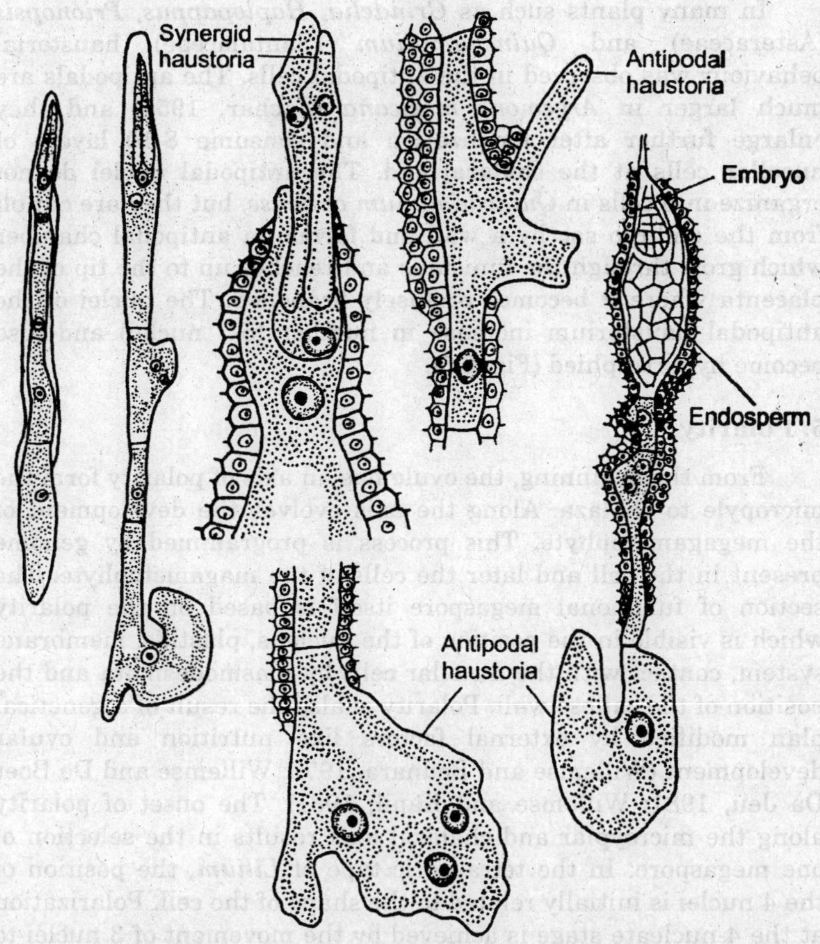

Fig. 6.6. Synergid and antipodal haustoria in *Elephantopus scaber*. **A.** Young embro sac; **B.** Embro sac showing antipodal haustoria. **C.** Synergid haustoria; **D-E.** Antipodal haustoria; **F.** Embryo, endosperm and antipodal haustoria (after Pullaiah, 1979)

lateral pouch like extension, which finally develop into hausioria. The micropylar haustorium becomes branched (Fig. 6.6D), the branches ramify and invade the sides of the integument. The chalazal haustorium, which becomes a bulbous foot like structure or a 'U' shaped structure, crushes and absorbs the chalazal tissue (Fig. 6.6 B,E,F). Haustorial remnants persist even after very well developed cotyledons have been formed in embryo.

In many plants such as *Grindelia, Haplopappus, Prionopsis* (Asteraceae) and *Quinchamalium* (Santalaceae) haustorial behaviour was observed in the antipodal cells. The antipodals are much larger in *Argemone mexicana* (Sachar, 1955) and they enlarge further after fertilization and consume 8-10 layers of nucellar cells at the chalazal end. The antipodal nuclei do not organize into cells in *Quinchamalium chilense,* but they are cut off from the embryo sac by a wall and forms an antipodal chamber which grow through the funiculus and reaches up to the tip of the placenta where it becomes profusely branched. The nuclei of the antipodal haustorium increase in number (5-7 nuclei) and also become hypertrophied (Fig. 6.5).

5. Polarity

From the beginning, the ovule has an axis of polarity form the micropyle to chalaza. Along the axis evolves the development of the megagametophyte. This process is programmed by genome present in the cell and later the cells of the magametophyte. The section of functional megaspore itself is based on the polarity which is visible in the position of the nucleus, plastids, membrane system, contact with the nucellar cells by plasmodesmata and the position of the callose wall. Polarity is also the result of a genetical plan modified by external factors like nutrition and ovular development (Willemse and Bednara, 1979; Willemse and De Boer Da Jeu, 1981; Willemse and Kapil, 1981). The onset of polarity along the microyplar and chalazal axis results in the selection of one megaspore. In the tetrasporic type of *Lilium*, the position of the 4 nuclei is initially related to the shape of the cell. Polarization at the 4-nucleate stage is achieved by the movement of 3 nuclei to the chalazal pole. Polarity exists also in the position of the plastids and the endomembrane system. In *Allium* the plastids prefer a chalazal position. In *Impatiens,* a bisporic type has plasmodesmatal contact with the chalazal nucellar cells and there is no clear polarized position of cell organelles at the onset of the 4-nucleate stage (De Boer De Jeu, 1978).

Polarity in the coenocytic megagametophyte is determined by the distribution of the cytoplasm as well as in its organization. The position of the vacuole or vacuoles is an indication for the cell polarity. In some Commelinaceae and Zingiberaceae the functional megaspore has only small vacuoles and an increase of cytoplasm takes place during its enlargement.

Histochemical tests for RNA and DNA in *Nicotiana* (Bannikova, 1971) reveal that at the 8-nucleate stage the micropylar cytoplasm has more RNA than the chalazal. A low RNA reaction has been observed in the area of the central cell. The nuclei in the micropylar part are larger than the chalazal nuclei and the largest nucleus in the micropylar zone is reserved for the future egg cell.

The enlargement of the megametophyte is completed before the cellular organization in the embryo sac. Studies of *Ammannia* (Smith and Herr, 1971) show that the maximum size is reached at 8-nucleate stage, while the greatest increase in growth along the micropylar chalazal axis mainly occurs at the 4-nucleate stage. The division of the nuclei of the coenocyte, are not always synchronous. The nuclei in the chalazal part divides, somewhat earlier (Mogensen, 1970). After the eight nucleate stage, the coenocytic megametophyte becomes cellular. This process can start at first in the chalazal or it is a simultaneous process (Newcomb, 1973).

6. Naked Embryo Sac

In most of the angiosperms the embryo sac resides inside the ovule, enclosed by the thick nucellus and the integuments. However, in some genera, such as *Galium, Philadelphus, Thesium, Torenia, Utricularia* and *Vandelia,* the embryo sac protrudes out through the micropyle. In *Torenia fournieri,* a member of the family Scrophulariaceae, the egg apparatus and the micropylar half of the central cell are located outside the micropyle (Fig. 6.7). The living naked embryo sac can be easily observed in intact ovules by differential interference microscopy, making it as suitable material to study the entire process of double fertilization, including pollen tube entry into the embryo sac (Fig. 6.7) delivery of the sperms and fusion of the male and female gametes.

Extra-ovular extension of the embryo sac at the micropylar end is quite common in the Santalaceae. In *Mida, Quinchamalium* and *Santalum* the embryo sac grows beyond the ovule at the 4-nucleate stage, whereas in *Leptomeria* it occurs at the mature embryo sac stage.

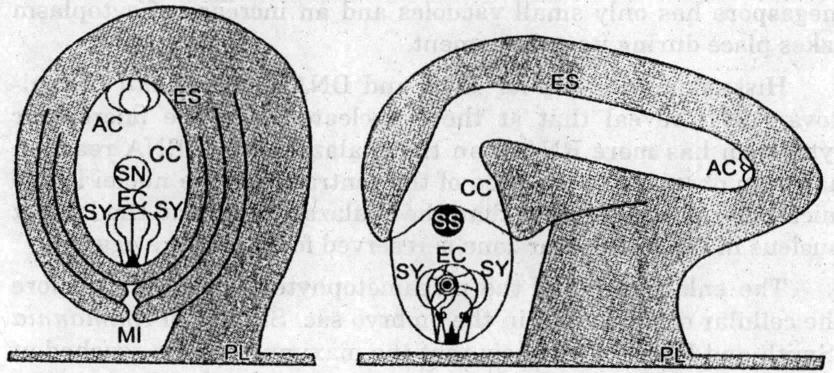

Fig. 6.7. Diagrams of a typical angiospermous ovule and that of *Torenia fournieri* with exposed egg apparatus and micropylar half of the central cell (after Higashiyama, 1999). **SY** : Synergid; **AC** : Antipodal Cells; **EC** : Egg Cell; **SN** : Secondary nucleus; **CC** : Central Cell; **ES** : Embryo sac; **MI** : Micropyle; **PL** : Placenta.

7. Embryo Sac haustoria

In some plants the entire embryo sac may grow beyond the ovular tissue. It may also form micropylar or chalazal extension or caeca from the Central cell which grows into various carpellary tissues and function as haustoria.

Most of the Santalaceous members show the lateral extension of the chalazal end of the embryo sac in the form of a lateral caecum leaving antipodals *in situ, e.g. Santalum album* (Fig. 6.8). In *Comandra* a caecum arises laterally on the funicular side of the gametophyte at the level of egg apparatus, and grows beyond the ovule into the placenta (Fig. 6.9) (Ram, 1957). In *Leptomeria cunninghamii* (Ram, 1957) the tip of the mature embryo sac grows beyond the ovule, bends upwards and extends up to the base of the style. Meanwhile caecum develops just above the level of the antipodal cells and after traversing the ovule it bends and enters the placenta. Thus, the outline of the embryo sac looks like an inverted N.

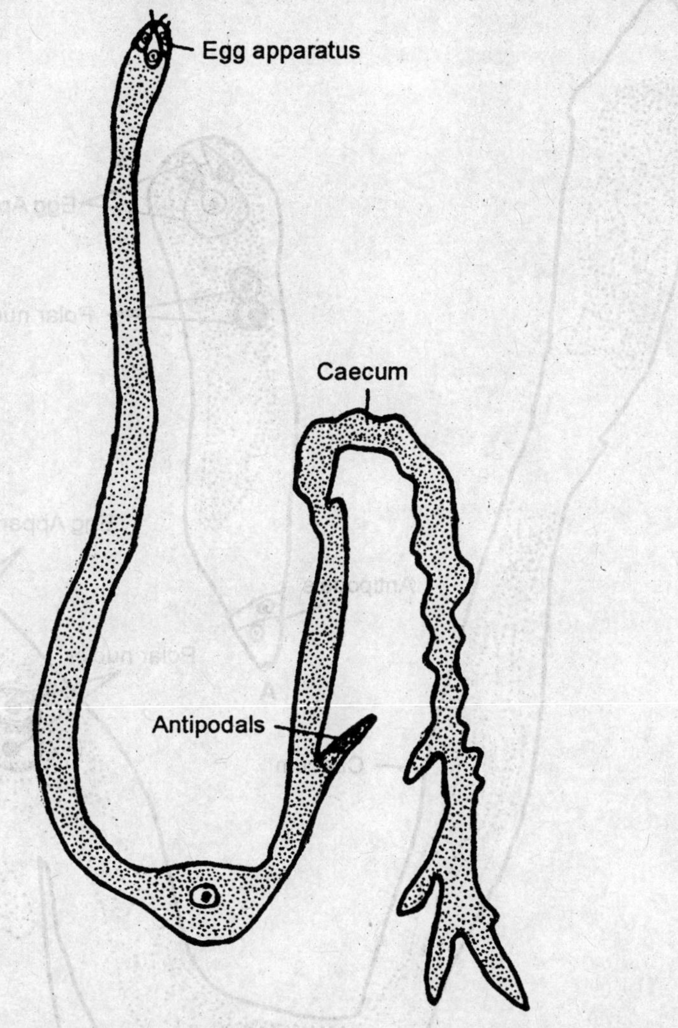

Fig. 6.8. Embryo sac haustorium in *Santalum album* (after Paliwal, 1956)

In *Nuytsia* the micropylar caecum may branched (Fig. 6.9A). In *Exocarpos strictus* numerous fingerlike outgrowths arise from the middle of the embryo sac and hang over the lower portion like a skirt (Fig. 6.10). They provide additional absorptive surface. In *E. menziesii* tubular haustroial processes arise from the tip of the embryo sac and extend beyond the placenta.

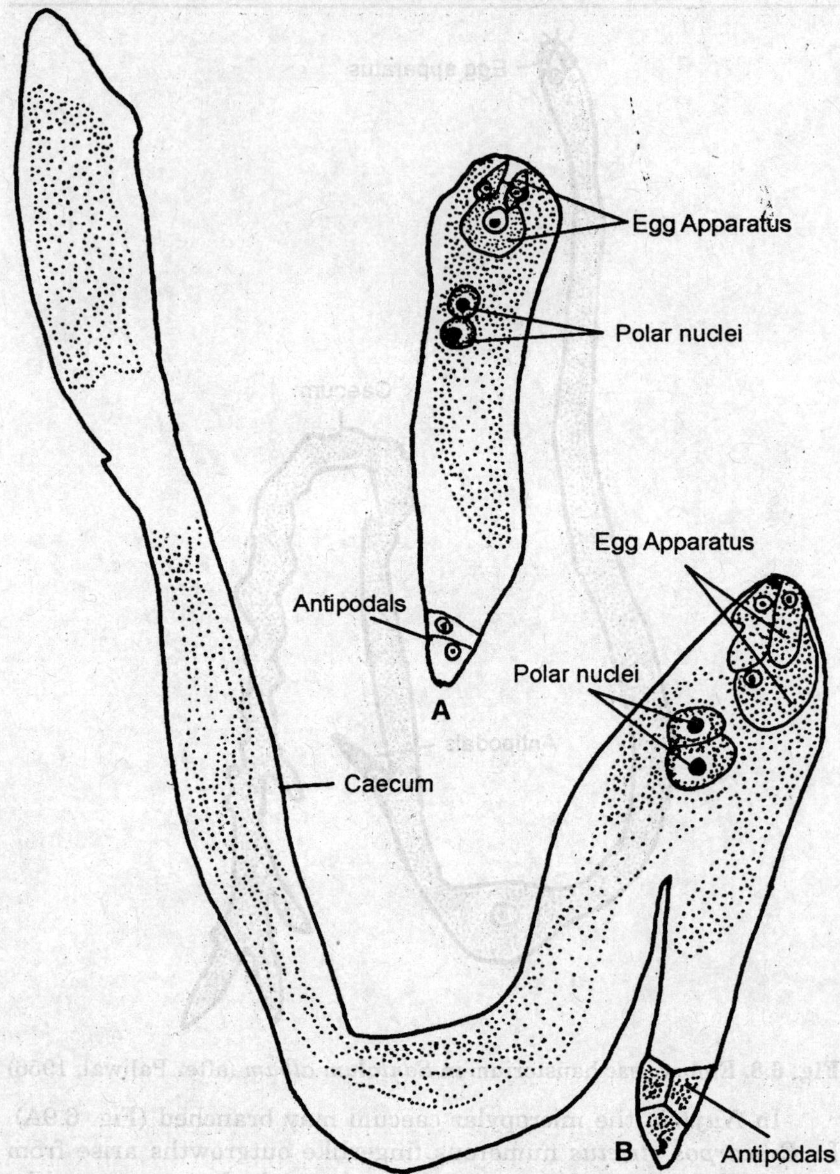

Fig. 6.9. *Comandra umbellata.* **A.** Young organised embryo sac; **B.** Mature embryo sac with prominent lateral caecum (after Ram, 1957).

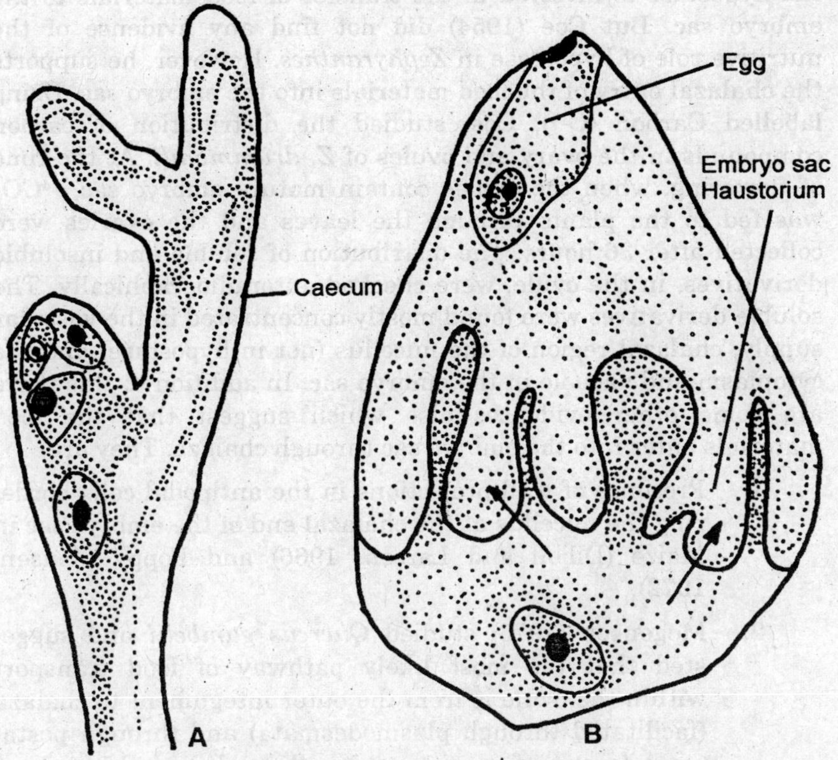

Fig. 6.10. Embryo sac haustorium. **A.** *Nuytsia floribunda*; **B.** *Exocarpus strictus*

Micropylar caecum is also seen in *Nuytsia* and *Atkinsonia.* In *Nuytsia* the micropylar caecum may branch.

Nutrition of the Embryo Sac

The embryo sac derives its nourishment mainly through the nucellus or integument, because they surround it on all sides. The ovular morphology suggests that, the nutrients enter the ovule mainly through the chalaza. The funicular vascular supply which is the normal channel for the supply of food materials to the ovule, terminates at the base of the integuments. From there the nutrients pass into the embryo sac through the nucellar tissue present between the vascular supply and the chalazal end of the embryo sac. Some authors like Venkata Rao (1953) suggest that the hypostase is involved in the transfer of food materials to the embryo sac. But Coe (1954) did not find any evidence of the nutritive role of hypostase in *Zephyranthes*. However, he supports the chalazal entry of the food materials into the embryo sac. Using labelled Carbon (C^{14}), Coe, studied the distribution of carbon compounds in the ovary and ovules of *Z. drummondi*. At the time of flowering, when the ovules contain mature embryo sac, $^{14}CO_2$ was fed to the plants through the leaves and the ovaries were collected after 36 hours. The distribution of soluble and insoluble derivatives, in the ovule, were checked autoradiographically. The soluble derivatives were found mostly concentrated in the vascular supply, chalazal region of the nucellus (not in hypostase) and the cytoplasm and vacuole in the embryo sac. In addition to this there are some other evidences also which suggest that the food materials enter into the embryo sac through chalaza. They are:

1. Presence of wall projections in the antipodal cells bordering the nucellus at the chalazal end of the embryo sac in Maize (Diboll and Larson, 1966) and Poppy (Jensen, 1972).
2. Mogensen (1973) studied *Quercus gambelii* and suggested that the most likely pathway of food transport within the ovule is from the outer integument to chalaza (facilitated through plasmodesmata) and through postament (group of persistent nucellar cells) to the embryo sac.
3. In fertilized ovules, the concentration of free amino acids is highest at the chalazal end of the embryo sac and there is a definite chalazal micropylar gradient (Ryczkowski, 1971). All these evidences support the chalazal entry of the food materials into the embryo sac.

Sometimes, the entire surface of the female gametopyte absorbs nutrients from the surrounding nucellar cells. In such cases, the nucellus is completely consumed by the developing embryo sac and the embryo sac comes in direct contact with the integument. In such case, the inner epidermis of the integument forms a glandular endothelium around the embryo sac. The endothelium has many similarities with the anther tapetum, and absorb the nutrients from the other tissues and supplies to the embryo sac. The embryo sac or the various components of it, of many plants become extraovular and produce conspicuous haustoria which absorb nourishment from the ovarian, plancental and stylar tissue. Examples of this type are found in the families Lentibulariaceae, Loranthaceae, Santalaceae etc.

The nucellar cells are found to store synthesized food and also seem to have high metabolic activity so that even new substances are synthesized. Poddubnaja-Arnoldi and Zinger (1961) recorded the presence of fat in the living ovules. In *Cyprepedium,* it appears at the time of fertilization, when the starch disappears. Fat is also brought into the embryo sac by the pollen tube and it concentrates around the egg apparatus and polar fusion nucleus. However, in *Calanthe* and *Dendrobium* fat is present in the ovules from the earlier stage. It accumulates at both poles of the embryo sac particularly in the micropylar portion. It is consumed after the fertilization when accumulation occurs in the embryo. The physiological and biochemical changes in the female gametophyte are of great importance since they control subsequent development leading to the production of fertile seeds.

Kallarackal and Bhatnagar (1980) studied the cytochemistry of the ovule of *Linaria bipartita.* They suggested that before fertilization, synergids are involved in the nutrition of the egg and the antipodals in the nutrition of the central cell, which later develops into the endosperm. The nutrients pass from the chalaza to micropylar end through the integumentary tapetum which in *Linaria* persists and is metabolically highly active. After fertilization, the micropylar endosperm haustorium become active and takes up the function of the synergids and the chalazal endosperm haustorium acts in the place of antipodals in providing nutrition to the developing endosperm (Bhatnagar and Kallarackal, 1980; Kallarackal and Bhatnagar, 1981). The possible pathways of the nutrient into the embryo sac in *L. bipartita* before and after fertilization are depicted diagrammatically in Figures 6. 11 A and B.

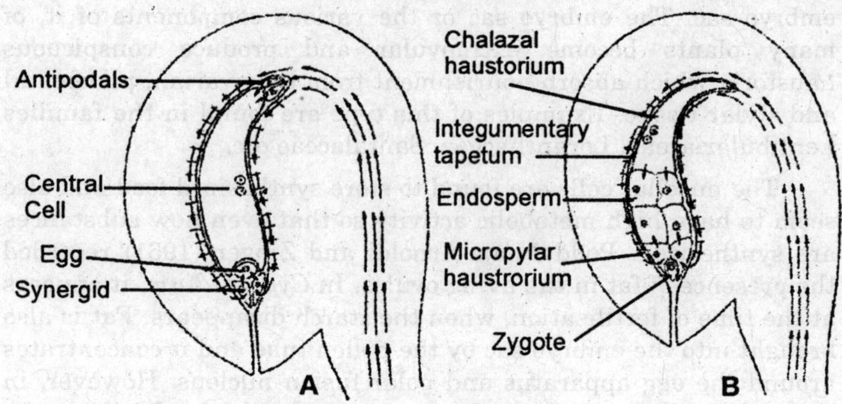

Fig. 6.11 Schematic diagram of the possible flow of nutrients in the ovule of *Linaria bipartita* before and after fertilization. **A.** Before fertilization; **B.** After fertilization

References

Berger, Ch. and O. Erdelska 1973. Ultrastructural aspects of the embryo sac of *Jasione montana* L. Cell Walls. *Caryologia*, **25**: 109-120.

Diboll, A.G. and D.A. Larson 1966. An electron microscopic study of the mature megagametophyte in *Zea mays*. *Am. J. Bot.* **53**: 391-402.

Kapil, R.N. and A.K. Bhatnagar 1976. The Antipodes. *Phyta*, **1**: 54-75.

Kapil R.N. and A.K. Bhatnagar 1981. Ultrastructure and biology of female gametophyte in flowering plants. *Int. Rev. Cytol.* **70**: 353-403.

Masand, P. and R.N. Kapil 1966. Nutrition of the embryo sac and embryo-a morphological approach. *Phytomorphology*. **16**: 158-175.

Mogensen, H.L. 1973. Some histochemical, ultrastructural and nutritional aspects of ovule of *Quercus gambelii*. *Am. J. Bot.* **60**: 48-54.

Newcomb, W. 1973. The development of the embryo sac of sunflower *Helianthus annuus* before fertilization. *Can. J. Bot.* **51**: 863-878.

Philipson, M.N. 1981. The haustroial synergids of *Cortaderia* (Gramineae) at maturity. *Acta Soc. Bot. Polon.* **50**: 151-160.

Swamy, B.G.L. and K.V. Krishnamurthy 1975. Embryo sac ontogenies in angiosperms. An elucidation. *Phytomorphology* **25**: 12-18.

Vijayaraghavan, M.R., W.A. Jensen and M.E. Ashton 1972. Synergids of *Aquilegia formosa*: Their histochemistry and ultrastructure. *Phytomorphology*, **22**: 144-159.

CHAPTER – VII

FLORAL CHARACTERS, POLLINATION MECHANISMS AND VECTORS

A. Janaki Bai

Department of Botany, Andhra University, Visakhapatnam - 530 003

In Angiosperms, pollination typically consists of three phases (1) release of pollen from the 'male' part of flower, (2) transfer from the paternal to the maternal part and (3) successful placing of pollen on the recipient surface of maternal part followed by germination of the pollen grain. Though the plant is usually a passive partner which remains static and displays its essential organs, in some plants there are moments of the style and stigma making a suitable position for contact with the visitor. The animal is the active partner and its importance is determined by the extent to which it effects cross-pollination in the plant population.

FLORAL CHARACTERS

The angiosperm flower structure can be fully understood when considered in relation to pollination ecology, *i.e.*, as a functional unit. The most frequent function of the calyx is that of protecting the young flower before it opens. The corolla generally attracts pollinators by long-distance advertising of the flower and its attractants. Anthers produce and present the pollen, and the pistil contains the ovule, the stigma constitutes the receptive structure and the style guides the growth of the pollen tube towards the ovules. These are certain morphological units with their most common function, but any of these functions may be missing in the flower under observation, and some of them may be taken over by other organs. A flower can be autogamous (self-pollinated) when it is pollinated by its own pollen. A flower pollinated by pollen from another flower is allogamous (cross-pollinated). Allogamy is of two kinds. It can be geitonogamy if the pollen comes from another flower of the same plant or it can be xenogamy when the pollen comes from another plant. Cross-pollination may be made within the individual flower if the stamens and stigmas are widely separated in space (herkogamy) or if they ripen at different times (dichogamy). Both features may be present in the same flower. When stigmas and anthers mature simultaneously, the flower is homogamous and these flowers can be self-fertilized. When stigmas and anthers mature at different times the flowers are dichogamous. If the anthers ripen first the flowers are protandrous and if the stigmas become receptive first the flowers are protogynous. All the flowers which open and expose their stamens and stigmas to the air are chasmogamic and majority of angiosperms belong to this category. Flowers which remain closed and produce good seed are necessarily self-fertilized and are called cleistogamous. These flowers often have the appearance of buds whose development has been arrested.

POLLINATION MECHANISMS AND VECTORS

Pollination can be brought about by the abiotic (natural) agencies or biotic (animal) agencies. The natural agencies of pollination include wind, water and rain. The animal agencies of pollination include insects, birds and bats.

I. ABIOTIC POLLINATION

Wind (Anemophily)

Pollination by wind occurs in 18 percent of all angiosperm families. Although conifers are thought to have been the direct ancestors of the wind-pollinated angiosperms, it is currently interpreted that these angiosperms originated independently from several different ancestral stocks. This mode of pollination is prevalent in such families as Poaceae, Cyperaceae, Juncaceae, and within the orders of Amentiferae and Urticales. In some plants anemophily seems to be active together with entomophily and the two modes of pollination have different relative importance. Wind pollinated flowers do not possess any attractive features, but are chiefly composed of the essential organs alone, and are exposed before leaves come out or exposed outside the leaf mass. The flowers are frequently arranged in 'catkins' and are usually unisexual. The male catkins are often pendulous and flexible so that they are easily shaken by the wind. *Casuarina* has the 'lamb tail' kind. The female inflorescence of this tree species is stiff erect catkin and become cone like in fruit. Other trees with catkins include Alder, Hazal, Birch, Poplar, Oak etc. The other major anemophilous group is Poaceae. The grasses have spreading paniculate inflorescences with wiry rachises which move freely in the lightest air currents. The flowers are with insignificant perianth or altogether lacking and the inflorescence is raised above the level of leaf mass. The anthers are versatile, *i.e.,* attached midway along their length so that they swing easily. The pollen usually pours out as soon as the anthers dehisce and leaves the anthers empty. The stigmas are large, characteristically exposed; they have feathery outgrowths adapted for intercepting wind-borne pollen. In other anemophilous flowers, if they are not inserted on a flexible axis, are often positioned on long slender filaments. Anemophilous plants in general produce greater amounts of pollen. For example male plants of *Mercurialis annuua* produce 39,30,00,000 and *Zea mays* 1,85,00,000 pollen grains per plant. This cannot be taken as a rigid rule. Many grasses have relatively small number of pollen per anther. The pollen are smooth coated, dry, have no oil coating and shed singly. The buonancy of pollen grains increases as the size of the grains decreases. In anemophiles the pollen grains belong to smaller size classes with a typical diameter of 20-30 μm. The number of ovules per flower is rather low, frequently reduced to one, so that each flower produces one seed. Anthers generally open when the weather is warm and dry. In rain, pollen is rapidly washed out of the air. In anemophiles small populations are always at a disadvantage, which are dependent on massive pollen incidence. But they overcome disadvantage by shedding pollen under low turbulent conditions as in *Emblica officinalis, Cicca disticha* and in many grass species in the tropics. Though considered inefficient and pollen movement is random, wind-pollinated plants have evolved strategies for successful reproduction fostering a high degree of outcrossing.

Water (Hydrophily)

The transfer of pollen in water generally takes place within a smaller volume of medium. Water pollination may take place either on the surface of the water, ephydrophily, or in the water, hyphydrophily. Not all pollinations taking place in the water is hydrophily since many small, autogamous terrestrial plants may flower even when submerged and the autogamous mechanism function in an air bubble confined within the flower. In ephydrophily, in general pollen is released in the water and floats up to the surface where stigmas are exposed as in *Ruppia*. The pollen grains appear as small, yellow dots on surface of water and spread rapidly, probably assisted by the oily coating on the pollen surface. The pollen of *Neptunia* and *Aeschynomene* sp. form a foam on the water surface, female flowers emerge into this and are again withdrawn. In *Vallisneria*, the whole male flower is released instead of individual pollen grains. Small depressions in the surface membrane around the emerging female flowers cause floating male flowers to slide down and the anthers attach themselves to the stigmas. This type '*Vallisneria* mechanism', is also found in various members of Hydrocharitaceae, partly in combination with explosive anthers as in *Hydrilla*. *Elodea canadensis* also exhibits surface hydrophily where the pollen is shed into the surface of water. The exine is spiny and air bubbles imprisoned between the spines enable it to float and keep it dry and increase its chance to reach the stigma. Hyphydrophily has been described for a few plants like *Najas*, *Halophila* and *Ceratophyllum*. In *Najas* the slowly sinking pollen grains are caught by the stigmas. In *Ceratophyllum* the anther dehiscence occurs in the air and then the pollen returns to the water. These pollen grains sink gently on to the elongated bifid stigmas of the female flowers. In *Zostera* the pollen dispersal unit is 2500 μ long and is more like a pollen tube than a grain. It can twist itself any narrow object in its way such as around the stigma. Most aquatic plants such as *Nymphaea*, *Alisma* and *Hottonia* are entomophilous. *Potamogeton* is anemophilous and *Lobelia* operates autogamy. But some aquatic plants have taken the advantage of the water also for pollination. Hydrocharitaceae members show a gradation from the flowers present at or above the water surface, to the totally submerged and truly aquatic flower. *Ottelia cordata* flowers are partly submerged then follows the flowers of free floating *Hydrocharis morus-ranae*, then comes *Stratiotes aloides*, a submerged aquatic plant which rises to the surface at flowering time. *Vallisneria spiralis* is with surface flowers and lastly *Halophila*, a totally submerged perennial of tropical seas and its flowers have truly hydrophilous pollination. The individual pollen grains of *Halophila* are short, but they are united into thread like strings which catch against the filiform styles.

Water pollination is accompanied by a disappearance of the positive attributes of scent, colour and nectar from the flower. The floral envelops are small, often only a single whorl of sepals remaining or it may also be absent and the flower may be with a single naked ovary or stamen. This simplification is an overall reduction in size and a tendency for the flower to become unisexual. If the flowers are to be pollinated above water, they are completely protected by enveloping bracts or spathes for the whole of their underwater life. Specific gravity of the pollen also aids in the water pollination. In some of the submerged plants, oil globules in the pollen grains make them lighter than water and keep them suspended to reach the stigmas by water currents.

Rain (Ombrophily)

Hagerup has described rain pollination in a number of flowers in the Faroes. The principle is that, during rain, the flowers fill with water up to a certain level. Pollen grains float on the surface of this water, and reach the stigmas that are exposed at the level of the water. This mechanism is observed in *Ranunculus repens*. In *Narthecium ossifragum* abundant hairs present on the filaments of the stamens hold the water drops. The drop bridges the gap between the dehisced anther and the stigma, and the pollen floats over on the stigma. Both these mechanisms are successful since these flowers set abundant seed in this area where pollinating insects are virtually absent. However, Daumann doubts the idea of ombrophily. If at all rain pollination is operative it would be autogamy. But allogamy, though not very effective, would be conceivable with rain-splash.

II. BIOTIC POLLINATION

In biotic pollination there should be bodily contact between the pollinator and the organ of pollen deposition and reception. The blossoms of biotic pollinators possess an attractant and means of making its existence known, generally by a large and conspicuous perianth. Pollen grains are of variable size, sculptured, sticky or tied together by thin viscin strands or in pollinia. Anthesis and production of attractant should synchronize with the activity of the pollinators. Biotic pollination always depends on external circumstances which may cause the pollinators to be or not to be present or active, and also the blossom must be in a 'proper' state. Of the three groups (insects, birds and bats) of biotic pollinators, insects play a major role in pollinating many plants. This is called entomophily, and include many groups of insects as thrips, beetles, flies, ants, wasps, bees, moths and butterflies. Insects play a role in pollination while searching for food to keep up their activities going.

Thrips

The association of thrips with flowers is well established. They are known mostly as flower visitors and dwellers. Their role in pollination is reported in European Ericaceae and Malaysian dipterocarps. In India *Thrips hawaiiensis* and *Haplothrips tenuipennis* are found inside the flowers of *Lantana camara*. They utilize pollen grains and stigmatic exudates as food and are associated with the yellow phase of the flowers. The stigmas become receptive in yellow phase only. The flower colour changes after pollination from yellow to shades of red, scarlet and mauve. Many species of thrips associate with Asteraceae flowers and their relation is species specific.

Beetles (Cantherophily)

Many species of beetles visit flowers, but in the great majority of cases, the plants reap no benefit. It is generally held that the beetles are the least specialized flower visitors and accomplish pollination in a mess - and - soil manner. They visit flowers primarily for edible rewards (pollen and nectar), mating activities and thermogenesis. There are no flowers adapted to pollination by beetles, although there are cases of beetles effecting pollination.

Beetles were probably the earliest of the flower-visiting insects and beetle pollination is prevalent in primitive angiosperms such as *Magnolia*. The beetle-pollinated angiosperms are divided into four overlapping modes - bilabiate, brush, chamber blossoms and painted bowl. It is the chamber blossom flowers that have been traditionally used to illustrate beetle pollination in 'primitive angiosperms'. But the syndrome of cantherophily may not be uniform with all the cantherophiles. Even in Magnolidae, diversity in the syndrome was noted. Characters of cantherophily include limited flower number, aggregation of beetles, prolonged visits, mating within the flower and visitor constancy. *Myristica insipida* and *Drimys brasiliensis* pollinated by beetles show different syndromes and the term Microcantherophily include small inconspicuous, exposed flowers offering pollen as a reward and the small beetles foraging on pollen. In India, the scarabeid beetles are the primary pollinators of *Chrozophora rottleri* (Euphorbiaceae), a weed of the dry arid habitats. The chrozophora-scarabeid beetle system does not totally agree with either of the two syndromes recognized, some features correspond with cantherophily and some with microcantherophily. Hence it represents a different system. *Euphorbia geniculata* received beetles of the families Coccinellidae, Nitidulidae, Bruchidae and Curculinidae. These two plants belong to Euphorbiaceae, a phylogentically advanced family.

Flies (Myophily)

Flies are more or less omnivorous. They generally visit blossoms for nectar and/or pollen. They are suggested as pollinators of primitive angiosperms. The sub-order Nematocera of Diptera is considered to be the most primitive and their relation with flowers is rather poorly known. However, they are important in certain flower pollination, fungus gnats and flowers; mosquitoes and *Habenaria* orchids; various nematocerans and Araceae; and Ceratopogonidae and Cocoa. The second suborder, Brachycera consists of a wide diversity of flower visitors. Some possess short mouth parts making visits to many species bearing shallow flowers with easily accessible nectar like umbellifers and some possess long mouth parts visiting deep flowers with tubular corolla. The Bombylidae have long sucking mouth parts capable of handling tubular flowers and are prominent among the flower-visiting Dipterans. In the third suborder Cyclorrhapha, Syrphidae is the most important family of flower visitors and feed on nectar and pollen. Members of Calypterae are present in abundance, visit the flowers frequently and transport pollen effectively. They are proved to be exclusive pollinators of *Sterculia foetida* and also the major pollinators of *Euphorbia antiquorum* and *Ziziphus mauritiana.* The ideal fly flowers produce nectar in small aliquots and that nectar is exposed freely and evaporation causes nectar to be concentrated. As flies are unspecialized and unreliable pollinators due to their limited flight range and feeding habits, the fly flowers possess adaptations to achieve outcrossing through mechanisms of protogyny, protandry, dioecism and temporal dioecism or duodichogamy. *Sterculia foetida* is functionally dioecious and *Euphorbia antiquorum* displays temporal dioecism. In both these plant species *Chrysomya megacephala* is promoting cross pollination. The parasitic flowering plant *Rafflesia* known for its largest flowers is pollinated by *Chrysomya.* Carrion and dung flies have no adaptation for flower visits, but the basis for the visit is deceit (Sapromyophily). Typical carrion and dung flies are attracted by odour of the blossoms resembling decaying protein that activate the flies for feeding or oviposition. Sapromyophilous blossoms are found in various families like Asclepiadaceae, Aristolochiaceae, Sterculiaceae, Orchidaceae etc. Various species of *Aristolochia* are typical examples of sapromyophily.

Ants (Myrmecophily)

Ants are extremely abundant in the tropical habitats. They are so small that they can sneak in and out of many blossoms, without even touching the essential organs. Hence they are considered as nectar thieves. The rarity of ant pollination is attributed to the limited foraging areas, lack of wings on the worker ant and due to presence of ant deterring structures in some blossoms and also petals which are chemically conditioned to ant-repellent effect and also due to reduction in viability and germination of pollen exposed to myrmicacin secretions of the ant's metapleural gland. However, ant pollination is extremely common in a few kinds of habitats. In the tropics, particularly in the dry season, floral nectar is most attractive both as an energy and water source to the ants. In Sahara with burning hot soil and few insects, ants are important pollinators. Much of the vegetation in this area is Euphorbiaceae members which have extra-floral nectaries and ants feed on the nectar. In Western Cascades of Oregan *Polygonum cascadense* is pollinated by ants. The orchid *Loperella fimbriata* is pollinated by pseudocopulation by winged males of the ant *Myrmecia urens*. It is suggested that these ant pollinators do not possess metapleural glands or the plants pollen must possess resistant characters. Metapleural glands are absent in all castes of the genus *Camponotus*. *Euphorbia dracunculoides*, a weedy species of the dry, arid lands of India is adapted for pollination by the ant *Camponotus*. In this species the cyathia are small, possess no odour, and secrete nectar in traces, thus unlikely to attract pollinators of high-energy demand. The plants are small and occur in extremely dense populations, thereby reducing ant-foraging distances. When the populations are thin, there is excessive branching, thus interplant distances are minimized thus facilitating low energy requiring crawling ants to move pollen between plants. Worker ants intensively forage at the cyathia. Nectar foraging by winged males largely promote xenogamy. In *Euphorbia geniculata* ants numerically preponderate and are the only pollinators when the plants occur in *Sorghum*, *Pennisetum* and Maize crops. The syndrome of ant pollination include dry-hot habitat, nectaries small, quantity of nectar too small to interet larger visitors, blossoms sessile, exposed near the ground. Gregarious occurrence, of several individuals facilitate xenogamy.

Wasps

The wasps in general are treated as unreliable and unsteady pollinators because majority of the true wasps are primarily animals of prey and feed their larvae with insect larvae. They visit flowers only for nectar to meet their carbohydrate requirement. At least some wasp species feed their larvae with nectar, and this is essential to the diet of *Vespula* larvae. Agonial wasps (*Blastophaga*) appear to be markedly selective and visit species of Figs, earning them the name 'wasp flowers'. *Vespa* species repeatedly enter and pollinate the flowers of *Scrophularia nodosa* and *S. aquatica*. In some higher wasps like *Polistes*, the mouth parts are modified into tubular structures as against the flat tongue of many wasp species. Through these tubular structures nectar can be sucked up and the wasps pollinate the blossoms. Studies in India proved the role of wasps in the pollination of such plant species as *Premna latifolia, Santalum album, Ziziphus mauritiana, Z. oenoplia, Vitex negundo* etc. Though these species receive visits from many insect species, they are more attractive to wasps and are abundantly visited by diverse wasps. Some wasps resort to nectar robbing by making holes in the corolla as in case of *Vitex negundo* by *Rhynchium metallicum* and *Ropalidia* species. Other foraging wasps like *Bembix* also use these perforations for nectar uptake. Other tropical species such as *Celosia argentea, Jatropha curcas* are also wasp pollinated. At Timbucta forest, *Acacia tortilis* and *Ziziphus jujuba* are pollinated by a big, clumsy wasp *Eumenes tinctor*.

Bees (Melittophily)

Bees, whether social or solitary, totally depend on flowers and visit more flowers than all the other groups of anthophilous insects. The bees require enormous quantity of pollen and nectar for their nutrition and for their brood. Morphologically and behaviourally, they are well suited to collect pollen and nectar. The bees may collect pollen nototribically on its back or may be sternotribically on the ventral surface of the abdomen by direct contact with anthers. The colour vision of the bees is well developed and visit yellow, blue or purple flowers. Social bees are the most active and best known pollinators. The genus *Apis* is active year round and visit any flower from which it can obtain food (polytropic) and communicate odour easily. Big bees either solitary or social are able to push aside blossom parts barring entrance to the nectar. These bees make interplant movements. Long tubed flowers massed together into dense heads are attractive to long tongued bees and short tubed flowers massed into flat headed inflorescences are preferred by short tongued bees. The carpenter bee (*Xylocopa*) flowers are zygomorphic, strongly made, large with adequate facilities for landing and have deep-seated nectar which is protected by tough base of the corolla tube. The role of Xylocopas, and other anthophorid bees is important in the pollination of floral species with poricidal anthers. These bees sonicate the anthers using trains of audible buzzes generated by the thoracic muscles. The mechanism results in the buzz-pollination of nearly 20,000 species of plants the world over. The Cassias are well suited for the buzz-pollination by different species of *Xylocopa*. *Cochlospermum religiosum* is cross pollinated exclusively by *Xylocopa latipes* and *X. pubescens*. The Xylocopas are important pollinators of Cassias, Lamiaceae, Acanthaceae and Verbenaceae also. The sunhemp and the Indian pulses are effectively pollinated by Xylocopas. Another solitary bee *Amegilla* is capable of buzz collecting of pollen in *Solanum* and *Lycopersicon*. The stingless bee *Trigona* is also known to buzz pollination in crop species such as *Mangifera indica* and *Cocos nucifera*. Several other solitary bee species involved in the pollination of several agricultural crops include *Amegilla, Andrena, Megachillae, Halictus, Lasioglossum, Colletes* etc. Bee species especially *Apis cerana indica* and *Apis mellifera* are important in pollinating the crop species and improving their yields. In the hills particularly Himalayas, *Bombus* is the most important flower visiting insect. The bees are diurnal in their activity and are active during day light. But they can also work in moonlight nights. The rockbee *Apis dorsata* is the sole visitor and pollinator of the night blooming *Pterocarpus santalinus* in its natural habitat. The flowers open at around midnight and are foraged during moonlight hours. *Xylocopa pubescens* is the sole pollinator of night blooming *Antirrhinum majus*.

Moths (Phalaenophily)

The moths are nocturnal, but some are diurnal. The night-time moths (Noctuids) alight on a blossom and then collect pollen on the legs and use nocturnal flowers. The day-time flying species (sphingids) forage on the diurnal and also nocturnal flowers that stretch into the following day. The sphingids, mostly hover while taking nectar and thus collect pollen on proboscis and head. They have a high metabolic rate, especially when hovering and require much food, consequently they are relatively important pollinators. Most of the moth flowers are nocturnal and open in the evening. Scent is the foremost attractive attribute of the moth flowers. The flowers are pale in colour, white, cream and light yellow, and have very long tubes. There is no landing platform, hence the moths hover in front of the flower and probe the proboscis to the base of the corolla tube. The pollination of *Yucca* by *Tegiticula* is well known. The pollination by hawk moths may be more common in tropics. Hawk moths may be important pollinators in orchid species, *Capparis* species and *Datura meteloides*. The sphingid hawk moth, *Macroglossum gyrans*, which is diurnal in activity pollinates *Xeromphis spinosa* at dawn and dusk when the flowers contain more nectar. A similar activity pattern is also reported on *Martynia annua* and *Datura repens*. Other plant species with adaptations for moth-pollination include *Carissa* species, *Holarrhena antidysenterica, Morinda tinctoria, Nyctanthes* species and Oleaceae members. *Macroglossum gyrans* also forages on *Tecoma* species, *Lantana camara, Jasminum angustifolium* and *Sansevieria*. The humming bird hawk-moth, *Macroglossa stellatorum* flowers have the long tube characteristic of the group, but have 'day' colours red or pink and a small landing platform. *Gnetum* is pollinated in Malaysia by moths.

Butterflies (Psychophily)

Butterflies do not feed their brood and food collection is for their own consumption. Most butterflies visit flowers for nectar during day time. Nectar provides energy and amino acids for nutrition and may increase butterfly longevity and egg production and egg maturation. Though butterflies are considered to occupy a parasitic mode of life as adults feeding on the nectar of flowers without pollinating them, studies indicate that butterflies do mediate pollen transfer. The butterflies prefer to visit flowers with long narrow tubes and flat-topped corollas. They may be consistent in their visit to a particular species if nectar is available at the source, or they may be erratic in their flower visit. The butterfly seasons correspond with the food plants of their larvae rather than to the blooming of any species of flower. They usually have a long period of activity because more than one brood is reared. The transfer of pollen by butterflies depends on floral architecture and intrafloral behaviour of the butterfly. The papilionids hover directly in front of the flower while probing for nectar, whereas the other butterflies perch on the flower inflorescence, while probing for nectar. Contact of essential organs of a flower on the body of butterfly, may be a proboscis, head, antennae, legs and wings. Pollen deposition on proboscis and head is treated as an important trait of psychophily. Certain floral species possess elongated staminal filaments and style and orient their flowers horizontally. Such orientation of flowers, occur in *Caesalpinia pulcherrima, Clerodendron infortunatum, C. inerme* and *Cadaba fruticosa*. The first three species are mostly visited by Papilionids, and the latter by Pierids *Colotis* and *Anaphaeis*. In *Clerodendrum infortunatum* the big bodied *Papilio polymnestor*, *Pachliopta aristolochiae*, *P. hector* and *Popilio polytes* perform pollination. In *Cadaba fruticosa* and *Caesalpinia pulcherrima* nectar is present in a nectarial tube, and the spatial separation of this and essential organs facilitates correct use and placing of proboscis. These four species are exclusively pollinated by the butterflies. The drifting behaviour of butterflies to visit a few flowers on a plant and fly away contributes to xenogamous pollinations. Nectar concentration of approximately 40 percent is optimum for butterfly feeding. Butterflies with a larger body size can extract nectar at a faster rate than small body size butterflies, and this behaviour is reflected in the flower handling time. The papilionid butterflies normally have a larger body size, able to have fast nectar extracting and cover a large number of flowers in a foraging trip.

Birds (Ornithophily)

Birds possess good external prerequisites for becoming pollinators. Birds can be divided into two natural groups: the non specialized nectar feeders with licking type of bill and tongue, and specialized nectar feeders with suctorial mouthparts. *Chloropis aurifrons*, *C. cochinchinensis*, *Dicaeum erythrorhynchos*, five species of sunbirds and white eye are the specialized nectar feeders. The primary feature of bird flowers is copious production of nectar, which is supplied by many small flowers or large inflorescences or separate large flowers. Of the 8 families of the flower visiting birds of the world, four families Nectariniidae, Zosteropidae, Dicaeidae and Psittacidae are represented in India. In all 58 Indian bird species are reported to visit 93 flowering plant species and to this list *Brugiera* species and *Acanthus ilicifolius* of mangrove forest, and *Anisomeles indica* and *A.malabarica* occurring in semi rocky areas may be added. *Brugiera* with explosive opening of flowers is pollinated by white-eye and the other three plant species by sunbirds. Out of the 93 plant species the birds are named for only 78 species. Of these birds the two sunbird species *Nectarinia asiatica* and *N. zeylanica* are the versatile flower visiting bird species foraging at 73-74 percent of the 78 flowering plant species. The sunbirds are passerine (perching) birds which alight on the stems when feeding and exploit the flowers either from above or below. In

Loranthaceae (mistletoes) all dioecious species are entomophilous, and all hermaphroditic species are bird-pollinated. The mistletoe flowers normally require an external pressure to open and such pressure comes from birds, and ultimately help the flowers in pollination. Spontaneous flower opening occurs in *Helixanthera intermedia* and *Dendrophthoe memecybifolia*. Four bird species – flower pecker (*Dicaeum concolor*), White-eye (*Zosterops palpebrosus*), small sunbird (*Nectarinia minima*) and purple sunbird (*Nectarinia asiatica*) are the effective pollinators of seven species of mistletoes. *Erythrina indica* is visited by 12 species of birds of which *Sturnia pagodarum* (Black-headed myna) is the commonest. Seven species of birds of six families frequent flowers of *Butea monosperma*, but only the purple sunbird *Nectarinia asiatica* is the effective pollinator. In *Helicteres isora,* flower colour changes from bluish grey (day 1) to red (day 2) and deep red (day 3). The flowers are tubular with wide mouth and probed by insects mostly for pollen and birds for nectar. Two bird species *Turdoides affinis* and *Acridotheres tristis* while collecting nectar come in contact with essential organs and get powdered with pollen on billforehead invariabily effecting pollination. An obligate association of birds and floral species appears to be rather uncommon, since the diet of flower birds include, not only nectar as an energy source but also insects to satisfy their protein requirement. Most of the floral species identified as bird flowers (*Moringa oleifera, Bauhinia purpurea, Peltophorum pterocarpum, Cassia* species, *Calotropis gigantea, Tecoma stans, Adhatoda zeylanica, Acanthus ilicifolius, Anisomeles* species, *Gmelina arborea* and *Cochlospermum religiosum*) have been proved to be pollinated by carpenter bees. However, the associations: White eye – *Bruguiera*, and asiatic nectar feeding assemblage of birds – mistletoes, may be taken as a mutualistic obligate association.

Bats (Chiropterophily)

Bats are tropical in distribution and bat pollination is typically a tropical phenomenon. Bats belong to the order Chiroptera and are nocturnal. Some bats can eat insects (insectivorous) and they belong to sub-order Microptera. Some eat flower petals and fruits (frugivorous) and they belong to sub-order Magachiroptera. Both frugivorous and insectivorous bats visit flowers for pollen and nectar, and their opening is tuned to the nocturnal activity of bats. The flowers are either in bunches, racemes, or even singly, dangle at the end of long leafless inflorescence stalk. Since bat pollination is a nocturnal phenomenon, detailed studies of the supposedly bat pollinated floral species are almost lacking. The only earlier study was of McCann with sausage tree *Kigelia pinnata.* Short nosed fruit bat, *Cynopterus sphinx* comes in contact with the dehisced anthers and stigma while drinking the copious nectar. Recent studies, on *Ceiba pentandra* indicate close correspondence between floral and bat activity. The blossoms are of the shaving brush type opening at dusk, nectar production starts at flower opening and the bat activity closely corresponds with the nectar presentation pattern. The large fruit bat, *Pteropus giganteus* that roosts in large trees visits flowers clustered near the tips of branches, lap nectar and eat pollen. The small bats *Cynopteris sphinx giganteus* and *Taphozous longimanus* roosting in the thatched houses also visit the flowers during the same period. While foraging, the snout and abdomen of the bat brushes against the anthers and stigma, and transfer the pollen to the stigmas of other flowers during successive foraging visits. Large bats make interplant visits, promoting xenogamous pollinations.

Apart from the agents of biotic pollinations described above, the lemurs, squirrels, marsupials, and rodents have been reported to regularly visit nectar-rich flowers and lap their nectar as an important source of food, and hence are the likely pollinators of the flowers they visit.

Further Readings

Faegri, K. and L. van der Pijl. 1979. *The principles of pollination ecology.* Pergamon Press, New York.

Percival, M. 1965. *Floral Biology.* Pergamon press. London.

Jones, C.E. and R.J. Little (eds.) 1983. *Handbook of Experimental Biology.* Scientific and Academic editions. Van Nostard Reinhold Company Inc., New York.

Proctor, M. and P. Yeo. 1972. *The pollination of flowers.* Taplinger Publishing Co., New York.

Proctor, M., P. Yeo and A. Lack. 1996. *The Natural History of Pollination.* Collins, London.

CHAPTER – VIII

FERTILIZATION

The process of fertilization includes the fusion of the male and female gametes. This event occurs inside the female gametophyte or the embryo sac. Angiosperms are unique in plant kingdom in having double fertilization. The phenomenon of double fertilization was first observed by Nawaschin (1898) and Guignard (1899) independently in *Lilium martagon* and *Fritillaria tenella.* In Angiosperms, the gametophytes are extremely reduced and are represented by a fewer number of cells, which are at a very high level of differentiation. The microgametophyte produces two sperm cells, whereas the megagametophyte gives rise to only one egg cell. The development and differentiation of the gametophytes are influenced by the sporophyte. The nutrition of the gametophyte and its duration of development are also governed by the sporophyte. The condition of the pollen and stigma are the two important factors concerned with fertilization.

The female gametophyte of the angiosperms is quite deep seated and lies in the ovarian cavity, which is quite away from the stigma. The male gametophytes (pollen) are generally deposited on the stigma and they have to go a long way through the style in order to reach its fusing partner namely the egg. The pollen grains germinate on the stigma and produce the pollen tube, which carries the sperm cells to their destination. The sperms are usually released by the pollen tube near the egg, of which one of them fuses with the egg (Syngamy) and the other fuses with the secondary nucleus (triple fusion). Syngamy and triple fusion together is known as double fertilization. It is unique to angiosperms. In Gymnosperms only one of the two sperms is functional; one unites with the egg and the other degenerates.

The distance through which the pollen tube has to travel varies with the species and it depends on the length of the style.

Structure and Function of Stigma

The stigma is the receptive surface of the pollen and it consists of specialized cells. The morphology of the stigma and style show great diversity and to some extent depends on the type of pollination. The function of the stigma is to accept the pollen and permit its germination. The stigma also provides nutrients to the pollen and directs the pollen tube growth. Both the pollen coating and stigma coating are involved in recognition and receiving signals.

The morphology of the stigma shows great diversity. Based on the morphology of the stigmatic surface, the amount of secretion and the nature of the surface cells, a general classification is suggested by Heslop-Harrison and Shivanna (1977). They have divided the stigmas into two categories namely - Wet stigmas (with copious fluid secretion) *e.g., Petunia* and Dry stigmas (limited secretions). In dry stigmas, the receptive cells are dispersed on multiseriate branches (*Zea, Sorghum, Hordeum* etc.) or concentrated in distinct ridges or zones. However, in wet stigmas, the receptive surface may have papillae (Liliaceae, Orchidaceae etc.) or non-papillate surface (*Dracaena, Alpinia* etc.) Trinucleate pollen is mostly associated with the dry stigmas, whereas binucleate pollen is seen in both cases.

Structural differences have been noted in the stigmatic papillae of floral morphs in many heterostylous plants. Ghosh and Shivanna reported that in *Linum grandiflorum* the stigma of the pin morph (long styled) is of the 'dry' type and smooth and the papilla is covered by a distinct cuticle pellicle layer. The stigma of the thrum morph (short styled), on the other hand, is of the wet type, warty and uneven and the cuticle-pellicle layer is disrupted.

Dumas (1978) classified the stigmas based on the function and the nature of secreted products. Stigmas have an abundant secretion of lipophilic or polysachharide substances or in some cases, less abundant secretions with polyphenols, tannins, flavonoids etc. Some others have no liquid secretion at all, but stigmas are covered with waxes or proteins. This classification can also be combined with the stylar structure. Hollow styles of monocots produce mainly polysaccharides, whereas solid styles of dicots, produce a lipophytic exudate. Some families like Orchidaceae, Solanaceae, Scrophulariaceae excrete a lipopolysaccharide substances.

Nature of the Stigma covering: The receptive surface is formed by the surface cells which are glandular in nature. This is followed by a transition zone of stigmatoid tissue without glandular cells. The stigmatoid tissue is also known as the neck of the transmitting tissue.

The dry stigma has a cuticle covered with a pellicle. The cuticle has some discontinuities for the water. Both the pellicle and the surface of the stigma remain hydrated. The pellicle is an extracellular product originated from the proteins of the stigmatic cells.

The exudate of wet stigmas mainly consists of amino acids, lipids, antioxidants, proteins and alkaloids. During pollination, the exudate serves as a nutritive for birds, bees, beetles, bats, moths etc. The exudate also had boron and serve as a nutrient for pollen. The exudate of wet stigmas varies in quality and composition and can be considered as a complex fluid. In *Trifolium*, it contains glucose, sucrose and traces of galactose, arabinose, some proteins and esterases. The exudate functions in the hydration of the pollen grain, stimulation or inhibition in pollen germination as well as in nutrition.

Fresh exudate of *Phaseolus* contains lipids, proteins, amino acids, sugars namely glucose, reducing acids and phenols. Alkaloids are absent. Sugars stimulate pollen germination; lipids prevent desiccation and also attract insects. Proteins function in recognition and the amino acids as nutrient for insects. Phenols function as protection against chewing insects and pathogens. Thus, the nature of the stigma covering is very complex and it is not possible to make any generalizations.

Structure and Function of Style

The morphology and anatomy of the style are variable. Three different types of styles are seen in angiosperms namely open, solid and half closed.

(i) Open type (Hollow Style): This type is more common in monocotyledons, *e.g.*, *Lilium, Amaryllis, Gladiolus*. In the hollow style there may be one or more canals; the number of canals normally corresponds to the number of carpels. The stylar canal is lined with canal cells. Pollen tubes creep on the surface of these canal cells. The canal cells are secretary in nature. The canal cells show dense cytoplasmic contents and are rich in cell organelles. The secretion from the canal cells contains protein, carbohydrates, lipids and shows activity for esterases and acid phosphatases.

It has been suggested that in *Lilium* the parenchyma cells underneath the canal cells synthesise some mucilaginous substances and transport them to the canal cells, which, in turn, secrete them into stylar canal. After pollination the canal is filled with secretions which serve as nutrition for the growing pollen tubes.

In the open style, canal cells provide elements for pollen tube growth and functionally similar to the intercellular substances of the solid style.

(ii) Solid type: Mostly seen in dicots, *e.g., Datura, Gossypium.* It has a solid core of transmitting tissue. The transmitting tissue of the solid style has either loosely arranged tissue or it may be more compact. This depends on the amount and quality of the intracellular substance. In general, the cells of the transmitting tissue show a thick cell wall with a small lumen or a thin cell wall with large amount of intracellular substance. The cells of the transmitting tissue contain numerous mitochondria, plastids, rough endoplasmic reticulum, dictyosomes and ribosomes. Along the transmitting tissue, carbohydrates and isoenzymes, such as peroxidase and acid phosphatase are found. The function of the intercellular substance is aiding in the pollen tube growth. In *Oryza sativa* the solid style is devoid of any transmitting tissue.

(iii) Half closed type: It is present in a few taxa in which the transmitting tissue can be limited to one side of the stylar canal. It is seen in family Cactaceae.

POLLEN GERMINATION

Sticking and Hydration

Pollen grains are transported to stigma by various agents such as wind, water, animals etc. and to some extent this depends on the pollen size, mass and sculpture. The first interaction between pollen and stigma is sticking. Adhesion results from the high viscosity and stickiness of the stigmatic exudate. The other factors involved are surface tension, wind force, electrostatic force, gravity etc. In *Brassica*, the incompatible pollen shows a less firm bends in sticking than compatible pollen.

The pollen grains swell in the presence of water indicating its rapid uptake. The water passes through the exine and is taken up by the intine and the cell. The full hydration of pollen is not necessarily a prerequisite for germination.

Pollen Tube Formation and Growth

After sticking, the pollen germination starts with imbibition. If the pollen is fully developed (trinucleate pollen) germination starts quickly. For pollen germination, water or a high humidity is necessary. In some cases, nutrients - mostly carbohydrates and minerals should also be present. Normally, from a pollen grain only one tube develops. However, in polysiphonous pollen grains, more than one tube may emerge from a grain; up to 10 pollen tubes have been observed in *Althaea rosea* and 14 in *Malva neglecta*. Only one of them makes further growth. The pollen tubes generally emerge from the germ pores of the pollen grains. All the contents of the pollen grain move into the pollen tube. Active growth of the tube is generally restricted to the tip region. Most of the cytoplasm of the grain is seen in the tip region and a large vacuole fills the grain. Callose plugs are formed in the basal region of the tube and they restrict the backward movement of the cytoplasm from the tip. Since there is formation of callose plugs, pollen tube of the mature grain is subdivided into many compartments. The callose plug starts originating from the wall of the tube, grows towards the centre and reduces the lumen and finally seals the tube. Small amount of the protoplasm, left in the grain, gradually disappears.

The pollen tube tip region, when observed under high magnification appears hemispherical and transparent. This transparent, apical zone is known as the 'cap block'. It is present only when the tube is growing, but later disappears. At this stage 4 distinct zones can be demarcated in the pollen tube. They are: (1) Apical zone or growth zone (2-4 μm long) which contains number of vesicles, but no cell organelles. (2) Sub-apical zone which is rich in granular cytoplasm and organelles. This region produces vesicles. (3) Nuclear zone, in which are present the vegetative nucleus and generative cell or sperms and (4) Vacuolation zone which forms a transition between the active and the inactive cytoplasm of the tube. This is highly vacuolated and ends with callose plugs. The region of callose plug formation is marked by electron dense ground plasm, absence of plastids and dictyosomes and presence of lipid bodies. The plug formation seals off part of the pollen tube. Pollen germination and pollen tube growth are processes of a free gametophytic phase.

Pollen Tube Wall

This consists of cellulose and pectin. In tobacco, the pollen tube consists of two layers. The outer primary wall is pecto-cellulosic in nature. The inner secondary wall consists of pectin, cellulose and callose.

Factors affecting Pollen Germination and Pollen Tube Growth

The factors affecting the pollen germination and pollen tube growth can be studied in two ways. (1) By culturing the pollen grains in a nutrient medium under the controlled conditions and to check their response to various substances (*in vitro* studies) or by observing the pollen germination and tube growth as it occurs in nature *i.e.,* in the stigma, style and ovary (*in vivo*).

In vitro Studies

1. Relative humidity: High relative humidity is very essential for pollen germination. It has been observed that the germination of the pollen occurs quickly in the saturated atmosphere.

2. Carbohydrates: A source of sugar is also essential for the germination of the pollen grains in a nutrient medium. The sugar helps in two ways. (a) for the control of osmotic pressure, (b) to serve as a respiratory substrate. Sucrose is found to be the best sugar for the promotion of the pollen germination and tube growth.

3. Boron: Boron in the form of Boric acid greatly helps in the pollen germination and pollen tube growth. Generally the mature pollen are deficient of boron, hence its addition enhances pollen germination. In nature, the boron deficiency is compensated by the stigma and style. The presence of boron, in the medium prevents the bursting of the pollen and also helps in the biosynthesis of carbohydrates.

4. Calcium: In nature, pollen grains contain small amounts of calcium. Even that small amount of calcium comes out in the form of Ca^{++} ions and leaves very little inside the pollen, which is normally insufficient for the pollen germination. Hence addition of calcium into the medium greatly helps in pollen germination and tube growth. In the *in vitro* studies, it has also been observed that the pollen germination is better when they are in groups, which is known as "Crowding effect", 'Population effect' or 'mutual effect'. This effect is due to the Ca^{++} ions brought into the medium by diffusion and getting trapped in between pollen grains.

5. Enzymes: Addition of cellulase and pectinase into medium helps in increasing the rate of tube elongation. Cellulase and pectinase are assumed to affect tube elongation through an increase in the plasticity of the tip region.

6. Hormones: Auxins and gibberellins to some extent promote the pollen germination and pollen tube growth.

7. Physical factors: Temperature is important among the physical factors. Increase in the temperature helps in the growth rate of the pollen tube. The optimum temperature required for pollen germination and tube growth is 20-30°C. Very rarely in some incompatible self pollinations, higher temperatures retard pollen tube growth.

In vivo Studies

In nature, the stigmatic surface is highly specialized and is capable of capturing and holding the pollen. In *Petunia hybrida*, the stigma development is studied by Konar and Linskens (1966). The stigma consists of a columnar tissue with a small depression at the centre. The stigmatic tissue is divisible into two zones namely an upper secretory zone and a lower storage zone. Some of the epidermal cells produce 2-celled stigmatic papillae.

The stigma mainly provides the water necessary for pollen germination. It also provides the other nutrients for germination in the form of an exudate.

Path of the Pollen tube

After pollen germination and formation of the pollen tube, for the growth of the pollen tube a continuous slimy passage is required. This mucilage originates below the cuticle of the outer wall of the epidermis (ectotrophic growth) or the inter-cellular substances gets filled with mucilage from the transmitting tissue (endotrophic growth) and helps in the free movement of the pollen tube.

The time required between pollination and fertilization depends on the rate of growth of the pollen tube and the distance it has to travel. For example in sugarbeet it has to grow only a few millimeters, whereas in maize it grows as much as 450 mm. The growth rate of the pollen tube is highly variable. It is 3 mm per hour in *Datura* and 35 mm per hour in *Taraxacum*. This growth is considerably influenced by the temperature and also the degree of compatibility of the parents.

Generally, the time interval between pollination and fertilization is 12-48 hrs. Sometimes, this may be as short as 15-45 minutes as in *Taraxacum* or 19-20 weeks as in the orchid *Paphiopedilum*.

After passing through the stigma and style, the pollen tube reaches the ovary. Basing on the mode of entry of the pollen tube into the ovule, three different types of fertilization was recognized.

1. **Porogamy:** This is the most common type. In this the pollen tube enters into the ovule through the micropyle (Fig. 8.1A).

2. **Chalazogamy:** In this, the pollen tube enters through the chalaza and it is seen in *Casuarina* and other Amentiferae (Fig. 8.1B).

3. **Mesogamy:** In this the entry of the pollen tube occurs through the funiculus as in *Pistacia* or through the integuments as in *Cucurbita* (Fig. 8.1C).

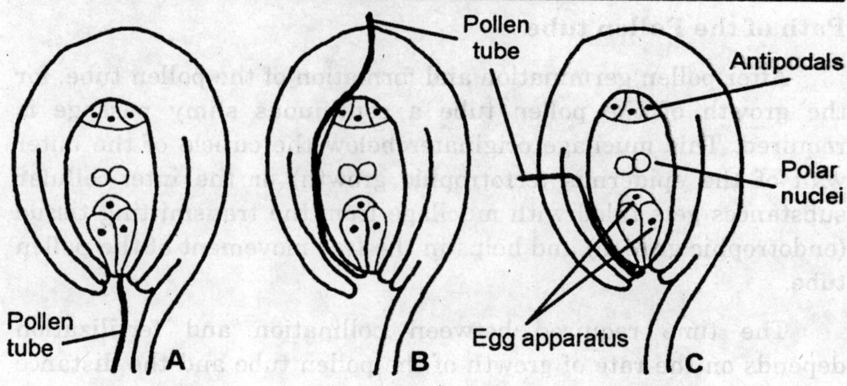

Fig. 8.1. Modes of pollen entry into the ovule.
A. Porogamy; B. Chalazogamy; C. Mesogamy.

With respect to porogamy, there is strong evidence from recent studies on *Arabidopsis*, maize and *Torenia fournieri* that low molecular weight secretary molecules from the female gametophyte, particularly synergids, play major role in short range attraction of pollen tube of the micropyle and the female gametophyte.

Higashiyama *et al.* (1991, 1998) and Higashiyama and Hamamura (2008) made a detailed study of the kinetics of double fertilization including the guidance of pollen tube towards the micropylar end of the embryo sac by direct observations of the live naked embryo sac in *Torenia fournieri* (Fig. 8.2). They developed an *in vitro* system in which the ovules with protruding embryo sac and semi-*in vitro* pollen tubes growing through the cut basal end of the style were co-cultured on thin layer of solid medium. Some of the pollen tubes grew precisely towards the central region of the filiform apparatus between the two synergids, the site of their entry into the embryo sac. The attraction of pollen tubes towards the micropyle is lost after the entry of a pollen tube into the embryo sac. The pollen tubes were required to grow through a cut style to become competent to be attracted by the guiding cues in the ovary, suggesting that female sporophytic tissue promotes sensitivity of the tube to respond to the attractants. If a pollen tube attracted by the micropyle is moved using a micromanipulator the tube precisely chases the micropylar end of the embryo sac (Higashiyama and Hamamura, 2008).

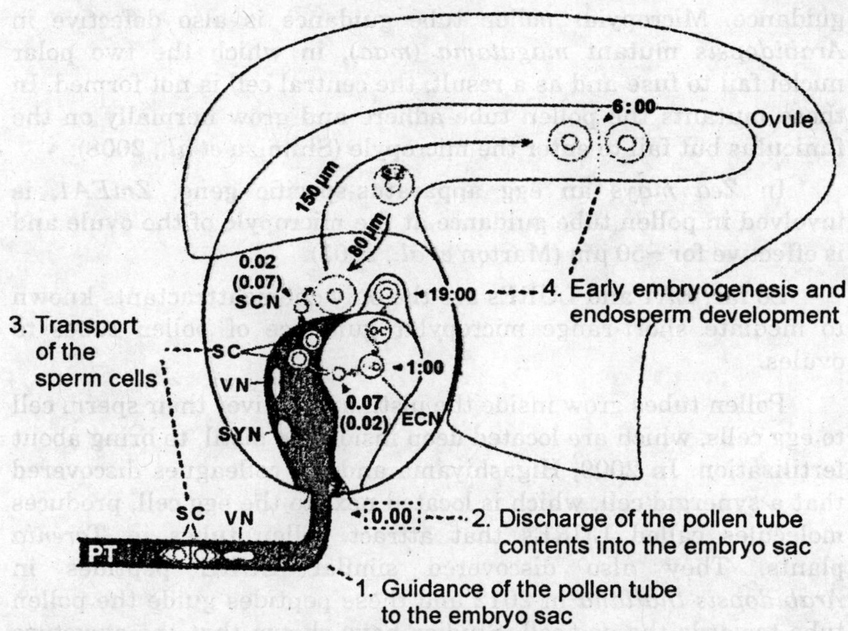

Fig. 8.2. A schematic representation of *in vitro* fertilization in *Torenia fournieri* (after Higashiyama, 1999); **ECN** : Egg Cell Nucleus; **SCN** : Sperm Cell Nucleus; **VN** : Vegetative Nucleus; **SYN** : Synerdid; SC : Sperm Cells

Thus *in vitro* system of *Torenia* suggested that some diffusible signal is derived from the micropylar end of the embryo sac. Pollen tubes do not enter the ovules with the synergid cells either laser ablated (Higashiyama *et al.*, 2001) or with abnormal filiform apparatus (Kasahara *et al.*, 2001). The distance of pollen tube attraction by synergid is 100-200 µm (Kasahara *et al.*, 2005; Punwani *et al.*, 2007). LUREs (LURE1, LURE2), ~9 kDa CRPs with 6 cysteines, were identified as the attraction molecules (Okuda *et al.*, 2009). These proteins are secreted by the synergid towards the micropyle and detected in the filiform apparatus (Higashiyama, 2010).

In *Arabidopsis*, a synergid cell-specific transcription factor gene, *MYB98*, is involved in micropylar guidance, which is effective for a distance of ~100 µm (Kashahara *et al.*, 2005; Punwani *et al.*, 2007). The *Arabidopsis* mutant *myb98*, impaired in filiform apparatus development, is impaired in the pollen tube guidance. Micropylar pollen tube guidance is also defective in *Arabidopsis* mutant *magatama* (*maa*), in which the two polar nuclei fail to fuse and as a result, the central cell is not formed. In these mutants the pollen tube adhere and grow normally on the funiculus but fail to enter the micropyle (Shimizu *et al.*, 2008).

In *Zea mays* an egg apparatus-specific gene, *ZmEA1*, is involved in pollen tube guidance at the micropyle of the ovule and is effective for ~50 μm (Marton *et al.*, 2005).

So far, EA1 and LUREs are the only chemoattractants known to mediate short-range micropylar guidance of pollen tubes to ovules.

Pollen tubes grow inside the pistil and deliver their sperm cell to egg cells, which are located deep inside the pistil, to bring about fertilization. In 2009, Higashiyama and his colleagues discovered that a synergid cell, which is located next to the egg cell, produces molecules called LUREs that attract pollen tubes in *Torenia* plants. They also discovered similar LURE peptides in *Arabidopsis thaliana* in 2012 and these peptides guide the pollen tube towards the egg cell. Studies have shown that the structure of LURE differs for each plant species and is specific for each plant's pollen tube, *i.e.* each LURE peptide preferentially attracts the pollen tube of the same plant species. Takeuchi and Higashiyama (2016) have discovered a receptor that is required for detection of LURE at the tip of the pollen tube for the model plant, *Arabidopsis thaliana.* They also found that this receptor works with multiple receptors that have a similar structure, in order to precisely detect the signals transmitted from the pistil. By accepting the various signals sent from the pistil, the kinase receptors enable the pollen tubes to grow to a position inside the pistil where they can detect LURE. Subsequently, the pollen tubes are guided to reach the egg cell and pass on their sperm cells for fertilization (Fig. 8.3).

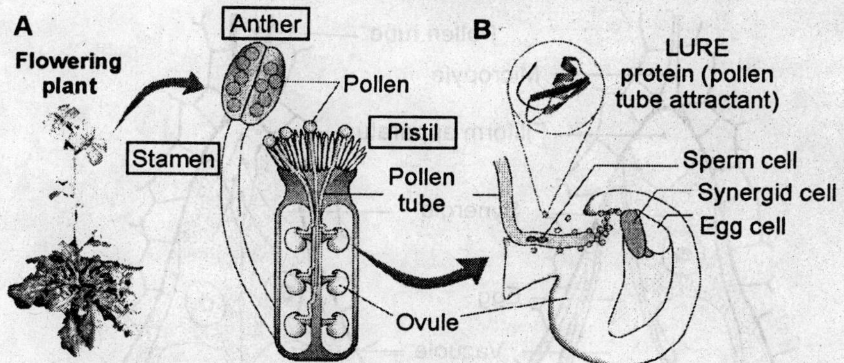

Fig. 8.3. Pollen tube growth and guidance by the LURE peptide in *Arabidopsis* (Takeuchi and Higashiyama, 2016)

Pollen Tube entry into the Embryo Sac

After the passage of the filiform apparatus the pollen tube grows into one of the two synergids (Fig. 8.4). The penetrated synergid shows signs of degeneration before the entry of the pollen tube, but after pollination (Fig. 8.5). Increase in the concentration of the GA_3 is supposed to be the possible cause of the breakdown of the synergid in plants like cotton.

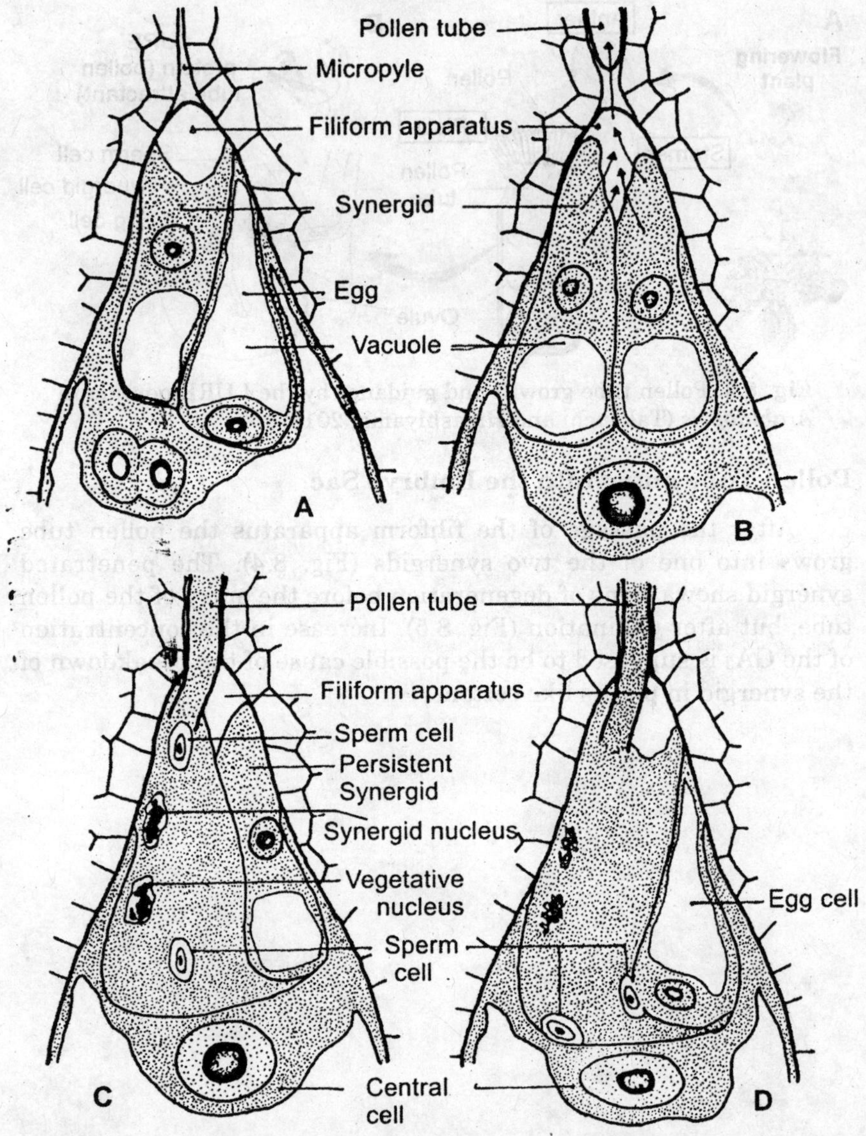

Fig. 8.4. A-D. Diagrammatic summary of pollen tube entrance, tube discharge and gametic transfer in angiosperms. **A.** Micropylar part of embryo sac; **B.** Proposed chemotropic activity of synergids and directed growth of pollen tube; **C.** Discharge of pollen tube; **D.** Proposed way of gametic transfer and fusion (after van Went

and Willemse, 1984)

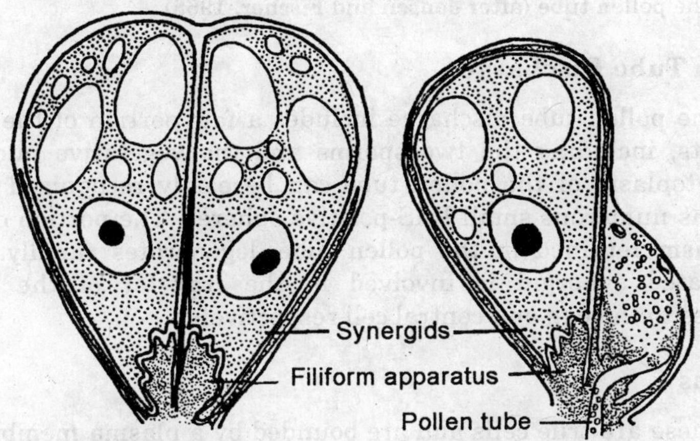

Fig. 8.5. Diagrammatic summary of the changes in the synergids after pollen tube discharge. **A.** Synergids in a pollinated flower; **B.** Synergids after pollen tube discharge (after Jensen and Fischer, 1968)

The pollen tube discharges its contents in the synergid. The discharge occurs through a pore on the pollen tube, which may be terminal or sub-terminal in position (Fig. 8.6). The discharge occurs in seconds. In some plants, the pollen tube is forked and one branch is directed towards the egg and the other towards the secondary nucleus.

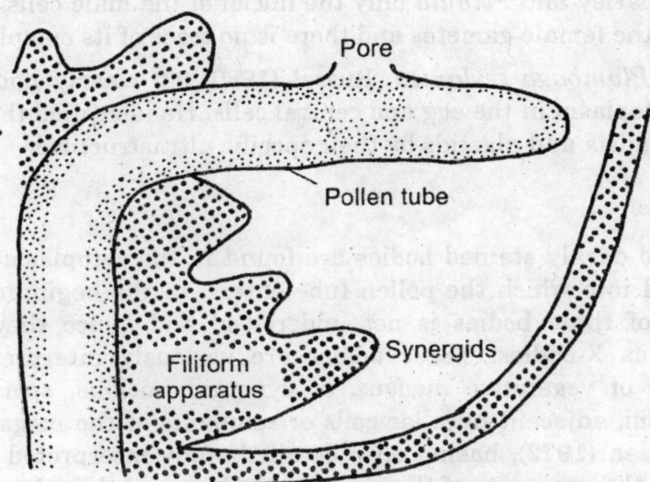

Fig. 8.6. A portion of synergid enlarged to show subterminal pore on the pollen tube (after Jensen and Fischer, 1968)

Pollen Tube Discharge

The pollen tube discharge includes a fair portion of the tube contents, including the two sperms and the vegetative nucleus. The cytoplasm of the pollen tube can be easily recognized as it contains numerous small PAS-positive spheres. The portion of the cytoplasm retained in the pollen tube degenerates rapidly. The penetrated synergid is involved in the transfer of the male gametes to the egg and central cell respectively.

Sperms

These are true cells and are bounded by a plasma membrane. After their release from the pollen tube, the two sperms remain together for some time and later separate. The cytoplasm of the sperm cell contains mitochondria, endoplasmic reticulum, ribosomes, dictyosomes and microtubules. Plastids are usually absent. The nucleus contains a distinct nucleolus. The microtubules present are believed to be associated with the motility of the sperms.

Our knowledge of the transfer of the sperm cells from the penetrated synergid into the egg and central cell is limited. In cotton, barley and *Petunia* only the nuclei of the male cells can be seen in the female gametes and there is no trace of its cytoplasm.

In *Plumbago zeylanica*, Russel (1980, 83) clearly noted the male cytoplasm in the egg and central cells. He identified the male mitochondria and plastids by their specific ultrastructure.

X-bodies

Two darkly stained bodies are found in the cytoplasm of the synergid into which the pollen tube enters. In the beginning, the nature of these bodies is not understood and hence they were named as X-bodies. These bodies are variously interpreted as remains of vegetative nucleus, or synergid nucleus, sperm cell cytoplasm, adjacent nucellar cells or remnants of the megaspores etc. Jensen (1972), basing on a detailed study interpreted one of them as the remnants of the synergid nucleus and the other as the remains of the vegetative nucleus.

POLLEN – PISTIL INTERACTION

In Angiosperms, the phenomenon of fertilization is very important and unique. It involves two fusions – fusion of egg with the male gamete (syngamy) and fusion between the secondary nucleus and second male gamete (triple fusion). Following fertilization, seeds and fruits develop.

Actually, the fertilization process is initiated on the stigma: After pollination, the pistil makes a crucial decision of recognition and this is followed by the acceptance or rejection of the male partner. A chain of reactions that occur on the stigma regulate the post-pollination behaviour of the pollen. In case of compatible pollination, the pollen grains germinate on the stigma, the pollen tube that emerged pass through the stigma and style and ultimately reach the embryo sac where it discharges the male gametes. If the pollination is incompatible, the pistil tissue prevents fertilization either through the inhibition of pollen germination, entry of pollen tube into the stigma or growth of the pollen tube in the style. Thus, the pollen – stigma interaction is very important in the biology of sexual reproduction, because the important function of the selection of the male gamete is not made by the egg, but by the pistillate tissue.

Pollination brings out many physiological changes in the pistil. They are increased in the respiratory activity, changes in RNA and protein synthesis, and increase in several enzymatic activities. In some taxa like cotton, one of the synergids degenerates even before the entry of the pollen tube into the ovule.

The first change that occurs in the pollen after it lands on the stigma is hydration. Along with hydration, the exine and intine proteins are released on the stigmatic surface. This flow of pollen wall proteins can be detected by immunoflorescence and histochemical techniques.

In wet stigmas, after pollination, the pollen wall products mix with stigmatic exudate. In dry stigmas, the pollen sticks on the pellicle which makes contact with the pollen covering. The proteins released by the pollen wall are involved in immunological reactions. These proteins act as antigens and are important in recognition. In the recognition system some complimentary products may also be involved.

The recognition is necessary for pollen acceptance and germination. The recognition of, an acceptable pollen is followed by pollen activation. The pollen starts to germinate and produce enzymes like cutinase for penetration. In rejection, the pollen activation, germination and tube growth is blocked. Many factors are involved in the recognition reaction. The various steps and events that occur in the stigma and style during recognition/ rejection reaction are summarized as follows:

Various steps and events in pollen-pistil interaction after pollination

	Steps	Events
Pollen-stigma interactions	Contact occurs between coat and stigma coat	out flow of exine components
	Attachment of pollen to stigma	Signals of recognition such as glycoproteins, lipoproteins
	Acceptance of pollen by stigma	Imbibition & hydration of pollen. Outflow of pollen wall components. Signals for acceptance, such as polypeptides
	Rejection of pollen	No hydration of pollen. Callose formation in stigma surface cells.
Germination	Activation of pollen	Activation of metabolism, excretion of enzymes affecting the stigmatic surface, as cutinase
	Germination of pollen	Formation of pollen tube, closure of non-functional pores, production and excretion of enzymes

	Steps	Events
	Rejection of pollen tube	Pollen tube growth stops after contact with stigmatic surface
Pollen tube growth	Penetration in the stigma	Dissolution of intercellular substance and cells walls, use of products of stigma
	Directed growth of the pollen tube	Changes in stylar cells of the transmitting tissue. Stylar cell degradation. Attraction of pollen tube. Signal of acceptance
	Growth of the pollen tube stops	Pollen tube growth stops after rejection and lacks nutrition

DOUBLE FERTILIZATION

Double fertilization was first observed by Nawaschin (1898). He found that both the sperms released by the pollen tube are involved in fertilization and this phenomenon is unique to angiosperms. The most important part of the fertilization is the fusion of the male and female nuclei (syngamy). The second sperm is involved in the fertilization of the secondary nucleus, which is formed by the fusion of two polar nuclei. Since the fusion of 3 nuclei takes place this is called triple fusion. Syngamy and triple fusion together is known as Double fertilization.

Light microscopic studies have indicated various modes of transfer of genetic material and nuclear fusions (Gerassimova-Navashina, 1960). According to Jensen (1973) one of the sperms comes in contact with the plasma membrane of the egg cell, while the other contacts the plasma membrane of the central cell (Fig. 8.7 A,B). At the points of contact the membranes dissolve and the sperm nuclei are released, one in the egg and the other in the central cell (Fig. 8.7 C). The male gametes are then passively carried along the cytoplasmic stream to the egg nucleus or the polar nuclei as the case may be (Fig. 8.7 D). The first sperm nucleus reaches the egg nucleus before the second sperm nucleus reaches the polars. However, the fusion of egg nucleus and sperm nucleus takes much longer than the fusion of the male nucleus and the polars.

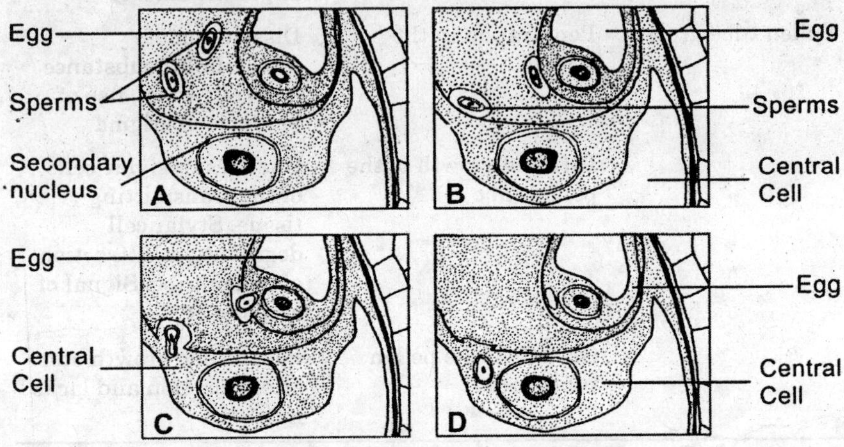

Fig. 8.7. Diagrammatic representation showing the proposed way of gametic fusion (after van Went and Willemse, 1984)

Basing on the condition of the sexual nuclei at the time of fertilization, three types of Karyogamy have been recognized - premitotic, postmitotic and intermediate (Fig. 8.8). In the first type, the fusion of the gametic nuclei will be completed before the zygotic mitosis, *e.g.*, Poaceae and Asteraceae. In the second category, i.e., post-mitotic type, the fusion of the gametic nuclei is completed after the initiation of the zygotic mitosis during prophase - metaphase, *e.g.*, Liliaceae. The third type is characterized by the incomplete mixing of the sexual nuclei. The chromosomes of the gametic nuclei remain unseparated and can be seen in the prophase of the zygotic nucleus. The above observations were also confirmed by Hu and Zhu (1979) through their studies.

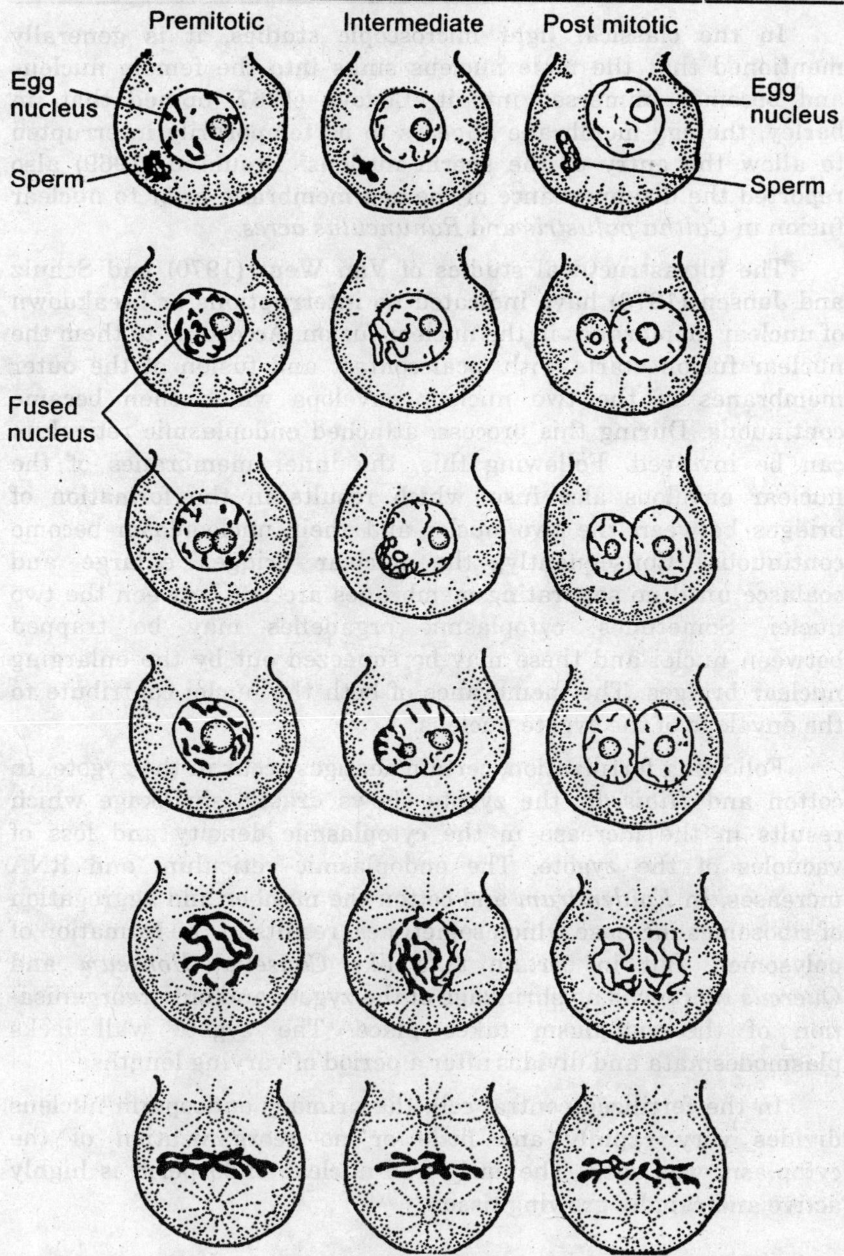

Fig. 8.8. Type of Syngamy (after Gerassimova-Navashina, 1960)

In the classical light microscopic studies, it is generally mentioned that the male nucleus sinks into the female nucleus and becomes immersed into it. Luxova (1967) opined that, in barley, the egg membrane appears to be temporarily interrupted to allow the entry of the sperm nucleus. Brouland (1969) also reported the disappearance of the egg membrane prior to nuclear fusion in *Caltha palustris* and *Ranunculus acres.*

The ultrastructural studies of Van Went (1970) and Schulz and Jensen (1973) have indicated no interruptions or breakdown of nuclear membranes in the nuclear fusion. According to them the nuclear fusion starts with local contact and fusion of the outer membranes of the two nuclear envelops which then became continuous. During this process, attached endoplasmic reticulum can be involved. Following this, the inner membranes of the nuclear envelops also fuse, which results in the formation of bridges between the two nuclei and their nucleoplasm become continuous. Consequently, the nuclear bridges enlarge and coalasce until no separating membranes are left between the two nuclei. Sometimes, cytoplasmic organelles may be trapped between nuclei and these may be squeezed out by the enlarging nuclear bridges. The membranes of both the nuclei contribute to the envelope of the zygote nucleus.

Following fertilization, certain changes occur in the zygote. In cotton and *Hibiscus,* the zygote shows drastic shrinkage which results in the increase in the cytoplasmic density and loss of vacuoles of the zygote. The endoplasmic reticulum and RNA increases. In *Epidendrum* and cotton the number and aggregation of ribosomes increase which sometimes, results in the formation of polysomes. But in certain taxa like *Capsella, Hordeum* and *Quercus* there was no shrinkage of the zygote and only reorganisation of the cytoplasm takes place. The zygote wall lacks plasmodesmata and divides after a period of varying lengths.

In the fertilized central cell, the primary endosperm nucleus divides very rapidly and little or no reorganization of the cytoplasm was seen. The early free nuclear endosperm is highly active and rapidly growing tissue.

POLLEN STORAGE

In nature, effective pollination occurs through various agencies, such as wind, water, animals, insects etc. In majority of cases, the pollen loses its viability very fast. The life of pollen is very short in grasses; for example pollen of *Paspalum* loses its capacity to germinate within 30 minutes of its shedding. Rarely in some fruit trees and gymnosperms the pollen remains viable for 24 hours or more. Methods of increasing the viability of pollen for longer periods is very important because it enables the transport of pollen to longer distances thereby helps in crossing the geographically isolated plants with desirable characters. It also helps in bridging over the undesirable interval in the blooming periods of the two parents.

The earliest reference concerning the handling and storage of pollen was that of the date palm. The male inflorescence of the date palm was considered an important item of commerce in the Middle-east about 400 years ago. But the methods used for storing the pollen at that time was not clearly understood. The methods used for the storage of pollen at present are described below.

Dry and cold storage: The longevity of pollen is governed by several factors, such as flowering time, temperature, relative humidity and light. In the last few decades, several attempts were made to increase the longevity of the pollen by controlling the temperature and relative humidity. Generally the subfreezing temperatures (-5 to -10^{0}C) and 25-50 per cent relative humidity extend the life of the pollen. Apple pollen, stored at -15^{0}C for 9 months showed 95 percent germination. By storing them at -190^{0}C the apple pollen longevity can be prolonged up to 2 years. Similarly, the life of alfalfa pollen can be increased from 8-15 days to 183 days by storing at low temperatures.

Like the low temperature, low relative humidity also prolongs the life of pollen of apple, apricot, peach, pear, plum cherry, grapes etc. But many of the grass pollen (except *Pennisetum typhoideum*) can be stored better at comparatively high relative humidity (70-100%). Low RH was found harmful to many graminaceous pollen.

For minimizing the deterioration of pollen during storage, especially due to desiccation, some pulverized anhydrous materials called dilutents are mixed. Out of about 30 dilutants used, *Lycopodium* powder, egg albumen and talc prolong the longevity of the pollen substantially.

The collection and storage of pollen has been reviewed by Stanley and Linskens (1974). Pollen must be dried over a desiccator before storage in vials at a low temperature. 2-celled pollen types retain their viability longer in storage than 3-celled pollen. Temperature and relative humidity are the major factors controlling successful storage.

For long distance transport, storing pollen under controlled conditions of temperature and relative humidity is inconvenient and expensive. Therefore, another method called 'Freeze Drying method' has been devised by King (1959). In this method, water from the pollen was removed by freezing them (sublimation) and later the pollen is stored in vacuum or in inert gas like helium. By this method, the pollen can be stored under controlled conditions and the shipment of pollen is inexpensive and also convenient package is possible.

Cryogenic storage (Cryopreservation): The pollen grains can be stored for longer periods without the loss of viability by keeping them in liquid nitrogen (-196°C). By using this method pollen grains of onion are found viable and fertile for 360 days. Similarly, pollen grains of *Vitis* (Ganesh, 1985, 1986) and papaya were stored for 448 days and 485 days, respectively in liquid nitrogen without the loss of viability.

The pollen grains can also be stored by keeping them in some organic solvents like acetone, benzene, ethanol, chloroform, phenol etc. These organic solvents are generally considered as toxic, but Iwanami *et al.* (1988) demonstrated that the pollen grains stored in them can germinate and effect fertilization. Hirata stored pollen grains of pear for one year and used them for pollination. Surprisingly it was found that fruits obtained in this pollination were bigger in size, sweeter in taste than those obtained by fertilization with pollen stored under dry, cold conditions. Similarly, lily pollen stored in ethanol were found effective in affecting fertilization after 10 years.

Pollen grains can also be stored in oils (Jain and Shivanna, 1990). In all the studies on the pollen storage conducted so far, viability was determined by germination. Pollen grains capable of germination are considered viable and capable for producing normal seeds. But there are very few studies on the effect of storage on pollen vigour. The studies of Nanda Kumar *et al* (1988) have demonstrated that, the stored pollen shows reduction in vigour in terms of germination and pollen tube growth. These pollen grains were not able to compete effectively with other pollen during pollen-pistil interaction. Sedgely (1981) tested the pollen quality by observing the ability of the pollen to penetrate ovules. He reported that for the pollen stored at -196^{0}C, this ability is very high. And for the pollen stored in other temperatures for longer than one month, though the pollen germinate, but failed to penetrate the stigmas. Storage greatly weakens the pollen tube growth. Thawing of pollen even once, during storage resulted in complete inviability. It is desirable to develop standard techniques for storing pollen not only for longer periods but also for maintaining their vigour.

Pollen Viability

Pollen viability is the ability of pollen grain to germinate and cause fertilization. Heslop-Harrison *et al.* (1984) defined this "as the competence of individuals of a given pollen population to deliver the male gametes to the embryo sac". There are certain tests to determine the efficacy storage method and to determine male fertility of the plant. These pollen viability tests can be divided into two categories, *viz.,* germinability tests and histochemical tests.

1. *Seed Set*: It is the most reliable test to determine the pollen viability. If the seed set occurs after pollination with the stored pollen, the pollen can be regarded as viable. But, sometimes, especially when pistil screening mechanisms operate in the taxa, this method can't be useful. Another disadvantage in this method is, one has to wait for a few days, *i.e.,* until the seed-set occurs to establish the viability of the pollen.

2. *Pollen germination and Pollen tube growth in living pistils*: This is one of the most satisfactory methods to test the pollen viability. In this method, pollinated pistils are taken and treated with 8N sodium hydroxide until the tissues become softer. Later the pistils are mounted

in 0.005 percent decolourized aniline blue and the tissues were gently pressed until they become flattened when observed under flourescent microscope, the pollen tubes show bright yellow flourescence in those regions where callose plugs are present.

3. *In vitro pollen germination*: This is one of the simplest methods to determine the pollen viability. The pollen grains were allowed to germinate in the laboratory with the help of suitable nutrient media (Brewbaker's medium). The percentage of germination and the rate of pollen tube growth can also be measured easily.
4. *Tetrazolium test*: The pollen viability can also be tested by using tetrazolium salts like 2,3,5 triphenyl tetrazolium chloride (TTC). A drop of the medium consisting of TTC salt in sucrose solution is added to a group of pollen grains present on the microslide and covered with a cover glass. The slide is kept in an incubator at 35-60°C in dark for ½ - 3 hours. The viable pollen grains turn red due to the reduction of TTC to red dye formazon.

Pollen Germination

Though the life span of the pollen in many species can be extended by artificial means of storage, it is not sure that the stored pollen can retain its ability for germination and the potency for fertilization, just like the fresh pollen. The probable reason for this may be loss of moisture, depletion of food reserves, decrease/ inactivation of the enzymes or other key substances during storage.

It was demonstrated that the stored pollen require for its germination a high humidity and high level of nutrients. The pine pollen stored for a short period showed optimum germination in 2 percent sucrose while stored pollen, stored for longer period require as much as 20 percent sucrose. Similar observations were made by Vasil (1962) with the pollen of *Arachis* and *Pennisetum*. Grobeg obtained normal sized tomato fruits when 4 day old stored pollen was used in fertilization, while with older pollen, the fruits obtained are smaller in size. Studies of Olmo (1942) on grape and

Hagiaya (1949) on tobacco indicate that the stored pollen though fail to germinate *in vitro*, can cause a satisfactory seed set when used in the field. Similarly, the pollen of Cacao stored over calcium chloride at -20°C to -30°C for 1-4 weeks showed only lower germination, but field pollinations showed 40-50 percent fruit set (Soria and Denys, 1961). This can be interpreted that stigmatic and stylar tissues help to make up some of the deficiencies that occur during storage and provide a natural environment for germination of the pollen and growth of the pollen tube.

Thus, the failure of germination of stored pollen in cultures does not necessarily testify that the pollen is dead or useless. The converse is also true, *i.e.,* the mere germination of stored pollen in vitro does not certify that the pollen will ensure fertilization. This implies that, there are also other factors which control germination of pollen. The prerequisites for pollen germination are: 1. Adequate supply of water, 2. Inorganic salts including Boron, 3. A carbon source – usually a sugar. Some of these requirements may be met through the pollen reserves itself, but sometimes one or more of these factors is limiting.

UNUSUAL FEATURES

Polyspermy

Normally, two sperms are released into the embryo sac. But sometimes, more than two sperms are also found and such a situation is usually referred as Polyspermy. This may be due to the formation and release of more than two sperms by the same pollen tube or due to the entry of more than one pollen tube into the same embryo sac. This occurs only as an abnormality. Polyspermy may result in the fertilization of the egg by more than one sperm or the additional sperms present in the embryo sac may be involved in the fertilization of the cells of the embryo sac other than the egg *i.e.,* synergids or antipodals. If the embryo sac receives more than one pollen tube, the egg and secondary nucleus may be fertilized by the sperms released by the different pollen tubes and this is known as the heterofertilization.

References

Kapil, R.N. and A.K. Bhatnagar 1975. A fresh look at the process of double fertilization in angiosperms. *Phytomorphology* 25: 334-368.

Takeuchi, H. and T. Higashiyama 2016. Tip-localized receptors control pollen tube growth and LURE sensing in *Arabidopsis*. *Nature*. DOI: 10.1038/nature17413 (http://dx.doi.org/10.1038/nature17413)

CHAPTER – IX

INCOMPATIBILITY

The inability of functional male and female gametes to effect fertilization in particular combinations is known as Incompatibility. This is due to the arrest of post-pollination events at different levels. Incompatibility occurs between species (interspecific incompatibility) as well as within the species (intraspecific incompatibility). Whereas interspecific incompatibility prevents fertilization between gametes of distantly related species, intraspecific incompatibility prevents fertilization between gametes of the same or other individuals of the same species. Intraspecific incompatibility is also called as Self-incompatibility.

Interspecific incompatibility is heterogenic *i.e.,* controlled by more than one gene at different loci on the chromosomes (Linskens, 1975). It prevents free cross pollination. In incompatible interspecific crosses either fertilization does not occur or syngamy is followed by abortion of hybrid embryo due to inadequate development of endosperm or embryo-endosperm incompatibility.

In nature different floral adaptations, such as dichogamy, herkogamy and unisexuality have evolved to prevent self pollination, but the most widespread and effective natural device to enforce outbreeding is self-incompatibility.

Dichogamy

Maturation of male and female sex organs at different times is called dichogamy. It ensures cross pollination and prevents self-pollination. Dichogamy is of two types (a) Protandry: Here anthers dehisce much before the stigma of the same flower attains receptivity, *e.g.,* Maize, *Saxifraga, Helianthus, Phlox, Impatiens.* (b) Protogyny: If the stigma loses its receptivity, by the time the anthers dehisce, it is called protogyny, *e.g., Pennisetum, Aristolochia, Helleborus, Annona, Prosopis, Solanum.*

Herkogamy

In a bisexual flower, if both sex organs mature at the same time, the self-pollination is prevented by the arrangement of male and female reproductive organs at different heights (*Hibiscus*) or projected in different directions (*Gloriosa*).

Unisexuality

Unisexuality or dicliny is a condition in which the flowers are either staminate (male) or pistillate (female).

(a) Monoecy

Staminate and pistillate flowers occur in the same plant, either in the same inflorescence, *e.g.*, castor, mango (*Mangifera indica*), banana (*Musa sapientum*), and coconut or in separate inflorescences, *e.g.*, maize. Other monoecious species are cucurbits (*Cucurbita* sp.), walnut (*Juglans regia*), chestnut, straw berries (*Fragaria* sp.), rubber (*Hevea brasiliensis*), grapes (*Vitis vinifera*) and Cassava (*Manihot utilissima*).

(b) Dioecy

The male and female flowers are present on different plants, the plant in such species are either male or female, *e.g.*, Papaya (*Carica papaya*), date, hemp, asparagus and spinach. In general sex is governed by a single gene, *e.g.*, asparagus and papaya. In some cases there are hermaphrodite flowers in addition to male and females, and a number of intermediate forms may also occur.

Self-Incompatibility

It refers to the failure of pollen from a flower to fertilize the same flower or other flowers on the same plant. More than 300 species belonging to 20 families of Angiosperms show self-incompatibility.

Intraspecific incompatibility is of two types (a) heteromorphic (heterostylous), and (b) homomorphic.

(a) Heteromorphic Incompatibility

When plants of the same species produce more than one morphological type of flowers, it is called heteromorphic incompatibility. Different plants of each species may produce two (dimorphic/distylous) or three (trimorphic/tristylous) types of flowers, but each individual produces only one type. The relative levels of the stigma and stamens in different forms, and their incompatibility relationships are diagramatically represented in Fig. 9.1. Distyly and tristyly together is referred to as Heterostyly.

Distyly:

In this category, two types of flowers are produced - long styled, also known as pin morphs and short-styled or thrum morphs (Fig. 9.1A). Pollination is successful between individuals of different morphs (intermorph). Self-pollination and pollination between plants of the same morph (intramorph) are incompatible. Distyly is common in several taxa of the families, Rubiaceae, Plumbagi-naceae, Linaceae and Boraginaceae (Bir Bahadur, 1978) and *Primula*.

Distyly is controlled by a single gene complex, S with two alleles S and s. The allele for short-style (S) is dominant over the allele for long-style (s). Short styled plants are heterozygous (Ss) and long styled are homozygous recessive (ss). Incompatibility in pollen is sporophytically determined. Pollen with 's' allele from short-styled form is compatible on long-styled pistil (ss), although incompatible on short-styled pistil (Ss).

Tristyly:

In this category, the plants produce three types of flowers - long-styled, mid-styled and short-styled (Fig. 9.1B). Each morph produces anthers at two different levels, which correspond to the level of stigma of other two forms. For a successful pollination pollen grains are required not only from flowers of other morphs, but also from anthers corresponding to the same level as that of the stigma (Fig. 9.1). For example, in short-styled morph pollen from long stamens are compatible on pistil of long-styled morph and not on mid-styled morph, those from mid-stamens are compatible on pistils of mid-styled morph and not on long-styled morph. Although most investigators limit the number of families with tristyly to four, Bir Bahadur (1978) considers tristyly to be present in 57 genera belonging to 14 families. Some of the examples are *Oxalis*, *Lythrum* and *Eichhornia*.

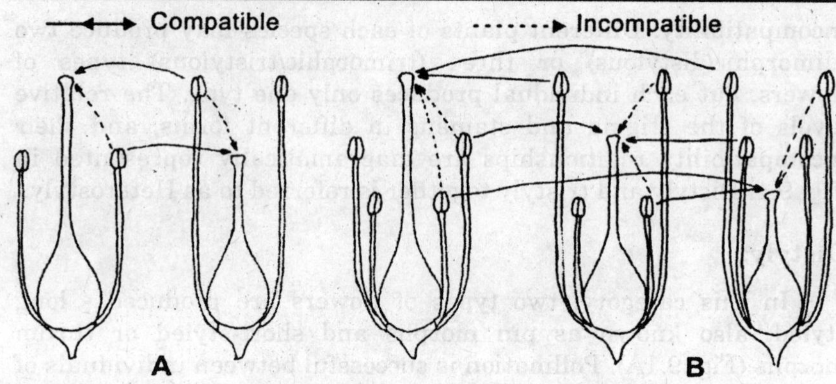

Fig. 9.1. Diagrammatic representation showing the proposed way of gametic fusion (after van Went and Willemse, 1984)

Tristyly is determined by two genes M and S, with two alleles each, S being epistatic to M. Long styled morphs are homozygous recessives for both genes (ss, mm), mid-styled morphs are homozygous recessive for s and homozygous dominant or heterozygous for M (ss MM/ss Mm), and short-styled morphs are heterozygous for S, while M may be in any form (Ss mm/Ss Mm/Ss MM).

In both distylous and tristylous plants pollen incompatibility is sporophytically determined.

(b) Homomorphic Incompatibility

In homomorphic incompatibility the flowers produced by different plants do not show morphological variations. Homomorphic incompatibility has been reported in more than 250 genera belonging to at least 71 families. In homomorphic cases, physiological barriers act in such a way that pollen grains do not germinate on a stigma of similar genetic constitution or their growth is so slow that by the time pollen tube reaches the embryo sac, the latter withers out.

Depending on the origin of factors determining the mating types on the pollen side, two categories of self-incompatibility have been recognized:

1. Gametophytic Self-Incompatibility (GSI): The incompatibility process is controlled by the genotype of the male gametophyte (pollen) itself, *e.g.*, Solanaceae, Onagraceae, Rosaceae, Poaceae, Fabaceae, Scrophulariaceae. In some species, *e.g.*,

Phalaris, Physalis etc., two loci (S and Z) govern incompatibility, while in some others, *e.g., Beta vulgaris* and *Papaver*, three loci are involved. In these cases, polyploidy does not affect the incompatibility reaction. Pollen tube grows very slowly in the style containing the same S allele as the pollen and fails to effect fertilization. Therefore, all the plants are heterozygous at the S locus.

2. Sporophytic Self-Incompatibility (SSI): The incompatibility process is determined by the genotype of the sporophytic tissue of the plant from which the pollen is derived, *e.g.*, Asteraceae, Brassicaceae.

Table 9.1. Differences between Sporophytic and Gametyophytic self incompatibility systems

Features	Sporophytic self incompatibility	Gametophytic self incompatibility
Pollen		
(a) Cytology	3-celled	2-celled
(b) *In vitro* germination	Difficult	Easy
(c) Viability	Short lived	Short lived
(d) Respiration rate	High	Low
(e) Pollen wall proteins	Present in the exine	Present in the intine
Pistil		
Stigma	Dry	Wet
Rejection reaction	On the stigma	In the style

Genetic Basis of Self-Incompatibility

Genetic basis of self-incompatibility was for the first time proposed by East and Mangelsdorf (1925) in *Nicotiana sanderae*. According to them, incompatibility reactions are controlled by a single gene, called S-gene, which has several alleles - S_1, S_2, S_3, S_4... S_n. Each plant has only two of these alleles. Cells of styles and stigmas are diploid and carry both the alleles of the plant. Pollen grains are haploid and, therefore, carry only one of the two alleles. A pollen grain carrying a self-sterility allele does not grow well on a female style carrying the same allele, but grows and fertilizes a plant that carries other alleles. A successful fertilization always denotes the presence of an allele in the pollen that is different from two alleles of the female plant. For example, a plant with the genotype S_1, S_2 produces two types of pollen grains, S_1 and S_2, none of which grows on a style with these alleles. Thus self-fertilization is ruled out. But if a plant with genotype S_2, S_3 is pollinated by such pollen grains, only the S_1 pollen grains germinates and fertilises, whereas S_2 pollen grains fail to grow. On the other hand, all the pollen grains of a S_1 S_2 plant are viable and active on a S_3 S_4 tissue (Fig. 9.2). Such a system leads to the production of heterozygous progeny, *e.g.*, *Trifolium, Nicotiana, Lycopersicon, Solanum, Petunia* etc.

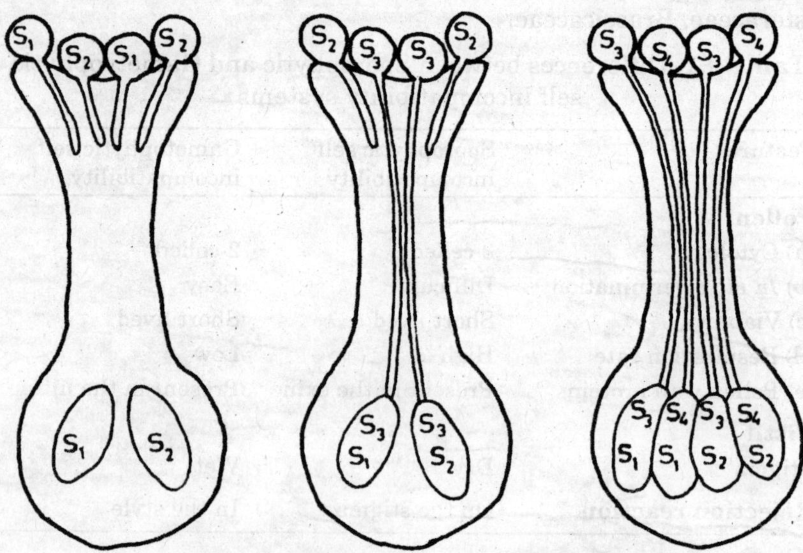

Fig. 9.2. Response of pollen on a pistil of S_1 S_2 plant, showing gametophytic self incompatibility

It should be emphasized here that in all the situations described so far, it is the S-allele of the pollen or the male gametophyte which determines the incompatibility reaction (GSI).

In the Sporophytic Self-Incompatibility (SSI) also, the self-incompatibility is governed by a single gene, S, with multiple alleles; more than 30 alleles are known in *Brassica oleracea*. In general, the number of S alleles is considerably large in the gametophytic than in the sporophytic system. The incompatibility reaction of pollen is governed by the genotype of the plant on which the pollen is produced and not by the genotype of the pollen. It was first reported by Hughes and Babcock in 1950 in *Crepis foetida* and by Gerstel in *Parthenium argentatum* in the same year. In the sporophytic system, the S allele may exhibit dominance, individual action (Codominance) or competition. Consequently there may be many complex incompatibility relationships.

In Sporophytic Self-Incompatibility (SSI) systems all the pollen of a plant behave similarly, irrespective of the S-allele they carry. For example, from a plant carrying S_1 S_2 alleles the pollen carrying S_1 or S_2 allele behaves as S_1 if S_1 dominant, or S_2 if S_2 is dominant; if there is no dominance both will behave as S_1 plus S_2. In other words, the presence of even one of the alleles of the stylar tissue in the sporophytic tissue of the male would render all the pollen of that non-functional with respect to that particular style. A S_1 S_2 plant, therefore, is completely incompatible to plants carrying S_1 S_2, S_1 S_4, S_1 S_5 or S_2 S_3, S_2 S_4, S_2 S_5 and so on, but shows 100 per cent compatibility with a plant carrying S_3 S_4, S_3 S_5 and so on (Fig. 9.3).

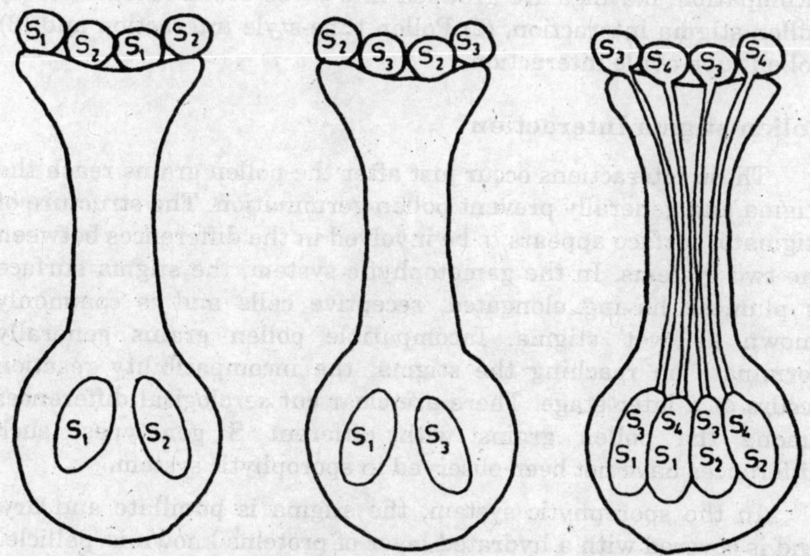

Fig. 9.3. Response of pollen on a pistil of S_1 S_2 plant showing sporophytic self incompatibility

For these two types of Incompatibility systems (GSI, SSI), Brewbaker (1957) observed, a correlation between pollen nuclear number and the place of incompatibility inhibition in homomorphic forms.

a. **Gametophytic:** Observed primarily in species having binucleate pollen with inhibition occurring during pollen tube growth.

b. **Sporophytic:** Linked with trinucleate pollen grains and inhibition occurring on the stigmatic surface or very early tube growth.

Although this correlation holds good in a large number of taxa, there are many exceptions. For example, members of Poaceae are characterized by 3-celled pollen with gametophytic type of incompatibility.

Mechanism of Self-Incompatibility

The mechanism of self-incompatibility is quite complex and is poorly understood. The various phenomena observed in self-incompatible matings are grouped into three broad categories: (1) Pollen-stigma interaction, (2) Pollen tube-style interaction and (3) Pollen tube-ovule interaction.

Pollen-stigma Interaction

These interactions occur just after the pollen grains reach the stigma and generally prevent pollen germination. The structure of stigmatic surface appears to be involved in the differences between the two systems. In the gametophytic system, the stigma surface is plumose having elongated, receptive cells and is commonly known as 'wet' stigma. Incompatible pollen grains generally germinate on reaching the stigma; the incompatibility reaction occurs at a later stage. There are clear cut serological differences among the pollen grains with different S genotypes; such differences have not been observed in sporophytic system.

In the sporophytic system, the stigma is papillate and dry, and is covered with a hydrated layer of proteins known is 'pellicle.' There is evidence that the pellicle is involved in incompatibility reaction. There are striking differences in the stigma antigens related to the S allele compositions. Within a few minutes of

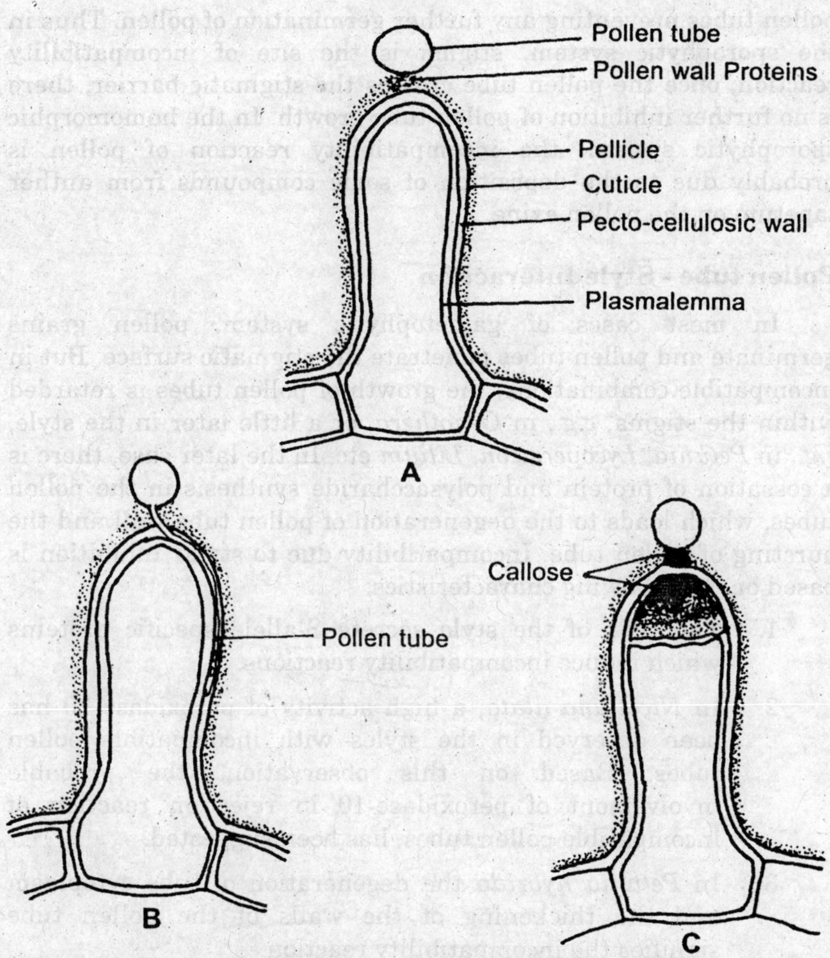

Fig. 9.4. Pollen stigma interaction in Brassicaceae and Asteraceae. **A.** Pollen wall proteins are released on to the pellicle, where recognition reaction occurs. **B.** The compatible pollen tube penetrates the cuticle and grows down the papilla. **C.** The incompatible pollen tube, although penetrates the cuticle, is inhibited from further growth (after Shivanna, 1977)

reaching the stigmatic surface, the pollen releases an exine exudate which is either protein or glycoprotein in nature. This exudate induces immediate callose formation in the papillae (which is in direct contact with the pollen) of incompatible stigma (Fig. 9.4). Often callose is also deposited on the young protruding pollen tubes preventing any further germination of pollen. Thus in the sporophytic system, stigma is the site of incompatibility reaction; once the pollen tube crosses the stigmatic barrier, there is no further inhibition of pollen tube growth. In the homomorphic sporophytic system, the incompatibility reaction of pollen is probably due to the deposition of some compounds from anther tapetum on the pollen exine.

Pollen tube - Style Interaction

In most cases of gametophytic system, pollen grains germinate and pollen tubes penetrate the stigmatic surface. But in incompatible combinations, the growth of pollen tubes is retarded within the stigma, *e.g.*, in *Oenothera*, or a little later in the style, *e.g.*, in *Petunia, Lycopersicon, Lilium* etc. In the later case, there is a cessation of protein and polysaccharide synthesis in the pollen tubes, which leads to the degeneration of pollen tube wall and the bursting of pollen tube. Incompatibility due to stylar inhibition is based on the following characteristics:

1. The cells of the style secrete S-allele specific proteins which induce incompatibility reactions.
2. In *Nicotiana alata*, a high activity of peroxidase-10 has been observed in the styles with incompatible pollen tubes. Based on this observation, the probable involvement of peroxidase-10 in rejection reaction of incompatible pollen tubes, has been suggested.
3. In *Petunia hybrida* the degeneration of tube cytoplasm and the thickening of the walls of the pollen tube signifies the incompatibility reaction.
4. In *Lilium longiflorum*, the tip of the pollen tube with incompatible styles show different structure and are unable to absorb the secretions of the pistil, hence exhibit retarded growth.
5. The styles of self incompatible species, such as *Lycopersicon peruvianum* exhibit high RNAase activity. It has been proposed that this RNAase of the style enter the pollen tube cytoplasm. This results in checking the growth of the pollen tube.

6. The pollen tube in an incompatible style may show various types of abnormal behaviour, such as branching of the tip of the pollen tube, development of callose plugs or restricted growth of pollen tube.

Pollen tube - Ovule Interaction

In some cases, *e.g., Theobroma cacao*, pollen tubes reach the ovule and effect fertilization. However, in incompatible combinations, embryos degenerate at an early stage of development.

Molecular Mechanisms of Self-Incompatibility

Self-incompatibility is widespread among angiosperms, but genes controlling it have been identified in only five families, *viz.*, Brassicaceae, Papaveraceae, Rosaceae, Solanaceae and Scrophulariaceae

In many species pollen recognition and rejection are controlled by a single multiallelic locus, the S-locus. It contains at least two genes, one encoding the male determinant that is carried by the pollen grain, and the other encoding the female determinant that is expressed in the pistil. Both the male and female determinants are polymorphic and inherited as one segregating unit. The variants of this gene complex are called S-haplotypes. The recognition of self/non-self pollen operates at the level of protein-protein interactions between the two determinants (Table 9.2) and an incompatible response occurs when both determinants are issued from the same S-haplotype.

Table 9.2. List of S-locus male and female determinant genes

Family	Type of SI	Male Determinant	Female Determinant
Brassicaceae	SSI	SP11/SCR	SRK
Solanaceae, Rosaceae, Scrophulariaceae	GSI	SLF	S-RNase
Papaveraceae	GSI	PrpS	PrsS

SSI: Sporophytic self-incompatibility; **GSI:** Gametophytic self-incompatibility; **SP11**, S-locus protein 11; **PrpS**, *Papaver rhoeas* Pollen S; **PrsS:** *P. rhoeas* Stigma S; **RNase:** Ribonuclease; **SRK:** S-locus Receptor Kinase; **SCR:** S-locus Cysteine Rich Protein; **SLF:** S-locus F-Box

Three main mechanisms of self-incompatibility have been identified: (i) Brassicaceae system, (ii) Solanaceae system and (iii) Papaveraceae system. The self-recognition system adopted by Brassicaceae and Papaveraceae depends on a specific interaction between male and female S-determinants derived from the same haplotype. The non-self recognition system, found in Solanaceae, depends on non-self (different haplotype)-specific interaction between male and female S-determinants, and the male S-determinant genes are duplicated to recognize diverse non-self female S-determinants.

(i) Brassicaceae System

The SI system of Brassicaceae, self-pollen are rejected at the surface of the stigma after discrimination of self-pollen from non-self pollen. This recognition and rejection is genetically controlled by a single S-locus, which contains three highly polymorphic genes, *SRK* (*S-locus Receptor Kinase*), *SP11 (S-locus Protein 11)/SCR (S-locus Cysteine Rich Proten)* and *SLG*. *SCR*/SP11 is pollen coat protein of tapetal origin, wheras SRK, the stigma SI determinant, is a plasma membrane spanning Receptor Kinse. Upon pollination SP11 penetrates the papilla cell wall and binds to SRK in an S-haplotype-specific manner. This binding induces the autophosphorylation of SRK, and the phosphorylated SRK interacts with MLPK (M-locus protein kinase) localized in papilla cell membrane. After subsequent signal transduction yet to be characterized, rejection of the self pollen occurs on the stigma. SLG is not required for recognition of pollen. It enhances the activity of SRK (Shimosato *et al.*, 2007; Watanabe *et al.*, 2012).

(ii) Solanaceae System

In this system, the S-locus encodes a single female and multiple male S-determinants, designated as S-RNase and SLFS, respectively. The female determinant is secreted in large amounts into the extracellular matrix of the style. In a pollinated pistil S-RNase is incorporated into the pollen tube and functions as cytotoxin. In cross pollination some members of SLFs interact with non-self S-RNase (from different S-haplotype), which degrades S-RNase and the pollen tube continues to grow. However, in self-pollination none of the SLFs interacts with S-RNase and, therefore, pollen tube RNA is broken down causing cessation of pollen tube growth.

S-RNase mediated self-incompatibility, first reported in Solanaceae by Anderson *et al.* (1986), also occurs in Rosaceae and Scrophulariaceae.

Papaveraceae System

The S-locus encodes female and male S-determinants, designated as PrsS (*Papaver rhoeas* stigma S) and PrpS (*Papaver rhoeas* pollen S), respectively (Wheeler *et al.*, 2009). In self-pollination the binding of the self-incompatibility protein (S protein) secreted by the stigma (PrsS) with its cognate pollen S-specific receptor protein (PrpS) on the pollen plasma membrane yet to be identified induces Ca^{2+} influx in the shank of the pollen tube. The increased Ca^{2+} level possibly triggers Ca^{2+}-dependent phosphorylation of a pollen protein P26 causing dramatic depolymerization of the pollen E actin cytoskeleton (Staiger and Franklin-Tong, 2003), leading to rapid growth inhibition and, ultimately, programmed cell death of the incompatible pollen tube (Iwano and Takayama, 2012).

Biological Significance of Incompatibility

In nature a balanced inbreeding and outbreeding of the plants is regulated by interspecific and intraspecific incompatibility. Repeated selfing of the plants leads to a highly homozygous individuals which have very low survival value. Self-incompatibility is one of the most effective outbreeding mechanisms, and is considered to be one of the main causes for the rapid evolution of Angiosperms. On the other hand, interspecific incompatibility brings about reproductive isolation and is, thus, responsible for the emergence of new races and species through the process of mutation and open hybridization.

In many crop plants such as maize, sorghum and knol-khol, hybrid vigour has been exploited by making use of male sterility. One of the major limitations to the exploitation of hybrid vigour in many other crop plants is the lack of male sterile lines and the lack of suitable chemicals to induce effective male sterility without affecting female fertility. However, self-incompatibility can be used equally effectively in producing hybrid seeds.

Methods to Overcome Incompatibility

For using hybrid vigour as a method of crop improvement, homozygous, true breeding cultivars are extremely important. Obtaining such cultivars through the conventional method of plant breeding, especially in outbreeding crops, requires many years of patient work. It involves selfing for many generations. Moreover, this technique cannot be applied to self-incompatible plants. Before the introduction of anther culture technique selfing was one of the chief methods to obtain homozygous individuals. For several plants where anther culture has failed to yield haploids, self-pollination continues to be an important approach to achieve this. In this context self-incompatibility is a serious problem.

Interspecific incompatibility is another major problem because it prevents distant hybridization and, thus production of better types by combining good characters from different plants.

There are many effective methods to overcome intra and interspecific incompatibility.

(i) Irradiation

Induction of mutations through X-ray irradiation of flower buds at pollen mother cell stage was one of the earliest successful techniques for the induction of self-compatibility. Lewis (1949) and Lewis and Crowe (1953) worked extensively on spontaneous and induced mutations at the S locus in *Oenothera organensis* and *Prunus avium*. Subsequently, mutations affecting the activity of the S-allele, in both pollen and style have been reported in *Trifolium*. The breakdown of intraspecific incompatibility following irradiation has been reported only in gametophytic systems (see Pandey, 1974). De Nettancourt and Ecochard (1968) could get higher seed-set per plant through gamma rays treatment of self-incompatible clones of *Lycopersicon peruvianum*. The enhanced seed set following selfing was due the prevention of premature abscission of flower.

Irradiation of the styles with X-rays (*Petunia* - Linskens *et al.*, 1960) or UV rays (*Oenothera* - Kumar and Hecht, 1965) also induces temporary break down of self-incompatibility. Apparently, irradiation damages the physiological mechanism of incompatibility in the style, so that the pollen tubes can pass through the style.

(ii) Bud Pollination

Bud pollination is by far, the most successful technique for overcoming self-incompatibility. It has been successful in both sporophytic and gametophytic systems such as *Brassica, Raphanus, Nicotiana* and *Petunia.* The degree of success varies from species to species, and the optimal stage of buds for overcoming self-incompatibility ranges from 2-7 days before anthesis. In *Nicotiana* (Pandey, 1963) and *Petunia* (Shivanna and Rangaswamy, 1969), the application of stigmatic exudate from the mature flower on the stigma of the bud before pollination, significantly increased seed set.

It is likely that in the style, the factor complimentary to the pollen factor for incompatibility reaction appears only at or just before anthesis. Therefore, if the stigma is self-pollinated at bud stage when this factor has not appeared the pollen tubes will grow normally and effect fertilization.

(iii) Stump Pollination

Some investigators have attempted to shorten the distance, through which the pollen tube has to grow in the pistil, by removing the style partially or fully and pollinating the cut-end. This has been particularly useful in crosses involving parents which differ in the length of the style. In crosses involving *Solanum pinnatisectum* x *S. canciforme* and *S. pinnatisectum* x *S. bulbocastanum,* Swaminathan (1955) removed the stigma together with a small portion of the style, and the cut-end was covered with a block of agar-sucrose-gelatin medium (which favoured pollen germination). Pollen grains were deposited on this medium and covered with a piece of moist cotton wool. Hybrid seeds were successfully obtained in both the crosses. A similar technique has also been successfully applied to crosses of *Nicotiana tabacum* x *N. rustica* and *N. tabacum* x *N. debney* (Swaminathan and Murthy, 1957). The length of the style in *Lathyrus hirsutus* and *L. odoratus* is 4 mm and 10 mm, respectively. In the cross *L. hirsutus* x *L. odoratus* fertilization occurs, whereas reciprocal cross in which the male parent has a short style the pollen tubes does not reach the ovary. Davies (1957) succeeded in overcoming this barrier (*L. odoratus* x *L. hirsutus*) by removing the stigma and a part or whole of the style and pollinating the cut-end of the style or the tip of the ovary.

(iv) Intra-ovarian Pollination

If the zone of incompatibility lies on the stigma or in the style, a direct introduction of pollen suspension into the ovary would be helpful. This technique involves injection of pollen grains (suspended in the suitable medium) directly into the ovary and achieving pollen germination, pollen tube entry into the ovule, fertilization and seed development. Viable seeds following intra-ovarian pollination have been obtained in several members of Papaveraceae, such as *Argemone mexicana, A. ochroleuca, Papaver somniferum, P. rhoeas* and *Eschcholtzia californica* by Kanta (1960) and Maheshwari and Kanta (1961, 1962). This technique has also been successfully used to achieve interspecific hybridization between *Argemone mexicana* and *A. ochroleuca* (Kanta and Maheshwari, 1963).

This technique has not been extended to any other systems. One of the major limitations is that it is not suitable in systems that do not have enough space in the ovary to contain the injected pollen suspension. Also, in taxa which require sugar for pollen germination, injection of pollen suspension with sugar makes the ovary prone to bacterial and fungal infection.

(v) Test tube Fertilization (*In vitro* Fertilization)

Basically, the technique involves culture of pollen grains and ovules together on a suitable nutrient medium and achieving pollen germination, pollen tube entry into the ovule, double fertilization and development of mature seeds. Kanta *et al.* (1962) reported success in *Papaver somniferum* by this technique. Subsequently, this technique was extended to *Argemone mexicana, Eschscholtzia californica, Nicotiana tabacum* (Kanta and Maheshwari, 1963; Maheshwari and Kanta, 1964), *Dianthus caryophyllatus* (Zenkteler, 1965) and *Dicranostigma franchetianum* (Rangaswamy and Shivanna, 1969). All these taxa are self-compatible and fertilization achieved *in vitro* was through compatible pollen. Rangaswamy and Shivanna (1967) attempted to overcome Self-incompatibility in *Petunia axillaris* by using the technique of test tube fertilization. Fertilization could not be achieved through culture of isolated ovules and pollen grains. Instead, the entire mass of ovules of an ovary, intact on the placentae together with a short length of pedicel, was cultured on the medium and the ovules dusted with pollen. This refined technique, termed placental pollination, helped to bring the pollen in direct contact with the ovules without disturbing the original arrangement of ovules and, thus prevented any injury to the ovules. Following placental pollination with either compatible or incompatible pollen, post-pollination events proceeded normally and resulted in viable seeds.

Subsequently, placental pollination has been successfully used in *Nicotiana tabacum* (Baltkova and Tupy, 1968), *Petunia hybrida* (Niimi, 1970), *Melandrium album, Agrostemma githago* (Zubkova and Shadky, 1975) and *Narcissus pseudonarcissus.*

Zenkteler (1967, 1970) and his associates (Zenkteler *et al.*, 1975) successfully applied the technique of test tube fertilization to raise interspecific and even intergeneric hybrids. They could successfully obtain hybrid progeny following the cross *Melandrium album* x *Melandrium rubrum*, *M. album* x *Silene schafta* and *M. album* x *Viscaria vulgaris*. Double fertilization and development of proembryo was achieved even in crosses involving members of Caryophyllaceae and Solanaceae (*Melandrium* x *Datura*).

(vi) Use of Mixed and Mentor pollen

In many taxa self-incompatibility has been overcome by using a mixture of self incompatible and compatible pollen (*Brassica* - Attia, 1950; *Theobroma* - Gledenning, 1969; Opeke and Jacob, 1969; *Cola* - Jacob, 1973). This method, however, requires an effective screening to distinguish the seeds/seedlings resulting from incompatible and compatible pollen.

The successful use of a mixture of mentor pollen (compatible pollen made ineffective for fertilization) and live incompatible pollen in overcoming interspecific incompatibility and the realization of the involvement of pollen-wall and stigma-surface proteins in incompatibility responses prompted several investigators to test the efficacy of mentor pollen in overcoming self-incompatibility. Mentor pollen is also referred to as recognition pollen and it is made ineffective for fertilization by chemicals or irradiation.

Self-incompatibility has been overcome by employing recognition pollen in *Cosmos* (Howlett *et al.*, 1975), *Petunia* (Sastri and Shivanna, 1976) and interspecific incompatibility in *Nicotiana*

(Pandey, 1977) and *Populus* (Knox *et al.*, 1972). While *Petunia* is an example for Gametophytic Self-Incompatibility, *Cosmos* shows Sporophytic Self-incompatibility. In a cross *Sesamum indicum* x *S. mulayanum* mentor pollen was effective in overcoming incompatibility at the level of the stigma, but not in the style; although incompatible pollen germinated and entered the stigma, their further growth was effectively inhibited (Sastri and Shivanna, 1976).

(vii) Hot Water and High Temperature Treatment

Temperature significantly affects the incompatibility reaction. In *Oenothera* the treatment of style and stigma with hot water at 50°C for 5 minutes before pollination, inactivates incompatibility reaction temporarily (Hecht, 1961). In *Lilium grandiflorum* also, a similar hot water treatment of either detached or intact styles at 50°C for 6 minutes prior to pollination was successful in overcoming self-incompatibility (Hopper *et al.*, 1967). Hot water treatment is considered to inactivate a protein which is involved in self-incompatible inhibition. Similar success in overcoming self-incompatibility by exposing plants to high temperature has been reported for *Brassica, Raphanus* (El Muraba, 1957), *Nemesia* (Campbell and Ascher, 1972), *Lilium* (Ascher, 1975) and *Tradescantia* (Qaraeen, 1980).

(viii) Application of Growth Substances and other Chemicals

Application of growth substances, such as NAA, IBA and GA_3, is one of the most commonly used methods to overcome incompatibility. This technique has been utilized to overcome incompatibility in *Petunia, Tagetus, Trifolium* and *Brassica* (Eyster, 1941), *Lilium* (Emswelles and Stuart, 1948) and *Lycopersicon* (de Nettancourt *et al.*, 1971). Generally it is presumed that the effect of growth substances is brought about by inhibiting floral abscission and, thus allowing the slow-growing pollen tubes to reach the ovary. Application of 2,4-D on ovaries, 24 h after pollination was successful in achieving interspecific crosses in *Solanum* (Dionne, 1958).

Use of a mixture of naphthalene acetamide and potassium gibberellate could overcome incompatibility in the cross involving

Phaseolus vulgaris and *P. acutifolius* (Al Yasiri and Coyne, 1964). Many chemical substances have been used to overcome self-incompatibility. In *Brassica oleracea,* the treatment of stigma with hexane, before pollination was effective (Ockendon, 1978). Hexane has been suggested to remove or inactivate incompatibility factors on the stigma.

Hormone application has been very successful in many graminaceous crosses. Kruse (1974) crossed *Hordeum* species with species of many other genera, such as *Dactylis, Triticum* and *Lolium* through application of growth substances.

(ix) Modification of Stigmatic Surface

The inter face of stigma surface pellicle with the pollen surface proteins resulting in incompatibility in plants with dry stigmas and plants with SSI-systems. In such cases, the application of lectin concanvalin A to the stigma would mask the pellicle and prevents the incompatibility. Other products used for this purpose are: detergents, such as Triton x-100 and hexane modify pellicle and inhibits the incompatibility reaction. However, this method is not of much use because these treatments show adverse effect on pollen adhesion and pollen germination. But application of NaCl increases pollen adhesion and germi-nation on the stigma and reduced callose formation in *Brassica campestris.*

(x) Increased CO_2 Level

It has been observed that an increase of atmospheric CO_2 level 100-fold, to 4-6 percent, while the relative humidity is 100 percent self-incompatible pollination would behave as compatible pollination in *Brassica.*

(xi) Somatic Hybridization

A new technique of isolating protoplasts and culturing them into complete plants has proved useful in overcoming all incompatibility barriers. In this method protoplasts of the plants to be crossed can be isolated and fused, and these fused protoplasts can be cultured to obtain complete plants. By using this technique both interspecific and intergeneric hybrids have been obtained. The first success in somatic hybridization was reported by Carlson *et al.* (1972) in *Nicotiana langsdorffii* + *N. glauca*. Later on several interspecific hybrids have been obtained which are sexually incompatible. These include *Nicotiana sylvestris* + *N. knightiana* (Maliga *et al.*, 1977). *N. tabacum* + *N. alata* (Nagao, 1979), *Datura innoxia* + *D. discolor* (Schider, 1978). Intergeneric hybrids have also been obtained applying this technique, *e.g.*, *Solanum tubero-sum* + *Lycopersicon esculentum* (Melchers *et al.*, 1978), *Daucus carota* + *Aegopodium podagraria* (Dudits *et al.*, 1979), *Nicotiana tabacum* + *Petunia hybrida* (Li *et al., 1982), Atropa belladonna* + *Nicotiana chinensis* (Gleba *et al.*, 1982).

References

Bhojwani, S.S. and S.P. Bhatnagar *1985. The Embryology of Angiosperms.* Vani Educational Books.

Shivanna, K.R. 1982. Pollen-Pistil interaction and control of Fertilization. In: B.M. Johri: *Experimental embryology of Vascular Plants.* Springer Verlag, Berlin.

Shivanna, K.R. and B.M. Johri 1985. The Angiosperm pollen - Structure and Function. Wiley Eastern, New Delhi.

Singh, B.D. 1994. *Plant Breeding - Principles and Methods.* Kalyani Publishers.

CHAPTER – X

ENDOSPERM

The process of double fertilization, which is unique to the flowering plants, triggers off a series of coordinated developmental sequences, in different parts of the ovule, among which the most important is the development of embryo and endosperm. The first fertilization or syngamy which results in embryo development is a feature shared by non-flowering plants also. The second fertilization, usually designated as triple fusion is an exclusive feature of flowering seed plants and its product, the endosperm is characteristic of only the angiosperms.

Historically the fate of second male gamete involved in triple fusion was traced initially in *Lilium martagon* and *Fritillaria tenella* (Nawaschin, 1898). Following this discovery, double fertilization was reported in a large number of Angiosperms. The fusion product of one sperm with the secondary nucleus results in the formation of the primary endosperm nucleus. The nuclei or cells derived through mitoses of this nucleus and daughter nuclei constitute the endosperm. The families Orchidaceae, Podostemaceae and Trapaceae are characterized by the absence of an endosperm.

The endosperm, being a pool of reserve food materials and regulatory chemicals is responsible for the nutrition, growth and germination of the embryo. The endosperm may either be consumed by developing embryo *e.g.*, peas, beans and pulses or it may persist in mature seeds, *e.g.*, cereals, castor bean and coconut.

DEVELOPMENT OF ENDOSPERM

On the basis of mode of development, the endosperm is classified into Nuclear, Cellular and Helobial types (Fig. 10.1).

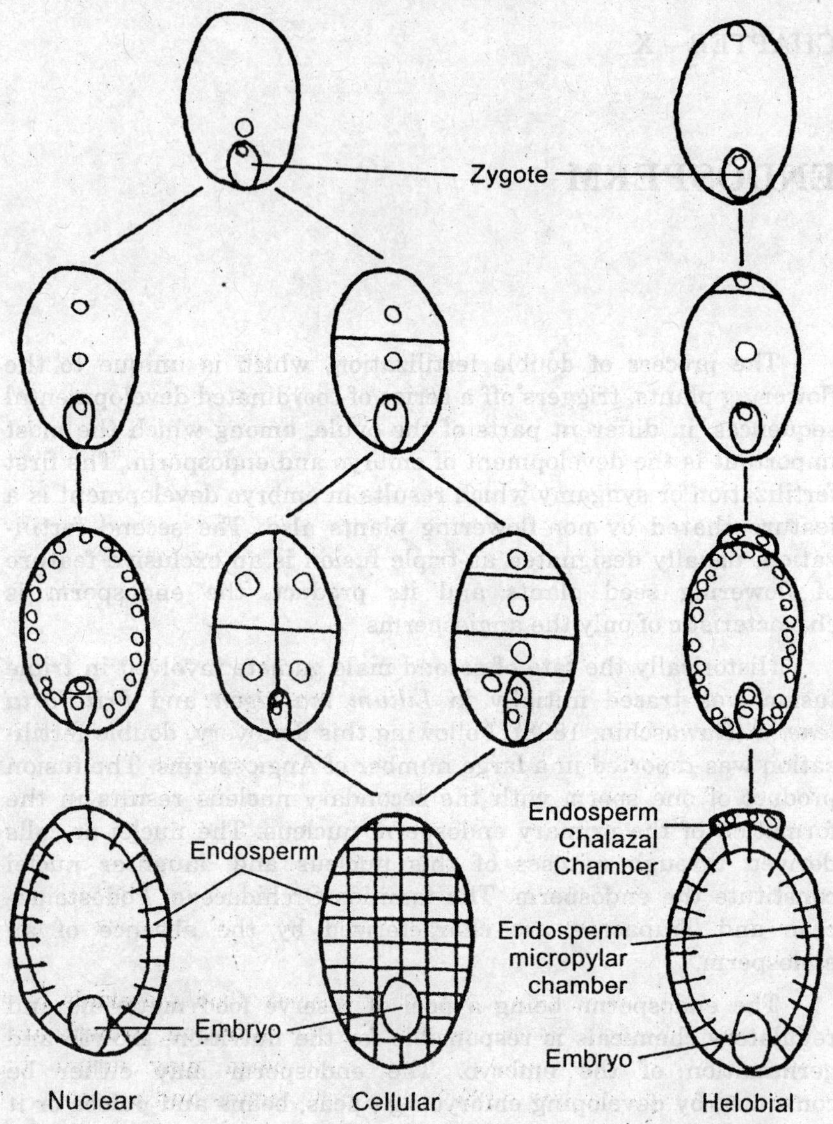

Fig. 10.1. Type of endosperm development patterns (after Palser, 1974)

I. Nuclear Type

In Nuclear type the first division of the primary endosperm nucleus and several of the following divisions are not followed by wall formation (Fig. 10.2. A,B). The free nuclei either aggregate in the micropylar or chalazal region or in both or may become distributed in a film of cytoplasm around the central vacuole. The endosperm remains free nuclear throughout till the endosperm is completely consumed by the developing embryo in *Oxyspora* (Subramanyam, 1951), *Limnanthes* (Mathur, 1956) and *Floerkea* (Maheshwari and Johri, 1956). In the majority of the plants, the endosperm eventually becomes cellular (Fig.10.2 C-E).

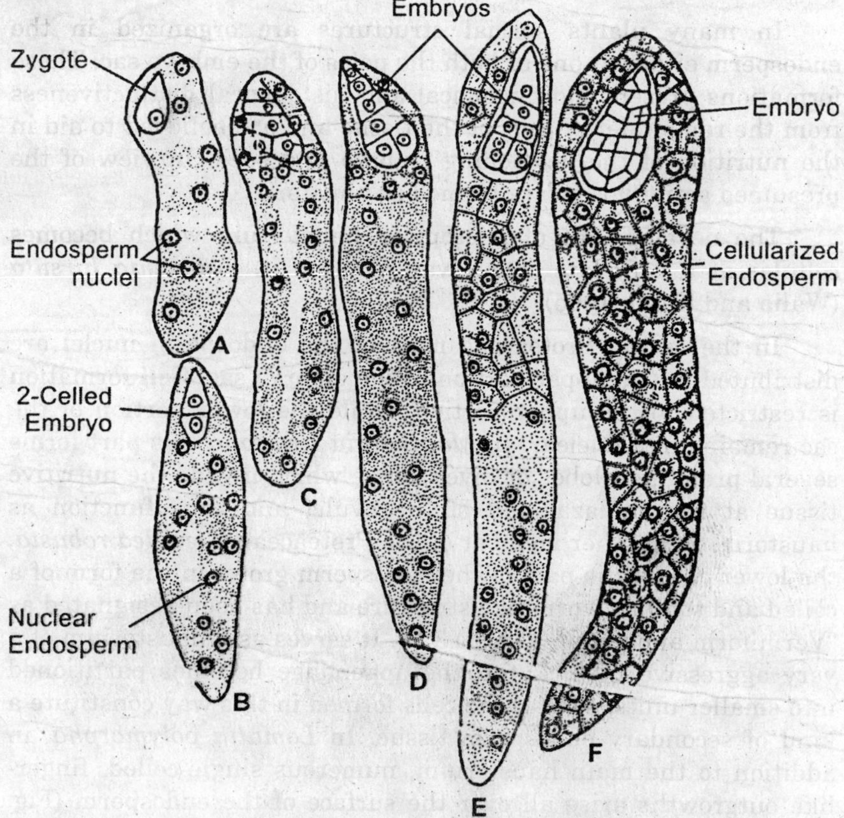

Fig. 10.2. Stages in the development of endosperm in *Chloris roxburghiana* (after Febulaus & Pullaiah, 1991).

Wall formation occurs at different stages of development in different species, as early as the four-nucleate endosperm in *Coffea arabica* (Mendes, 1941), *Galinsoga ciliata* (Harris, 1935), 8-16-nucleate stage in *Tithonia rotundifolia* (Pullaiah, 1978) or even after the formation of 1000 free nuclei as in *Anemone* (Bhandari, 1986). Wall formation is centripetal or it is initiated around the proembryo at the micropylar end and proceeds to the chalaza (Fig. 10.2 C-F). In most plants wall formation once initiated continues until there are no more free nuclei. The cells divide until the maturity of the seed and get filled with reserve food. In a few families like Brassicaceae, Cucurbitaceae, Fabaceae and Proteaceae wall formation does not proceed beyond the central region, after initiation. The chalazal region thus remains free nuclear.

In many plants special structures are organized in the endosperm either at one or both the poles of the embryo sac. These formations present morphological and histological distinctiveness from the remainder of cells of the tissue and are believed to aid in the nutrition of the developing embryo and seed. In view of the presumed activity they are termed as haustoria.

The development of micropylar haustorium which becomes cellular at a very late stage, was reported in *Frankenia hirsuta* (Walia and Kapil, 1965).

In the family Proteaceae most of the endosperm nuclei are distributed in the upper portion of the embryo sac. Cell formation is restricted to the upper portion, while the lower portion of the sac remains free nuclear. In *Macadamia ternifolia* this part forms several prominent lobes or diverticulae which invade the nutritive tissue at the chalazal end of the ovule, and thus function as haustoria. In another member of the Proteaceae *Grevillea robusta*, the lower coenocytic part of the endosperm grows in the form of a coiled and tubular worm-like structure and has been designated as "Vermiform appendix" (Fig. 10.3 B). It serves as a haustorium of a very aggressive type. Later, the appendage becomes partitioned into smaller units, so that the cells formed in this way constitute a kind of secondary endosperm tissue. In *Lomatia polymorpha*, in addition to the main haustorium, numerous single-celled, finger-like outgrowths arise all over the surface of the endosperm (Fig. 10.4) and thus, increase the absorptive surface (Venkata Rao, 1962).

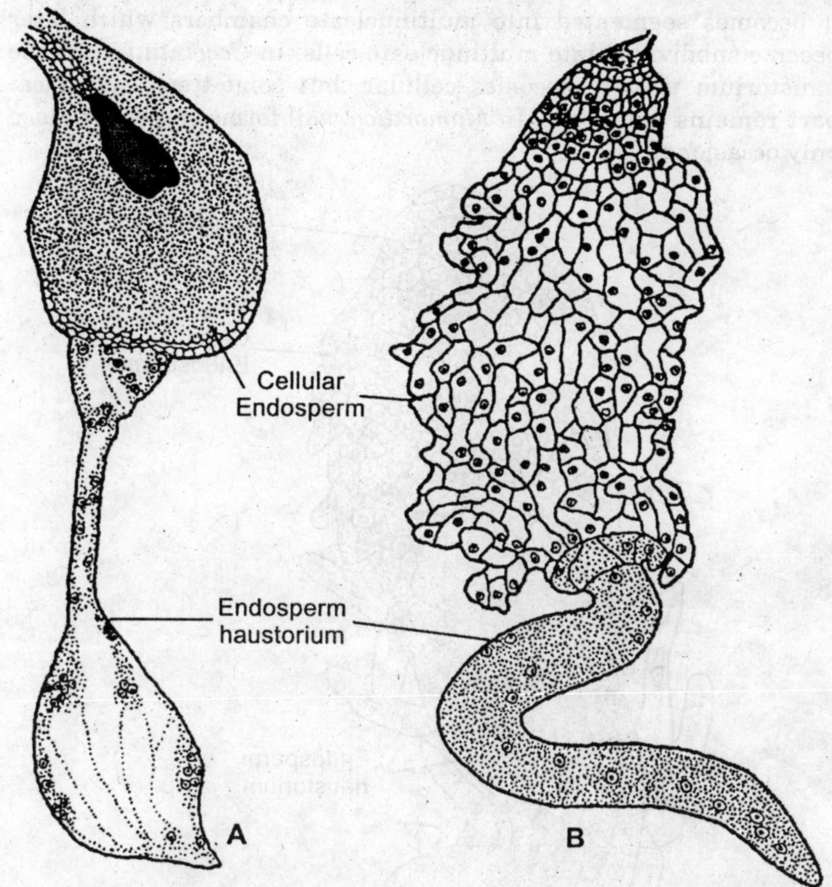

Fig. 10.3. Endosperm haustoria. **A.** *Crotalaria* - Leguminosae (after Rau, 1951); **B.** *Grevillea* Proteaceae (after Kaushik, 1938)

Dissections and microtome sections have revealed many interesting types of chalazal endosperm haustoria in the family Cucurbitaceae. Soon after fertilization, the lower part of the embryo sac grows downwards into a long tubular process. When the proembryo is at globular stage, the upper broader part of the endosperm has become cellular and is then designated as "endosperm proper". The lower free nuclear portion elongates still further. In some plants it frequently reaches down to the chalazal region. To begin with, the endosperm haustorium is coenocytic and in many cucurbits it remains so throughout. In *Citrullus fistulosus* it becomes segmented into multinucleate chambers which later become subdivided into multinucleate cells. In *Coccinia indica* the haustorium usually becomes cellular, but some-times its apical part remains coenocytic. In *Momordica* wall formation takes place only occasionally.

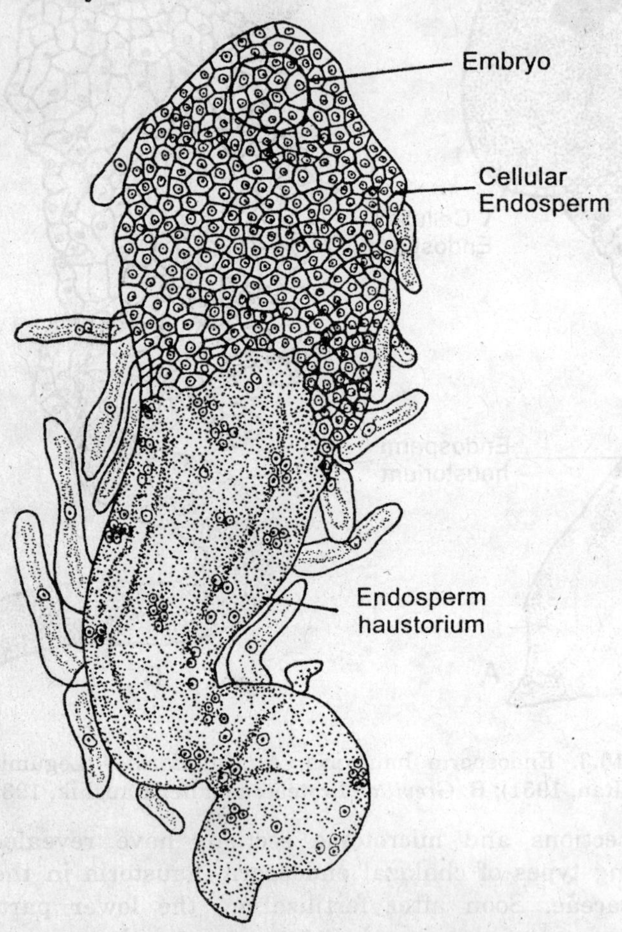

Fig. 10.4. Endosperm hautorium in *Lomatia polymorpha* (after Venkata Rao, 1962)

The tip of the haustorium in *Cucurbita pepo* and *Cucumis sativus* (Fig. 10.5 A,B) becomes swollen. The longest endosperm haustorium is reported in *Echinocystis lobata* where it measures up to 16 mm in length (Fig. 10.5C). Occasionally the peripheral cells of the lower region of endosperm proper bulge out during later stages of the development of seed. In *Luffa aegyptiaca* and *Melothria maderaspatana* such bulges are usually referred to as "secondary haustoria".

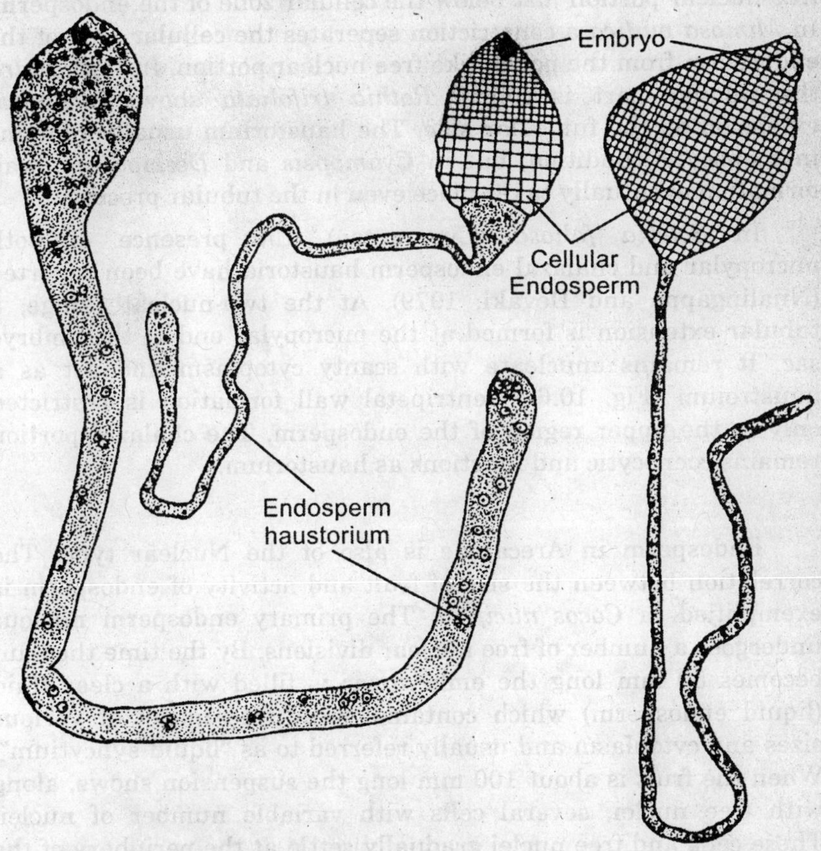

Fig. 10.5. Endosperm haustoria in Cucurbitaceae. A.B. *Cucumis sativus* (after Chopra, 1955). C. *Echinocysti lobata* (after Chopra and Seth, 1977)

Chalazal endosperm haustorium is also recorded in Fabaceae (Leguminosae). In *Stylosanthes mucronata* cell formation is absent. Cell formation is restricted to the region immediately around the embryo, while in others it is restricted to upper half to two thirds of the embryo sac. The lower half of the endosperm remains free nuclear, assumes vesicular contour, contains hypertrophied nuclei and functions as haustorium. The shape and size of the haustorium varies in many taxa. In *Dichrostachys, Leucaena, Mimosa hamata, Crotalaria* and some species of *Cassia* it assumes a tubular form with a vesicular tip (Fig. 10.3 A). In *Cassia mimosoides* and *C. occidentalis* there is another bulbous free nuclear portion just below the cellular zone of the endosperm. In *Mimosa pudica* a constriction seperates the cellular part of the endosperm from the pouch like free nuclear portion. In *Calliandra* the tubular part is coiled. *Rothia trifoliata* shows a lateral extension on the funicular side. The haustorium usually remains in coenocytic condition, but in *Cyamopsis* and *Desmodium,* wall formation eventually takes place even in the tubular process.

In *Scleria foliosa* (Cyperaceae), the presence of both micropylar and chalazal endosperm haustoria have been reported (Nijalingappa and Devaki, 1979). At the two-nucleate stage, a tubular extension is formed at the micropylar end of the embryo sac. It remains enucleate with scanty cytoplasm and act as a haustroium (Fig. 10.6). Centripetal wall formation is restricted only to the upper region of the endosperm. The chalazal portion remains coenocytic and functions as haustorium.

Endosperm in Arecaceae is also of the Nuclear type. The correlation between the size of fruit and activity of endosperm is exemplified in *Cocos nucifera.* The primary endosperm nucleus undergoes a number of free nuclear divisions. By the time the fruit becomes 50 mm long the embryo sac is filled with a clear fluid (liquid endosperm) which contains many free nuclei of various sizes and cytoplasm and usually referred to as "liquid syncytium". When the fruit is about 100 mm long the suspension shows, along with free nuclei, several cells with variable number of nuclei. These cells and free nuclei gradually settle at the periphery of the cavity and layers of cellular endosperm make their appearance which forms the coconut meat. Due to further division of these cells the quantity of the cellular endosperm increases. In mature coconuts the liquid endosperm becomes milky and devoid of any free nuclei or cells.

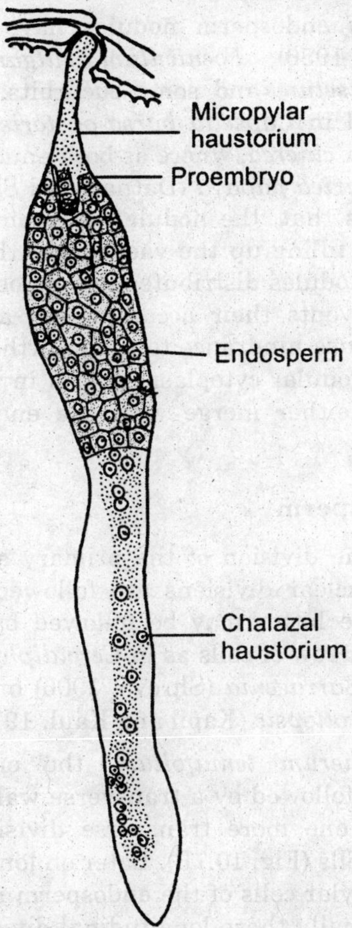

Fig. 10.6. Endosperm haustoria in *Scleria foliosa* (after Nijalingappa and Devaki, 1979)

The development of endosperm in *Areca catechu* is comparable to that in coconut. However, the embryo sac cavity is small and it becomes totally filled by the growth of the endosperm and becomes extremely hard later.

Endosperm Nodules

Another noteworthy feature in the Nuclear type of endosperm is the formation of enucleate and nucleate cytoplasmic nodules or vesicles. Enucleate endosperm nodules have been reported in *Passiflora* (Raju, 1956), *Foeniculum vulgare* (Gupta, 1964), *Stackhousia, Pennisetum* and some cucurbits. Nucleate nodules have been reported in *Capsella bursa-pastoris* (Maheshwari and Sachar, 1954), *Salix cinerea*, where as both enucleate and nucleate nodules occur in *Carica papaya* (Dathan and Singh, 1976). Dalbir Singh (1964) states that, the nodule development/formation is a possible method of filling up the vacuoles of the endosperm. The formation of such nodules distributes the endosperm nuclei to the periphery and prevents their accumulation around the zygote, which otherwise cause hindrance to the growth of the embryo (Al-Altcu, 1977). The nodular cytoplasm is rich in various organelles. The nodules may either merge with the endosperm proper or degenerate.

II. Cellular Endosperm

In this type the division of the primary endosperm nucleus and subsequent nuclear divisions are followed regularly by cell wall formation. The latter may be followed by transverse walls resulting in a linear row of cells as in *Cercidiphyllum* (Swamy and Bailey, 1949) and *Sarracenia* (Shreve, 1906) or the walls may be crosswise as in *Parrotiopsis* (Kapil and Kaul, 1972).

In *Hymenantherum tenuifolium,* the endosperm nucleus undergoes division followed by a transverse wall (Fig. 10.7A). The two cells undergo one more transverse division resulting in a linear row of four cells (Fig. 10.7B). Later on longitudinal divisions occur in the micropylar cells of the endosperm making it biseriate (Fig. 10.7 C). Gradually these longitudinal divisions are extended to the chalazal endosperm cells and the endosperm cells there also become biseriate. Further divisions in these cells occur in all planes resulting in a massive endosperm tissue (Pullaiah, 1976).

Formation of haustoria is a very common feature of this type of endosperm and it is more varied than in the Nuclear type. The haustoria may be micropylar or chalazal or both and are characteristic feature of majority of sympetalous families. Usually for obtaining nutrients and transporting them to the developing endosperm, the haustoria grow in the direction of the tissue rich in food reserves. The hypertrophied, often polyploid nuclei indicate high metabolic activity.

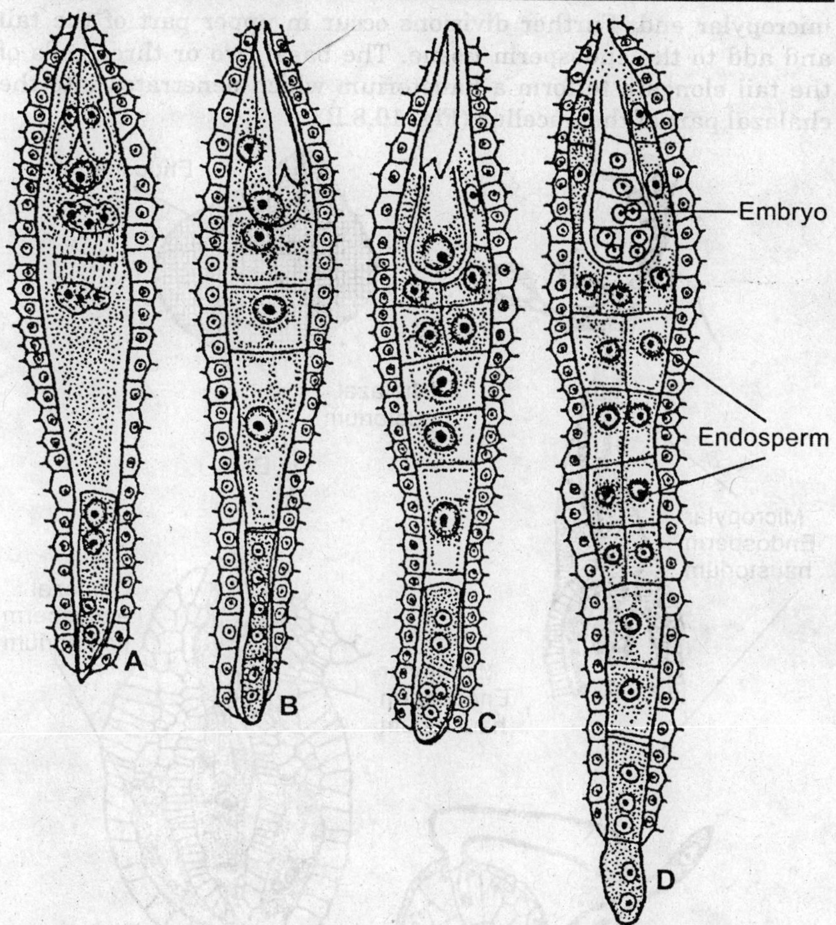

Fig. 10.7. Cellular endosperm development in *Hymenantherum tenuifolium* (after Pullaiah, 1976). **A-B.** Transverse division of Endosperm Cells; **C-D.** Vertical division of Endoperm Cells.

Micropylar haustoria are reported in *Impatiens roylei* and *Hydrocera triflora* (Fig. 10.8 A). A two-celled chalazal haustorium is developed in *Magnolia obovata.* The first division of the primary endosperm nucleus is followed by transverse wall resulting in two chambers of almost equal size. Divisions in the micropylar chamber are rapid in all directions. The chalazal chamber divides transversely and at comparatively slow rate. This results in a tail like chalazal part attached to the more massive tissue at the micropylar end. Further divisions occur in upper part of the tail and add to the endosperm tissue. The basal two or three cells of the tail elongate to form a haustorium which penetrates into the chalazal part of the nucellus (Fig. 10.8 B).

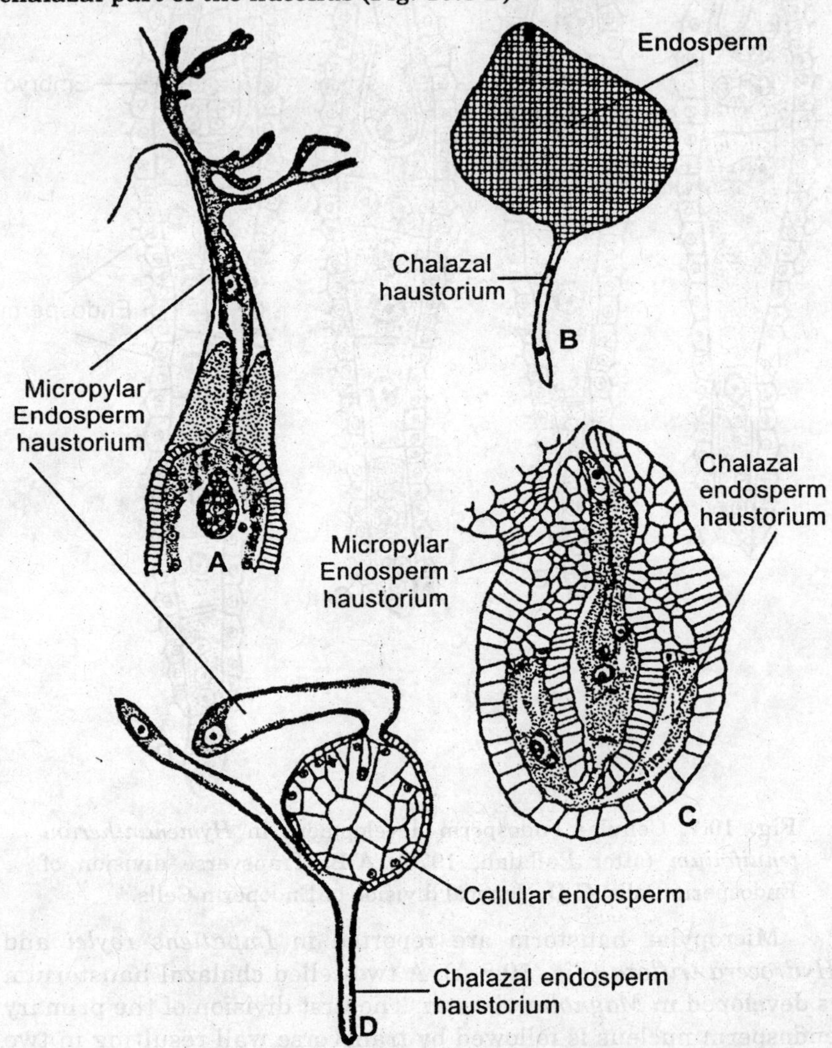

Fig. 10.8. Endoperm haustoria. **A.** *Impatiens roylei* (after Dahlgren, 1934); **B.** *Magnolia obovata* (ater Kapil and Bhandari, 1964); **C.** *Klugia notoniana* (after Arekal, 1961); **D.** *Nemophila aurtia* (after Svenen, 1925)

In *Exocarpus sparteus* the division of the endosperm nucleus is by a transverse wall. Of the two cells thus produced, the lower divides vertically, whereas the upper segments transversely. The two cells of the lower tier do not divide further, but elongate considerably and produce many tubular processes that serve as efficient haustoria.

In *Iodina rhombifolia* (Bhatnagar and Sabharwal, 1968), the chalazal haustorium is aggressive and profusely branched. The haustorium is actually formed before fertilization. The chalazal end of the fertilized embryo sac forms an extensive caecum, the lower end of which extends into the placenta and branches. After fertilization, the division of the primary endosperm nucleus is followed by transverse partitioning of the central cell, resulting in the formation of a micropylar chamber and a chalazal chamber. The endosperm proper is derived from the micropylar chamber alone. The chalazal chamber functions as an aggressive uninucleate haustorium. The nucleus migrates into the caecum and becomes much hypertrophied (about 600 um). Profuse branching at the free end gives the haustorium a corolloid appearance.

In *Melampyrum* (Arekal, 1963) both micropylar and chalazal haustroia are formed. The micropylar haustorium comprises of a single 4-nucleate cell with many tubular processes one of which enlarges and enters the funiculus during later stages. The chalazal haustorium is a binucleate cell broader above and narrow below.

In the family Acanthaceae endosperm development is of the cellular type. The first division of the primary endosperm nucleus cuts off a small chalazal haustorium. The micropylar cell divides transversely to give rise two daughter cells, of which the upper becomes binucleate and forms micropylar haustorium. The central cell which is the largest, forms a sac like out growth in which there are several free nuclear divisions. A large vacuole appears in the centre of the cell and the nuclei become distributed in the peripheral layer of the cytoplasm. The endosperm proper is formed chiefly in the central chamber. It may be nuclear or *ab initio* cellular. Wall formation takes place at the 32-nucleate stage in *Barleria cristata* and *Ruellia tuberosa*, at the 64-nucleate stage in *Dipteracanthus* and *Ruellia pulcherrima* or at the 128-nucleate stage in *Ruellia decaisniana* and *Adhatoda vasica*. If wall formation is not complete, a free nuclear "basal apparatus" is formed below the cellular endosperm. Such basal apparatus is seen in *Dipteracanthus* and *Adhatoda.* In *Asystasia gangetica,* it is very aggressive and forms many multinucleate and enucleate finger like processes. The endosperm is *ab initio* cellular in *Acanthus, Crossandra, Elytraria, Andrographis* and *Haplanthus.* In *Andrographis* some of the endosperm cells elongate to function as secondary haustoria.

Deviation from the general pattern of endosperm development in the Acanthaceae is followed by *Blepharis* and *Thunbergia.* The liner three-celled stage, present in all other members of the family, does not occur in these plants. They also lack the chalazal haustorium. After division of the primary endosperm nucleus a micropylar chamber and a chalazal chamber are formed. The former grows into a haustorium where as the latter gives rise to the endosperm proper. In *Thunbergia alata* the micropylar haustorium is a branched, coenocytic structure and the endosperm proper forms tube-like secondary haustroia (Mohan Ram and Wadhi, 1964).

Klugia notoniana shows both micropylar and chalazal haustoria (Fig. 10.8C). The chalazal haustroium is initially binucleate, but the two nuclei fuse forming uninucleate cell. The haustorium grows laterally and upward consuming subepidermal cells (Fig. 10.8C). The micropylar haustorium comprises two uninucleate cells.

Both micropylar and chalazal haustroia have also been recorded in *Nemophila* of Hydrophyllaceae. Of the three cells formed the micropylar and chalazal cells develop into haustoria, while cell divisions are confined to the central cell resulting in the endosperm proper (Fig. 10.8D).

Composite Endosperm

A unique type of endosperm development present in Loranthaceae. There being no true ovule, all the embryo sacs in an ovary lie close to each other. They develop concurrently and their tips ascend to various heights in the style and stigma. After fertilization, the primary endosperm nucleus moves to the basal part of the embryo sac where it divides. The endosperm forms simultaneously in several embryo sacs. During later development the endosperm of all the embryo sacs in an ovary fuse to form a composite endosperm (Fig. 10.9 A,B), a feature so far restricted only to this family. The only exception is *Struthanthus vulgaris,* where the endosperm develops from a single embryo sac and thus a composite structure is not formed (Venturelli, 1981).

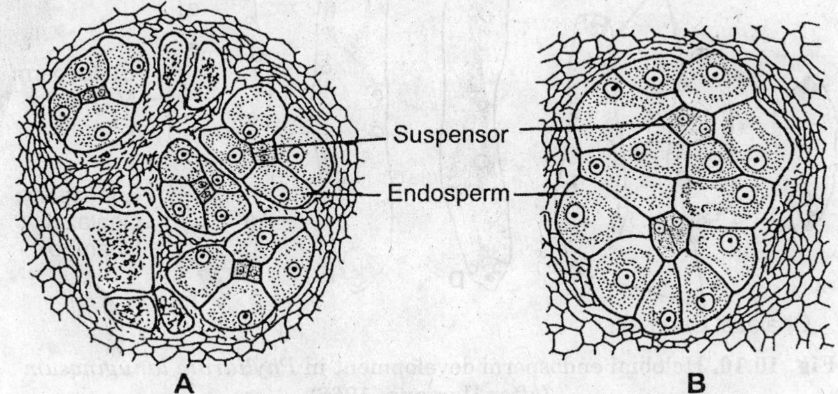

Fig. 10.9. Composite endosperm in *Tolypanthus* of Lorathaceae (after Dixit, 1961)

III. Helobial Endosperm

The Helobial type of endosperm is characterized by the division of the primary endosperm nucleus which moves to the chalazal end of the embryo sac, giving rise to two unequal chambers – the micropylar chamber usually larger than the chalazal. Free-nuclear divisions occur in the micropylar chamber before it undergoes cellularization. By contrast, the nucleus of the chalazal chamber either remains undivided, or divides only a few times. The chalazal chamber usually remains coenocytic, but in a few cases like *Phylidrum lanuginosum* (Kapil and Walia, 1965), *Juncus prismatocarpus* (Zaman, 1950), it gets partitioned into cells.

In *Sparganium* (Sparganiaceae) the primary endosperm nucleus lies near the antipodals and its division results in a large micropylar and a small chalazal chamber (Fig. 10.10). Wall formation in the micropylar chamber does not always precede that in the chalazal chamber (Asplund, 1973). The endosperm cells become multinucleate.

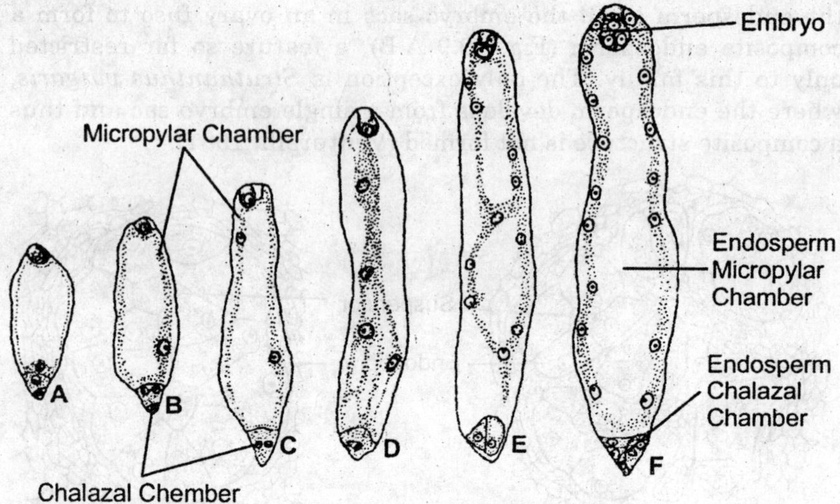

Fig. 10.10. Helobial endosperm development in *Phylidrum lanuginosum* (after Hamann, 1966)

In *Asphodelus* (Eunus, 1952) the first division is simultaneous in both the chambers (Fig. 10.11 A, B), but further divisions take place more rapidly in the micropylar chamber (Fig. 10.11 C). The nucleus of the chalazal chamber divides only once or twice, while that of the upper chamber undergoes several free nuclear divisions, the daughter nuclei becoming parietally disposed. After wall formation is completed (Fig. 10.11 D), centripetal growth occurs and results in a large mass of endosperm. At the same time the chalazal chamber shows dense cytoplasm and hypertrophied nuclei. Its basal portion grows in the form of a wedge-shaped structure and has a haustorial function (Fig. 10.11 E). The chalazal chamber persists even in fairly old seeds.

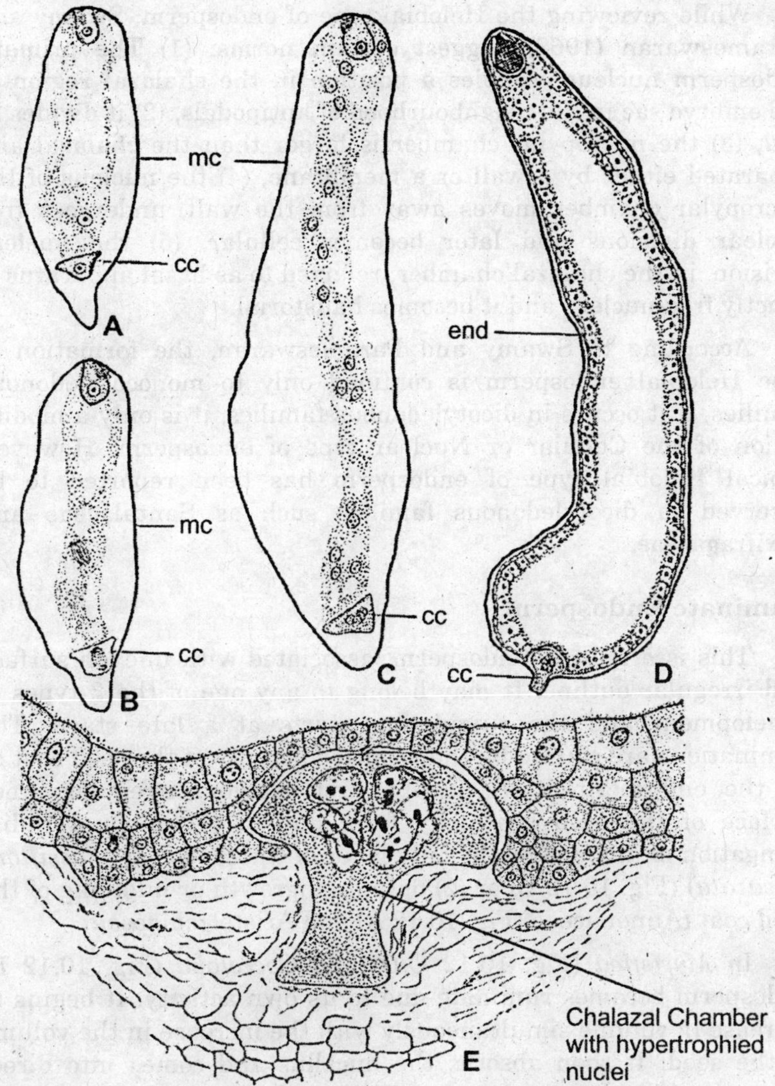

Fig. 10.11. Development of endosperm in **Asphodelus tenuifolius. A-B.** Large micropylar chamber and small chalazal chamber; **C.** Multinucleate micropylar chamber and quadric-nucleate chalazal chamber; **D.** After wall formation the upper chamber; **E.** Enlarged view of the lower part of seed, showing hypertrophied nuclei in the chalazzal chamber of the endosperm. **CC :** Chalazal chamber; **mc :** Micropylar chamber; **end :** endosperm.

While reviewing the Helobial type of endosperm, Swamy and Parameswaran (1963) suggest certain norms: (1) The primary endosperm nucleus occupies a position in the chalazal region of the embryo sac in the neighbourhood of antipodals, (2) it divides *in situ*, (3) the micropylar chamber is larger than the chalazal and separated either by a wall or a membrane, (4) the nucleus of the micropylar chamber moves away from the wall, undergoes free nuclear divisions and later becomes cellular, (5) the nuclear division, in the chalazal chamber, referred to as basal apparatus is strictly free-nuclear and it becomes haustorial.

According to Swamy and Parameswaran, the formation of true Helobial endosperm is confined only to monocotyledonous families, if it occurs in dicotyledonous families, it is only a modification of the Cellular or Nuclear type of endosperm. However, typical Helobial type of endosperm has been reported to be observed in dicotyledonous families such as Santalaceae and Saxifragaceae.

Ruminate Endosperm

This is a mature endosperm, associated with uneven surface and irregular outline. It may belong to any one of the 3 types of development and the rumination starts at a late stage. The ruminations are caused by the localized activity of the seed coat or by the endosperm itself. In the first category irregular inner surface of the seed coat may form due to (1) unequal radial elongation of any one or the only layer of the seed coat (*Passiflora calcarata*) (Fig. 10.12 B) or (2) definite ingrowth or infolding of the seed coat (Annonaceae (Fig. 10.12 A) and Aristolochiaceae).

In *Myristica* (Fig. 10.12 C,D) and *Coccoloba* (Fig. 10.12 E) endosperm becomes ruminate due to its own activity. It begins to increase in volume simultaneously with the increase in the volume of the seed. It soon absorbs the nucellus and comes into direct contact with the seed coat. During further growth of the endosperm the irregular inner surface of the seed coat makes it ruminate. On the other hand, in *Andrographis* and *Elytraria* the endosperm exhibits unequal peripheral activity during late stage of its development and causes the seed coat to attain an irregular outline.

On morphological basis, Periasamy (1962a) distinguishes seven types: *Annona*, *Passiflora*, *Myristica*, *Spigelia*, *Verbascum*, *Coccoloba* and *Elytraria*. In all these types, except *Elytraria*, irregularities occur in the growth of integument/s which bring about the rumination of endosperm. In the *Elytraria* type, during the development of seed, localized regions in the peripheral layers of cellular endosperm show active growth so that it becomes ruminate.

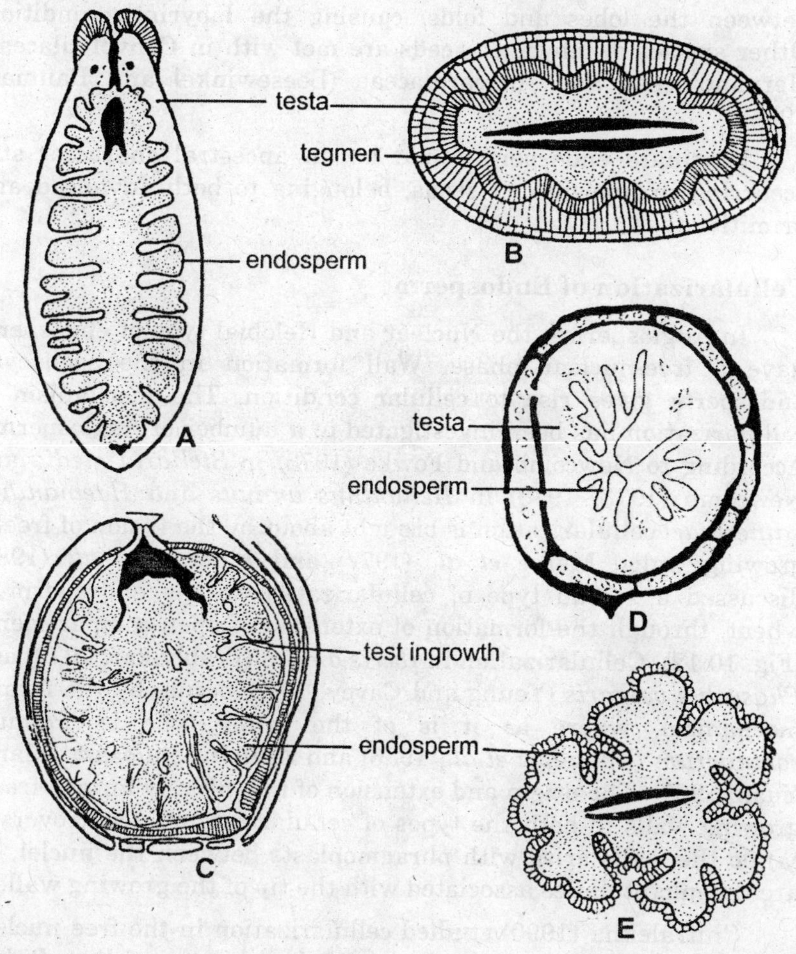

Fig. 10.12. Seeds with ruminate endosperm **A.** *Annona squamosa*, **B.** *Passiflora calcarata*, **C, D.** *Myristica fragrans* **E.** *Coccoloba uvifera* (A,B,E after Periasamy, **C.** after Voigt, 1988, **D.** after Mohana Rao, 1974)

Seeds which reveal a network of lobes in sections in any plane are termed 'labyrinth seeds' (Van Heel, 1970, 1971 a,b). This structure may be either due to rumination of the endosperm or lobing and folding of the cotyledons. In seeds, thus, when lobing and folding of the cotyledons occur together the rumination, the labyrinth structure is very prominent. In *Kingiodendron pinnatum*, *Erycibe griffithii* and *E. tomentosa* the seeds show excessive folding of cotyledons, followed by intrusion of the testa in between the lobes and folds, causing the labyrinth condition. Other species of labyrinth seeds are met with in Convolvulaceae, Hernandiaceae and Anacardiaceae (Boesewinkel and Bauman, 1984).

Ruminate endosperm could be an ancestral character still occurring in present-day seeds, belonging to both advanced and primitive taxa.

Cellularization of Endosperm

In Angiosperms, the Nuclear and Helobial type of endosperm have a free-nuclear phase. Wall formation in the coenocytic endosperm gives rise to cellular condition. The mechanism of cellularization has been investigated in a number of Angiosperms. According to Newcomb and Fowke (1973) in *Stellaria media* and Newcomb (1973, 1978) in *Helianthus annuus* and *Haemanthus katherinae*, cellularization is brought about by the fusion of freely-growing walls. Mares *et al.* (1977) and Fineran *et al.* (1982) discussed a second type of cellularization in the endosperm of wheat, through the formation of extension of open compartments (Fig. 10.12). Cellularization in *Iberis amara* (Prabhakar, 1979) and *Phaseolus vulgaris* (Yeung and Cavey, 1988) occurs by the former mechanism, where as it is of the latter type in *Papaver somniferum* (Bhandari *et al.*, 1986) and *Allium fistulosum* (Xiang-Yuan, 1987). The origin and extension of the initially formed freely growing walls in both the types of cellularization is controversial (a) do the walls arise with phragmoplasts between the nuclei, (b) are the microtubules associated with the tip of the growing wall.

Chitralekha (1990) studied cellularization in the free nuclear endosperm of *Argemone mexicana, Asphodelus tenuifolius, Bidens pilosa, Brassica campestris* and *Ranunculus scleratus,* and concluded that cellularization of free-nuclear endosperm conforms

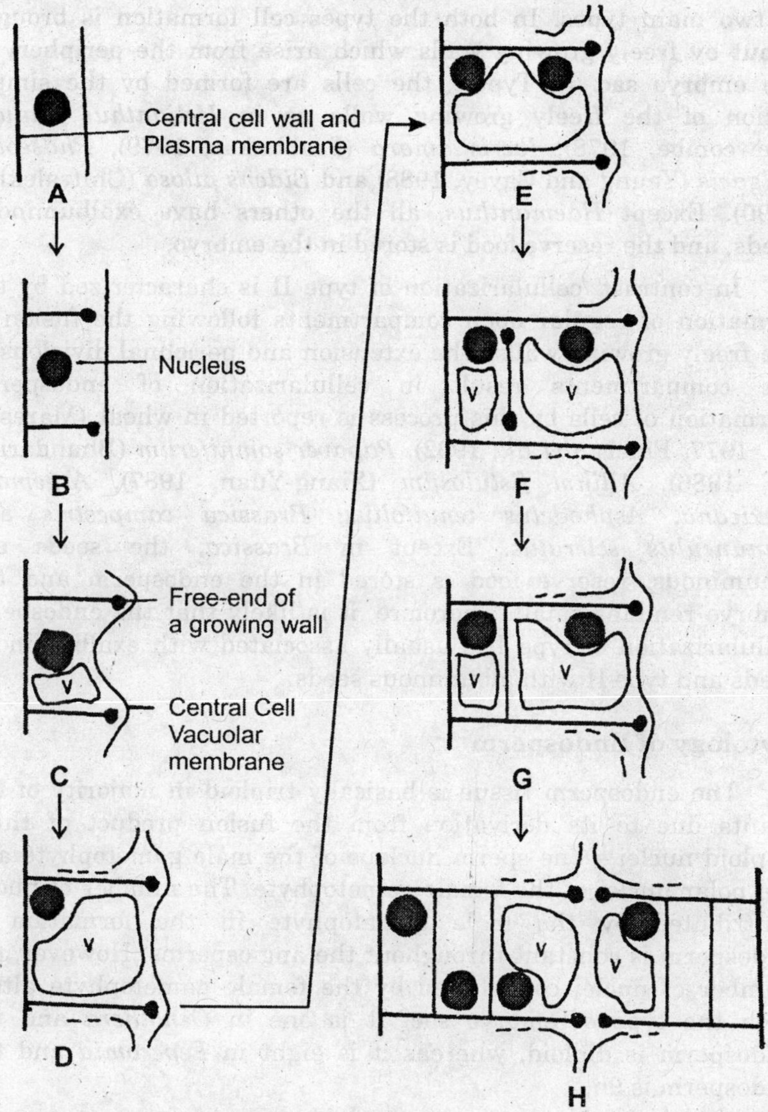

Fig. 10.13. *Triticum aestivum*, diagrammatic sketch of events leading to cell wall formation in free-nuclear endosperm (Adopted after Mares *et al.*, 1977)

to two main types. In both the types cell formation is brought about by freely growing walls which arise from the periphery of the embryo sac. In Type-I, the cells are formed by the simple fusion of the freely growing walls as in *Helianthus annuus* (Newcombe, 1978), *Iberis amara* (Prabhakar, 1979), *Phaseolus vulgaris* (Yeung and Cavey, 1988) and *Bidens pilosa* (Chitralekha, 1990). Except *Haemanthus,* all the others have exalbuminous seeds, and the reserve food is stored in the embryo.

In contrast, cellularization of type II is characterized by the formation of regular open compartments following the fusion of the freely growing walls. The extension and periclinal divisions of the compartments result in cellularization of endosperm. Formation of cells by this process is reported in wheat (Mares *et al*., 1977, Fineran *et al*., 1992), *Papaver somniferum* (Bhandari *et al*., 1986), *Allium fistulosum* (Xiang-Yuan, 1987), *Argemone mexicana, Asphodelus tenuifolius, Brassica campestris* and *Ranunculus scleratus*. Except in *Brassica,* the seeds are albuminous, reserve food is stored in the endosperm and the embryo remains small. Therefore, it is likely that the endosperm cellularization of type I is usually associated with exalbuminous seeds and type II with albuminous seeds.

Cytology of Endosperm

The endosperm tissue is basically triploid in majority of the plants due to its derivation from the fusion product of three haploid nuclei – one sperm nucleus of the male gametophyte and the polar nuclei of the female gametophyte. The number of nuclei contributed by the male gametophyte in the formation of endosperm is constant throughout the angiosperms. However, the number of nuclei contributed by the female gametophyte alters with the type of embryo sac. It is one in *Oenothera* and the endosperm is diploid, whereas it is eight in *Peperomia* and the endosperm is 9n.

The endosperm tissue exhibits high frequency of polyploidization of its cells during its development. While Tandon and Kapoor (1961 a,b, 1963) and Kapoor and Tandon (1963, 1964) worked out the cytology of species of *Nothoscordum* and *Zephyranthes.* Erbrich (1965) studied the endosperm cytology in several flowering plants. In *Thesium alpinum,* the ploidy of the endosperm haustorium is up to 384n and the highest ploidy is reported in *Arum maculatum* where it becomes 24567n. In *Melampyrum pratense* each nucleus of the chalazal haustroium attains a ploidy of 1536n. However in *Dendrophthoe falcata* all the cells of the mature endosperm show remarkable stability in their chromosome number and are triploid (Johri and Nag, 1974). Endomitosis and nuclear fusion are some of the methods of polyploidization of endosperm cells (Kapoor, 1962). An increase in the size of endosperm nuclei can result from polyploidy or polyteny as in *Pedicularis* and Maize, respectively.

The common mitotic abnormalities in the endosperm tissue are the chromosome bridges, lagging chromosomes, spontaneous breakage of the chromosome and fragmentation of nuclei (Kapoor, 1962). Size of the nuclei and the number of nucleoli per nucleus also vary greatly.

The endosperm is generally non-chlorophyllous. However chlorophyllous endosperm also occurs in *Crinum, Mathiola, Raphanus* and *Viscum*.

Aleurone Tissue

The cereal endosperm at maturity comprises two different tissues. The outer aleurone layer consists of living cells and is usually one cell layer thick in *Triticum aestivum, Zea mays, Triticale, Secale cereale* and 2-4 cell layers thick in *Hordeum vulgare*. The inner starchy endosperm is dead at maturity. In rice aleurone cells are of two types: (a) rectangular cells with dense cytoplasm surrounding the embryo and (b) cuboidal cells with dense cytoplasm around the starchy endosperm. The aleurone cells have thick walls and non-vacuolated cytoplasm. They are interconnected by plasmodesmata.

The most prominent organelles of the aleurone cells are aleurone grains, followed by sphaerosomes. The main components of aleurone grains are protein, phytin, phospholipids and carbohydrates. Structurally, the aleurone grains possess two kinds of inclusions, besides ground substance: (a) Globoids present within the globoidal cavities – they contain phytin and lipids; (b) protein-carbohydrate bodies which are 1-1.5 um in diameter. The ground substance also contains high concentration of protein, but it is less than that in the protein-carbohydrate bodies.

Improvement of the Nutritive value of endosperm by Genetic Engineering

Cereal endosperm is the major staple food of about one third of the world's population. It is a good source of calories, but lacks or is very poor in vitamins and proteins. Genetically engineered rice has been developed which synthesizes β-carotene (pro-vitamin A) in the endosperm, due to which the grain appears yellow and, therefore, called 'Golden Rice' (Ye *et al.*, 2000). The dietery β-carotene can be converted into vitamin A, which is important for normal development of humans. Deficiency of vitamin A causes night blindnesss or total blindness in children, besides other diseases.

Similarly, maize lines with 100-fold enhancement in lysine content of the endosperm have been developed by genetic engineering (Frizzi *et al.*, 2008).

Functions of Endosperm

1. In endospermous seeds, the persistent endosperm is rich in reserve food materials in the form of carbohydrates, fats and/or proteins. During seed germination these substances are digested and utilized for the growth of the seedling until the latter develops the green pigments and is able to manufacture its own food.
2. The more important contribution of endosperm is the nutrition of embryo during its early stages of development – from zygote to globular embryo (Maheshwari and Rangaswamy, 1965).
3. During early stages of endosperm, as a nurse tissue, does not show specificity for its own embryo. Thus liquid endosperm or endosperm extract from immature fruits not only supports growth of its own embryo but can also nourish the embryos of other Angiosperms, *e.g.*, coconut milk, corn milk, liquid endosperm of horse chestnut and walnut.
4. Young endosperm is rich in various growth hormones such as auxins, cytokinins and gibberellins whose concentration decreases after certain age of the tissue. Zeatin, a very potent cytokinin, is extracted from young endosperm of maize. Added to the basic nutrient medium coconut milk also induces the

differentiation of embryos (embryoids) from plantlets from various plant tissues. Immature endosperm is also potent for inducing divisions in highly differentiated and mature cells, such as those of secondary phloem in carrot.

5. Endosperm also regulates the precise mode of embryo development. Removed from the influences of the endosperm very young embryo fails to mature even in the richest artificial nutrient medium. Moreover, the isolated older embryos often skip the normal stages of embryogeny and show precocious germination in cultures.

   a. At the time of fertilization the embryo sac has very little nutritive material. After the endosperm advances in development it stores enough food substances for adequate supply for the developing embryo.
   b. In the majority of angiosperms the zygote divides only after the endosperm has reached a reasonable stage of development. Even in those cases where the zygote divides before or simultaneously with the endosperm the latter soon surpasses the embryo in growth.
   c. Generally the embryo grows only when the endosperm develops properly. If the endosperm aborts as happens in many incompatible crosses, the growth of embryo is adversely affected.
   d. In the absence of endosperm (Orchidaceae, Podostemaceae and Trapaceae) special provisions exists to ensure the nutrition of the developing embryo.
   e. During its growth the embryo depletes the surrounding cells of the endosperm of their contents. In many plants, such as legumes and cucurbits, the embryo consumes the endosperm completely before the seed attains maturity.

References

Olsen, O.A. 2007. Endosperm. Developmental and Molecular Biology series. Plant cell Monographs. Vol. 8. Springer, Berlin.

CHAPTER - XI

EMBRYO

The discovery of syngamy by Strasburger (1884) and double fertilization by Nawaschin (1898) are the two important events in the history of angiosperm embryology. Syngamy results in the formation of zygote. The zygote takes up a predetermined mode of development (embryogeny) and gives rise to an embryo, which, later develops, into a complete plant. The embryo may consist of one (monocots) or two (dicots) cotyledons depending up on the species. The embryonal axis above the cotyledon is termed as epicotyl and the portion below the cotyledons is called hypocotyl. The epicotyl ends up with the plumule (embryonic shoot) and the hypocotyl with radicle (embryonic root). The monocot embryo differs from that of the dicots in having a single cotyledon. The embryo is highly specialized in certain groups like grasses.

ZYGOTE

The zygote takes some rest before it divides. The resting period of the zygote is known as the dormancy and it varies from plant to plant. The resting period is very short in taxa belonging to Asteraceae and Poaceae, and is very long in *Colchicum*. This dormancy period may sometimes correlate with the type of endosperm-shorter in cellular type than the nuclear type. Striking changes occur in the zygote during this dormancy period and at the end of it, the zygote shows more pronounced polarity.

The unicellular zygote undergoes a series of divisions and develops into a well organized embryo with much differentiated organs. In angiosperms, the zygote is located at the micropylar pole of the embryo sac. Its lower end is attached to the embryo sac wall and the apical portion projects into the central cell.

After fertilization, the size of the zygote is reduced to almost half in taxa such as *Gossypium* (Jensen, 1968). This decrease in cell size results in large accumulation of cell organelles, like ER, ribosomes, plastids, mitochondria around. the zygote nucleus. The number of dictyosomes which were involved in the wall synthesis, increase in number around the zygote. The ribosomes are grouped together so as to form polysomes which indicates the beginning of the metabolic activity.

The zygote before its division is a highly polarized and isolated cell because plasmodesmatal connections of the zygote with other cells were lost. The zygote nucleus surrounded by a number of plastids and mitochondria, takes up the chalazal position, while the micropylar end is occupied by one or more vacuoles and limited number of cell organelles.

EMBRYOGENY

Early Embryogenesis

In the majority of the angiospermous taxa studied so far, the first division of the zygote is transverse and results in the formation of two superposed cells, which can be designated as apical cell '*ca*', and the basal cell '*cb*'. But rarely, as in Loranthaceae, Piperaceae and Balanophoraceae, the first division of the zygote is veritcal or oblique. The development of the embryo in the early stages is almost similar in both monocots and dicots, although differences arise during the formation of the cotyledons. The young embryo, starting from the 2-celled stage to the formation of organs is generally regarded as the proembryo.

DEVELOPMENT OF THE EMBRYO

I. Systems suggested by Schnarf, Johansen and Maheshwari

Basing on the variation in the orientation of wall in the zygote, plane of cell division in the apical cell of the 2-celled proembryo and the contribution made by the apical cell (*ca*), and basal cell (*cb*) to the body regions of the embryo proper, Schnarf (1929), Johansen (1945) and Maheshwari (1950) classified the embryogeny in the dicots into 5 categories as follows:

1. The apical cell (*ca*) of the 2-celled proembryo divides by a vertical wall resulting in the formation of an inverted T-shaped proembryo.

   2. The basal cell (*cb*) plays a minor part or none in the development of the embryo .... **- Crucifer or Onagrad type**

   2. Both the apical cell (ca) and basal cell (*cb*) contribute equally to the development of the embryo .... **- Asterad type**

1. The apical cell (*ca*) of the 2-celled proembryo divides transversely and form a linear 4-celled proembryo.

   3. The basal cell (*ca*) takes up a minor part or none in the subsequent development of an embryo.

      4. The basal cell (*ca*) forms a short suspensor of two or more cells - .... **- Solanad type**

      4. The basal cell (cb) does not divide further and the suspensor if any is also formed by the apical cell (*ca*) .... **- Caryophyllad type**

   3. Both the apical call (*ca*) and basal cell (*cb*) contribute equally to the embryo development .... **- Chenopodiad type**

Some examples of the families showing different types as noted in the Table

Embryogeny	Families
1. Onagrad or Crucifer type	Annonaceae, Onagraceae, Cruciferae (Brassicaceae), Magnoliaceae, Scrophulariaceae.
2. Asterad type	Asteraceae, Balsaminaceae, Violaceae
3. Solanad type	Aizoaceae, Campanulaceae, Linaceae, Solanaceae, Theaceae etc.
4. Caryophyllad type	Caryophyllaceae, Molluginaceae, Portulacaceae
5 Chenopodiad type	Amaranthaceae, Basellaceae, Chenopodiaceae

Johansen (1950) reported a number of variations in each type and designated them by the names of the taxa characteristic of the particular variation. Generally the type of embryogeny is constant throughout the family. But, however, rarely more than one type of embryogeny may be seen in a species. For example *Anemone rivularis* shows both Onagrad and Solanad types.

Fig. 11.1 gives a schematic representatin of the classificatory types based on Schnarf (1929) and Johansen (1950).

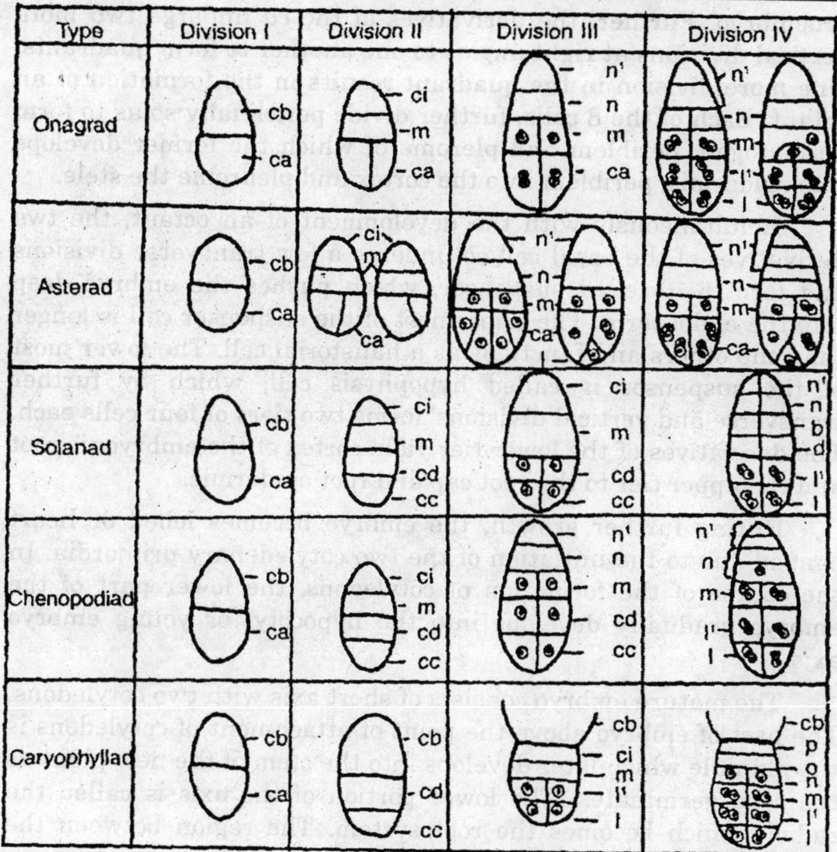

Fig. 11.1. Schematic representation of main types of embryogeny, based on Schnarf (1929), Johansen (1950) and Maheshwari's (1950) classification

(i) Onagrad type

To illustrate complete development of a dicot embryo in detail, Hanstein (1870) work on *Capsella bursa-pastoris* serves as a role model. It shows Onagrad or Crucifer type and is as follows.

The first division of the zygote is transverse and results in the formation of an apical cell (*ca*) and basal cell (*cb*). The basal cell is cut towards the micropyle and the apical cell towards the embryo sac cavity. Later, the *cb* divides transversely and *ca* divides vertically resulting in the formation of an inverted T-shaped proembryo. Further, the derivatives of the *ca* undergo two more vertical divisions at right angles to one another to form quadrants. One more division in the quadrant results in the formation of an octant. Each of the 8 cells, further divide periclinally so as to form dermatogen, periblem and plerome of which the former develops into epidermis, periblem into the cortex and pleurome the stele.

Simultaneously with the development of an octant, the two derivatives of the basal cell *cb* undergo a few transverse divisions and form 6-10-celled suspensor, which pushes the embryo deep into the endosperm. The uppermost of the suspensor cell is longer than the others and functions as a haustorial cell. The lower most of the suspensor is called hypophysis cell, which by further transverse and vertical divisions' forms two tiers of four cells each. The derivatives of the lower tier yield cortex of the embryonic root and the upper tier to the root cap and root epidermis.

During further growth, the embryo becomes lobed or heart shaped due to the initiation of the two cotyledenary primordia. In the course of the formation of cotyledons, the lower part of the embryo gradually develops into the hypocotyl or young embryo axis.

The mature embryo consists of short axis with two cotyledons. The part of embryo above the point of attachment of cotyledons is the plumule which later develops into the stem of the new plant as the seed germinates. The lower portion of the axis is called the radicle which becomes the root system. The region between the plumule and radicle is known as the hypocotyl. The cotyledons are normally food storage organs which digest and absorb food from the endosperm and which may also serve as first leaves after germination.

(ii) Asterad type

Soueges (1920) gave a detailed account of development of embryo in *Senecio vulgaris*. Since then several other members of this family have been studied and all of them follow uniform pattern of embryo development (see Pullaiah, 1984).

The zygote divides transversely to form a terminal cell *ca* and a basal cell *cb*. The terminal cell *ca* divides vertically and basal cell *cb* transversely to form four-celled T-shaped proembryo. The two derivatives of *cb* are designated as *m* and *ci*. The two juxtaposed cells of tier *ca* undergo one more vertical division at right angles to the previous one producing quadrants *q*. Due to oblique divisions in them octants are formed. The cell *m* divides into two juxtaposed cells. By another vertical division at right angles to the first one, quadrants are formed in this tier also. The cell *ci* divides transversely and gives rise to two superposed cells *n* and *n'*. Of these the former undergoes two vertical divisions at right angles to each other. The *n'* divides by a transverse wall and forms two superposed cells, namely *o* and *p* (Fig. 11.2).

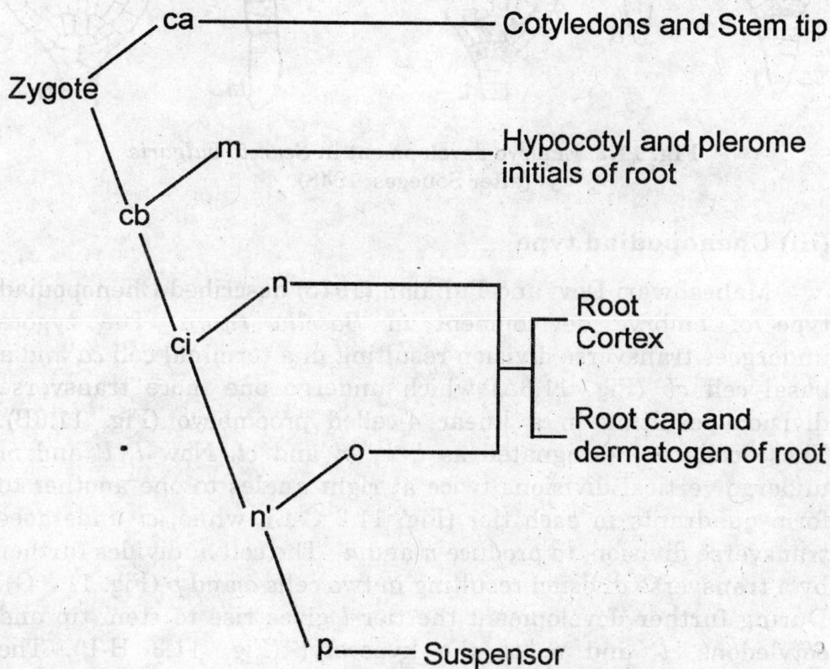

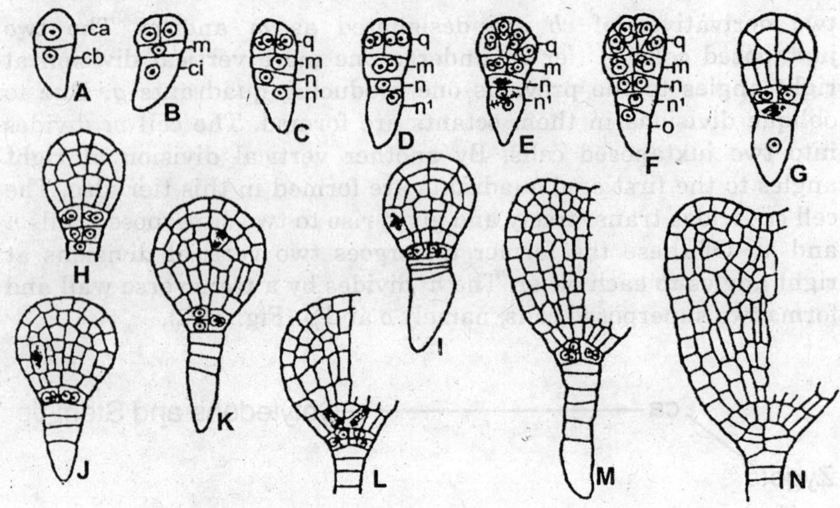

Fig. 11.2. Embryo development in *Senecio vulgaris* (after Soueges, 1948)

(iii) Chenopodiad type

Maheshwari Devi and Pullaiah (1975) described Chenopodiad type of embryo development in *Basella rubra*. The zygote undergoes transverse division resulting in a terminal cell *ca* and a basal cell *cb* (Fig. 11.3A) which undergo one more transverse division resulting in a linear 4-celled proembryo (Fig. 11.3B). These cells are designated as *l, l', m* and *ci*. Now *l, l'* and *m* undergo vertical divisions twice at right angles to one another to form quadrants in each tier (Fig. 11.3 C-E), while, *ci* undergoes transverse division, to produce *n* and *n'*. The cell *n'* divides further by a transverse division resulting in two cells *o* and *p* (Fig. 11.3 G). During further development the tier *l* gives rise to stem tip and cotyledons, *l'* and *m* to the hypocotyl (Fig. 11.3 H-P). The derivatives of *n* function as the hypophysial initials and contribute to the formation of root tip. The cells *o* and *p* undergo further divisions and forms suspensor (Fig. 11.3 J-P).

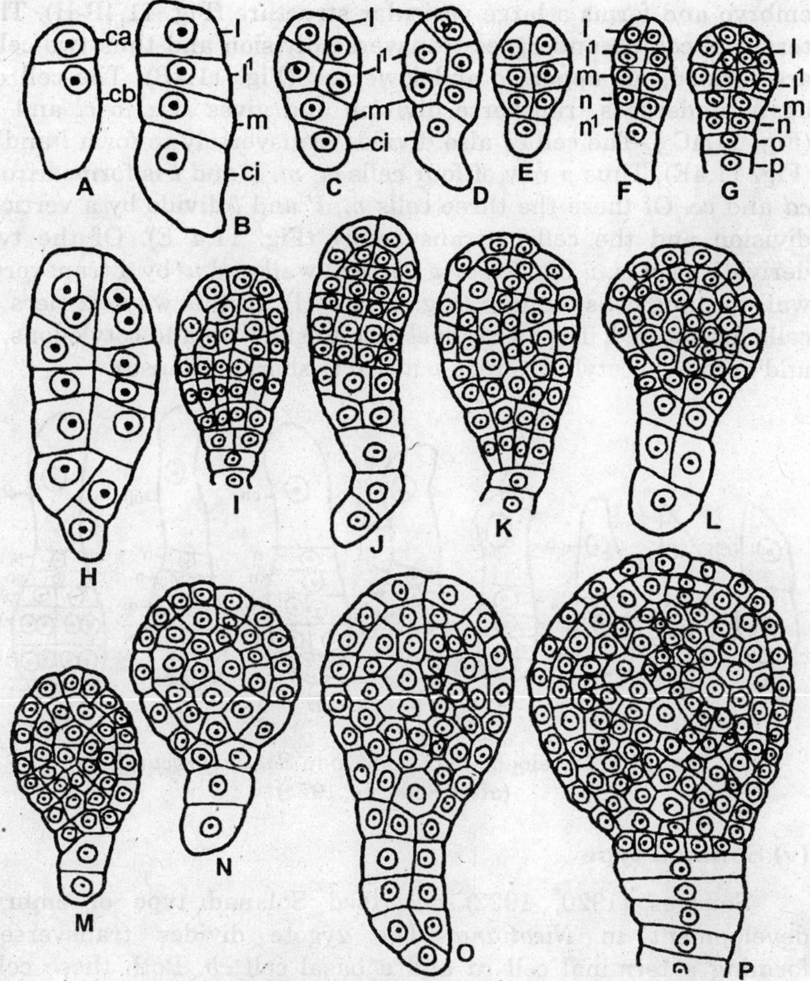

Fig. 11.3. Embryo development in *Basella rubra* (after Maheswari Devi and Pullaiah, 1974)

(iv) Caryophyllad type

Soueges (1924) described caryophyllad type in *Sagina procumbens.*

The zygote divides by a transverse division forming a terminal cell *ca* and basal cell *cb* (Fig. 11.4A). The basal cell *cb* remains undivided which does not take part in the development of embryo and forms a large vesicular structure (Fig. 11.4B-H). The terminal cell *ca* undergoes transverse division and thus two cells are formed, an upper *cc* and lower *cd* (Fig. 11.4B). The cell *cd* again undergoes transverse division and gives rise to *ci* and *m* (Fig. 11.4C). The cell *cc* also divides transversely to form *l* and *l'* (Fig. 11.4E). Thus a row of four cells *ci, m, l'* and *l* is formed from *cd* and *cc*. Of these the three cells *m, l'* and *l* divide by a vertical division and the cell *ci* transversely (Fig. 11.4 E). Of the two derivatives of *ci*, *n* divides by a vertical wall and *n'* by a transverse wall forming cells *o* and *p* (Fig. 11.4 F-H). In this way six tiers of cells are formed. The tier *l* gives rise to stem tip and cotyledons, *l'* and *m* to hypocotyl, *n* to root, *o* and *p* to short suspensor.

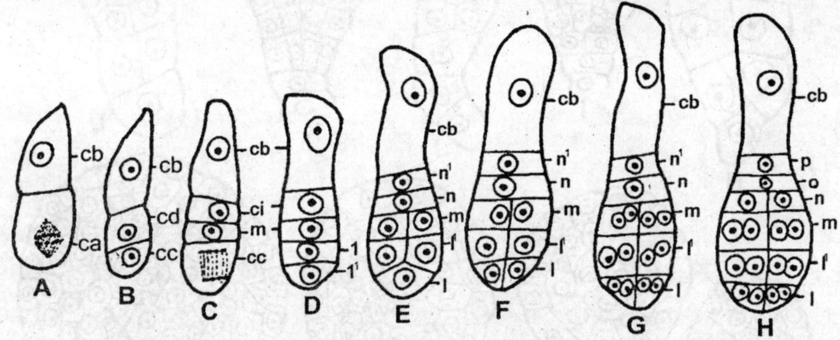

Fig. 11.4. Development of embryo in *Sagina procumbens* (after Soueges, 1924)

(v) Solanad type

Soueges (1920, 1922) described Solanad type of embryo development in *Nicotiana*. The zygote divides transversely forming a terminal cell *ca* and a basal cell *cb*. Both these cells undergo transversely resulting in a 4-celled linear proembryo. The cells are designated as *l, l', m* and *ci*. Now *l* and *l'* divide by vertical walls oriented at right angles to each other to give rise to quadrants in each tier, while *m* and *ci* divide transversely to produce *d, f, n* and *n'*. By subsequent divisions *l* gives rise to stem tip and cotyledons, *l'* to hypocotyl and to the periblem and plerome of the root, *d* to the root tip and root cap. The remaining cells *f, n* and *n'* produce suspensor (Fig. 11.5).

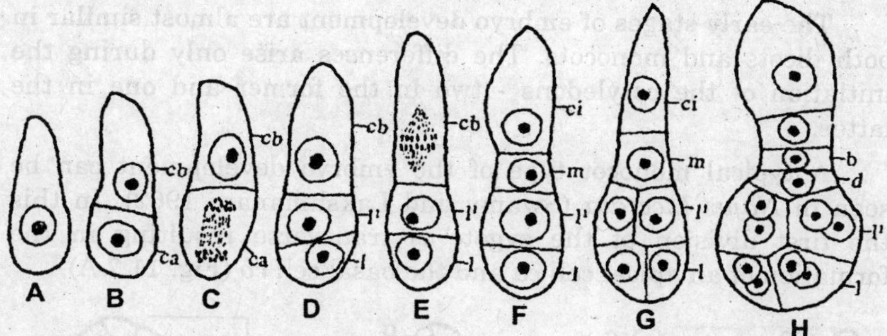

Fig. 11.5. Development of embryo in *Nicotiana* (after Soueges, 1922)

II. Embryo Development in Monocotyledons

The development of embryo up to the octant stage is almost similar in Monocotyledons and Dicotyledons. The differences appear later.

According to Lakshmanan (1972) the chief difference between embryos of Dicotyledons and Monocotyledons lies in the number of cells of the terminal quadrant which contribute to the formation of cotyledon(s) and epicotyl. In Dicotyledons it is the two opposite cells of terminal quadrant that give rise to the two cotyledons (Fig. 11.6A,B). In Monocotyledons the number of cells involved in cotyledon formation is variable, practically all the four cells (except a few cells derived from one of the quadrant cells) in the Philydraceae (Fig. 11.6C), three cells of the quadrant in Iridaceae, Sparganiaceae and Pontederiaceae (Fig. 11.6D) and two adjoining cells in the Potamogetonaceae, Hydrocharitaceae and Amaryllidaceae (Fig. 11.6E).

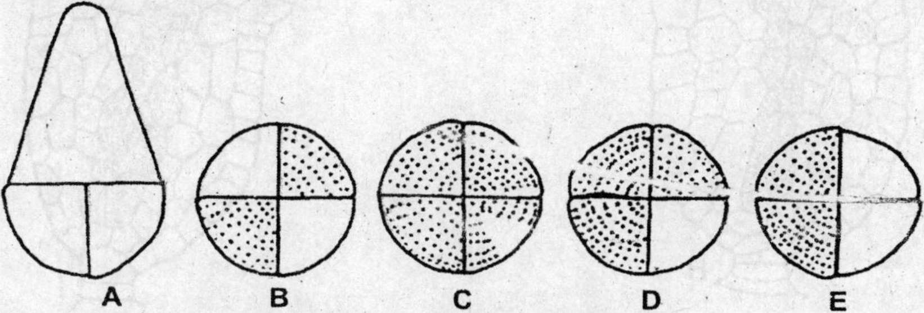

Fig. 11.6. Derivation of cotyledons in Dicotyledons and Monocotyledons (after Lakshmanan, 1972)

The early stages of embryo development are almost similar in both dicots and monocots. The differences arise only during the initiation of the cotyledons - two in the former and one in the latter.

A typical monocot type of the embryo development can be seen in *Najas lacerata* (Swamy and Lakshmanan, 1962). In this the first division of the zygote is transverse resulting in the formation of an apical cell *ca* and the basal cell *cb* (Fig. 11.7 A).

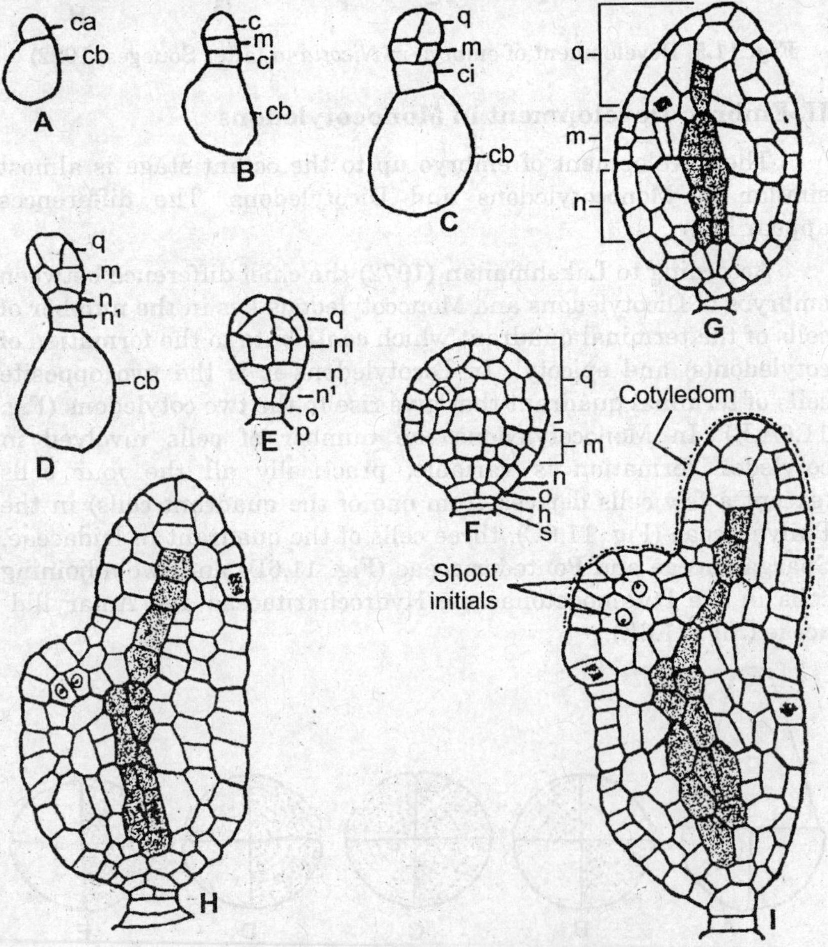

Fig. 11.7. Development of embryo in *Najas lacerata* (ater Swamy and Lakshmanan, 1962)

The basal cell *cb* does not divide further and transforms itself into a large vesicular structure, which forms a part of the suspensor (Fig. 11.7 A-D). Thus the entire embryo is derived from apical cell *ca*. During further development, the *ca* undergoes a transverse division and forms two cells namely *c* and *d*. Of these the cell *d* once again divides transversely. In this way a linear proembryo of four cells (*c, m, ci, cb*) is formed (Fig. 11.7 B). Two vertical divisions at right angles to each other in the two distal cells (*c* and *m*) lead to the formation of two superposed tiers (*q* and *m*) of four cells each (Fig. 11.7 C,D). In the mean time cell *ci* divides transversely to give rise to *n* and *n'* (Fig. 11.7 D). Whereas, cell *n* divides vertically, *n'* undergoes transverse division giving rise to two cells (*o* and *p* Fig. 11.7E). During further growth (Fig. 11.7 F-I) tier *q* gives rise to single cotyledon, *m* gives rise to the hypocotyls. The radicle is organized from the derivatives of *n*, while *o* gives rise to root cap. Derivatives of *p* form suspensor.

Differentiation in the Grass Embryo

The grasses exhibit some peculiarities in their embryogeny and also in the structure of the mature embryo and hence deserve special mention. In the caryopsis, the mature embryo is present towards the base on the dorsal side. It is comparatively smaller than the endosperm. The mature embryo has a single cotyledon designated as the scutellum. In median longitudinal section, the scutellum appears as a long, shield like vascularized structure laterally attached to the axis. The point of attachment of the scutellum to the axis is known as the **scutellar node** and it demarcates the axis into two regions. The upper region is the epicotyl which comprises the shoot apex and leaf primordia. The epicotyl is surrounded by a hallow sheath called **coleoptile**. At the time of germination, the coleoptile becomes green and photosyn-thetic and the first foliage breaks through it. The area between the base of the coleoptile and the point of attachment of the scutellum has been referred as the mesocotyl. In the region below the scutellar node the axis terminates in a radicle with a root cap and is ensheathed by a structure called **coleorhiza** (Fig. 11.8). During germination the root emerges through coleorhiza.

The embryo development in four species of *Triticum* was examined in great detail by Batygina (1969). She opines that the embryogeny in *Triticum* follows a fixed pattern, which is different from that of monocots, and dicots, and she named it as the "Graminad type". This type of embryogeny is also seen in several other members of Gramineae and is characterized by the regular occurrence of oblique divisions. Thus, the mature embryo of grasses with its special structures has proved of systematic value.

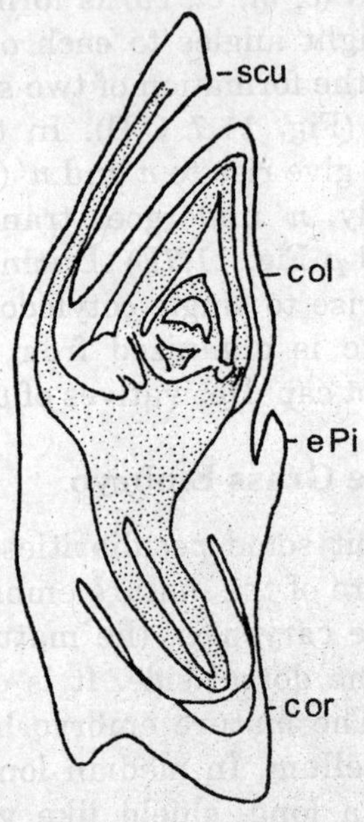

Fig. 11.8. Mature embryo of *Triticum* as seen in median longitudinal section (Batygina, 1969). *scu* - scutellum, *col:* coleptile, *epi:* epiblast, *cor:* coleorrhiza.

Embryo Development in *Paeonia*

In angiosperms first division of the zygote is followed by wall formation resulting in a two-celled proembryo. Yakovlev and Yoffe (1957) reported unique pattern of embryogeny in *Paeonia*. The zygote undergoes free nuclear divisions forming a coenocytic structure (Fig. 11.9 A-C). Later, the nuclei become lodged in the peripheral layer of cytoplasm around a large central vacuole (Fig. 11.9D). Now wall formation occurs, and the coenocyte becomes cellular, except in the central part (Fig. 11.9E). Some of the peripheral cells give rise to the embryo initials (Fig. 11.9F) of which one matures into adult embryo.

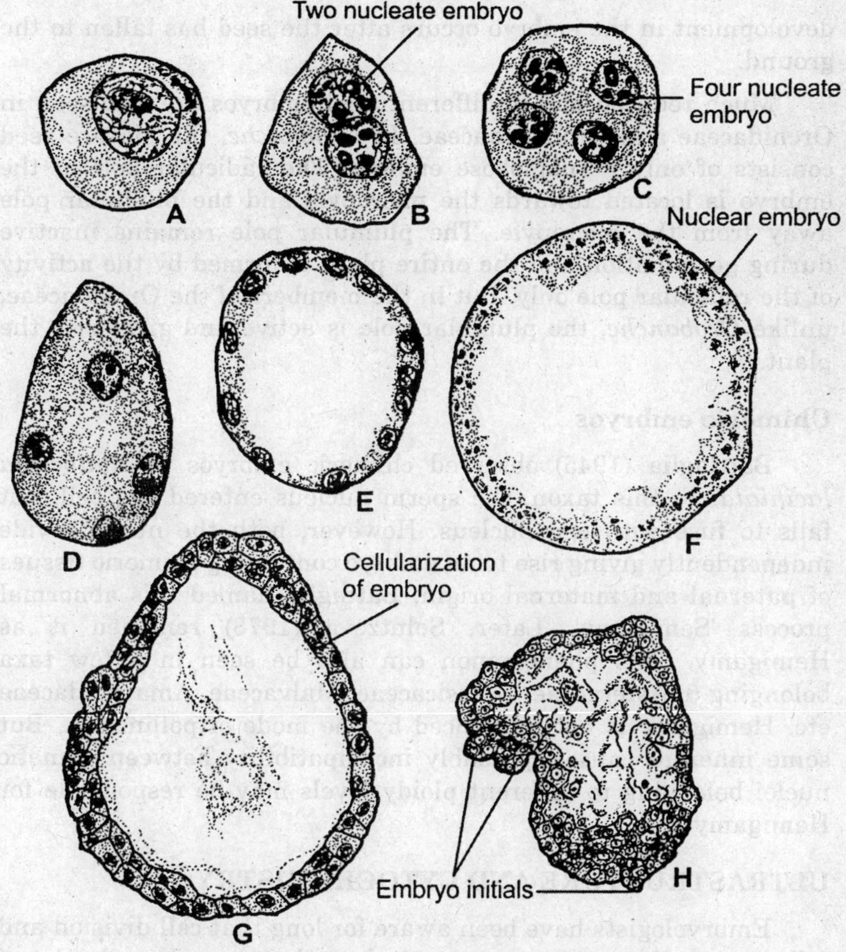

Fig. 11.9. Embryo development in *Paeonia* (after Yakovlev and Yoffe, 1965)

Reduced or Under-developed Embryos

In some parasitic or saprophytic taxa of the families Balanophoraceae, Burmanniaceae, Gentianaceae, Orchidaceae, Orobanchaceae, Rafflesiaceae etc. the mature embryo even at the shedding stage remains under-developed without any differentiation. In *Eranthis hiemalis* (Ranunculaceae) also the embryo at shedding is very much under-developed and is pear-shaped. But further development in the embryo occurs after the seed has fallen to the ground.

Much reduced and undifferentiated embryos are also seen in Orchidaceae and Orobanchaceae. In *Orobanche*, the mature seed consists of only a subglobose embryo. The radicular pole of the embryo is located towards the micropyle and the plumular pole away from the micropyle. The plumular pole remains inactive during germination and the entire plant is formed by the activity of the radicular pole only. But in the members of the Orchidaceae, unlike *Orobanche*, the plumular pole is active and gives rise the plant.

Chimeric embryos

Battlaglia (1945) observed chimeric embryos in *Rudbeckia laciniata*. In this taxon, the sperm nucleus entered egg cell, but fails to fuse with egg nucleus. However, both the nuclei divide independently giving rise to an embryo containing chimeric tissues of paternal and maternal origin. Battaglia named this abnormal process 'Semigamy'. Later, Solntzeva (1978) renamed it as Hemigamy. This phenomenon can also be seen in a few taxa belonging to Asteraceae, Brassicaceae, Malvaceae, Amaryllidaceae etc. Hemigamy is not influenced by the mode of pollination. But some inherent causes, probably incompatibility between gametic nuclei belonging to different ploidy levels may be responsible for Hemigamy.

ULTRASTRUCTURE AND CYTOCHEMISTRY

Embryologists have been aware for long that cell division and cell destiny play an important role in embryogenesis and a lot of information has been gathered about cell segmentation and organogenesis at light microscope level. However, interest in the subcellular and molecular changes underlying these processes is relatively recent and has resulted from the application of newer tools and techniques, such as EM, autoradiography, cytochemistry etc. But, the number of taxa studied is small, but enough to excite interest.

The apical cell and the basal cell resulting from the division of the zygote are cytologically as well as biochemically distinct from one another. These cells also exhibit a marked difference in their ultrastructure. The apical cell is generally more electron dense than the basal cell and contains more ribosomes per unit volume. Vacuoles are smaller and fewer in the apical cell, but are larger in the basal cell. In cotton, the plastids are perinuclear in the apical, but they are casual in the basal cell. Although, typically the zygote has plasmodesmatal connections with the synergids and central cell, these are lost in the 2-celled proembryo.

In an young proembryo, the cells of the embryo proper contain few dictyosomes and scanty endoplasmic reticulum, but a large amount of polysomes. Mitochondria and plastids are common, and the proembryo stains intensely for protein and nuclic acid. A sharp contrast is observed between the cells of the embryo proper and those of the suspensor. However, in *Hordeum*, there is little or no indication of cytological distinction between the cells of the suspensor and the embryo proper except for the progressive decrease in the number of vacuoles towards the suspensor in a young proembryo. The hypophysis bears ultrastructural resemblance to the suspensor rather than the embryonal cells. The cells of the hypophysial origin in *Capsella* and *Linum* have fewer ribosomes, more endoplasmic reticulum and larger vacuoles than those of embryo and frequently contain starch and lipid. Plasmodesmatal connections are present in the common walls separating adjacent suspensor or embryonal cells. In several taxa, the plastids in the embryonal cells differentiate into chloroplasts so that the embryo becomes green. Yakovlev and Zhukova (1980) surveyed 1,094 species belonging to 666 genera from 182 families of both dicots and monocots and noted that 428 species belonging to 224 genera and 72 families have green embryos. This character should be of some systematic value.

The ultrastructural features observed during early embryogenesis are closely associated with cytochemical changes. Generally, the apical cell stains more darkly than the basal cell for RNA and protein as in cotton, *Vanda* and *Linum*. However, in *Capsella*, *Quercus* etc. the basal cell contains more RNA and protein than the apical cell.

In a globular proembryo, all the cells show equal staining, but as in the ultrastructure, there is a notable difference between the suspensor and proembryonal cells. The hypophysis shows a cytochemical affinity with the suspensor cells.

In the heart shaped embryo an enhanced accumulation of nucleic acids and protein can be seen in the cotyledonary primorodia. Hypophysis stain lightly, and suspensor weakly. The radicular pole also records an increased staining intensity in *Capsella*.

Vallade *et al.* (1978) measured cytophotometrically the amount of DNA in the zygote as well as in the first two cell generations of the proembryo in *Petunia hybrida*. There was an increase in the DNA level. In cotton, nuclear DNA level remained more or less constant during embryogenesis with practically no cytoplasmic DNA. But earlier Yoo and Jenson (1966) had demonstrated the decrease in total DNA as well as the presence of cytoplasmic DNA. Woodward (1956) also noticed similar differences in the content of DNA in the embryos of different ages in *Tradescantia*.

Several investigators have also paid attention towards the localization of enzymes during embryogenesis. Enzymes are known to catalyze biochemical reactions in cell systems. Hydrolytic enzymes notably phosphatases and oxidative enzymes, chiefly oxido-reductases have been investigated. Phosphatases hydrolyze phosphate esters. Alkaline phosphatase has been localized in the embryos of *Stellaria media*. This enzyme occurs both in the embryo and suspensor. In the embryonal cells the concentration of enzyme is uniform. But acid phosphatase activity is minimum in the suspensor when compared with embryo proper. Within the embryo most of the enzyme occurred in the cells of the protoderm and the hypophysis did not show any enzyme activity.

Raghavan (1976) studied the distribution of acid phosphatase in *Capsella*. In the globular embryo the enzyme first appears and shows a uniform distribution. At heart-shaped embryo stage, the enzyme is mainly confined in the cotyledons and root apex. In older embryos, the enzyme, is mainly confined to the root and shoot apex and cotyledons.

Of the oxidative enzymes, succinic dehydrogenase, is a mitochondrial enzyme and generally indicate the sites of high respiratory activity. Forman and Jenson (1965) studied its distribution in the embryos of cotton.

Distribution of cytochrome oxidase, another key oxidative enzyme has been ivnestigated in *Stellaria* and its activity is higher in the meristematic regions. Bhalla *et al.* (1979) localized several respiratory enzymes in the globular, heart-shaped and mature embryos of *Tropaeolum majus.* They observed highest activity at the globular stage of development.

Another enzyme involved in oxidation-reduction reactions is peroxidase which is supposed to regulate auxin levels in the tissues. Alvarez and King (1969) observed highest activity of this enzyme in *Vanda* at younger stages of development and less during later stages. Using the technique of gel electrophoresis, they noticed that the number of isoenzyme bands also decreased during the development of the embryo. The results indicate that IAA levels are kept low during younger stages of development, whereas its concentration is allowed to increase during later stages. Thus the ultrastructural and histochemical studies on embryogenesis have revealed the differential morphogenetic potentialities of the apical and basal cell, already known from conventional LM studies.

SUSPENSOR

Structure and Function

In most of the angiosperms, the early proembryo is differentiated into two distinct regions namely, the suspensor and the embryo proper. The suspensor is an ephemeral structure and is present on the radicular part of the embryo. In the early stages of development, the suspensor grows much faster than the embryo and it reaches its maximum development by the globular or heart-shaped stage of the embryo. Only the remnants of the suspensor may be seen in the mature seed. The early embryologists are of the opinion that the suspensor is an organ, which merely pushes the embryo proper into a nutritionally rich environment, namely the endosperm. But recent studies have indicated that the suspensor by itself takes up the nutritive function in some cases. The suspensor shows lot of variation in its size, shape and in the number of cells.

In some taxa like *Lycopsis, Penaea, Tilia* etc. there is no suspensor at all. A reduced suspensor is seen in *Bryonia, Euphorbia, Ruta* etc. In the plants, which produce, the micropylar endosperm haustoria, the suspensor is not prominent or short lived. A long filamentous suspensor is characteristically present in Brassicaceae and certain taxa like *Pedicularis* of Scrophulariaceae. In certain families like Crassulaceae, Fumariaceae, Rubiaceae, etc. the suspensor becomes massive and haustorial. The suspensor shows wide variation in the family Fabaceae from its virtual absence to very massive structure, which may be filamentous or clustered like a bunch of grapes. In *Loranthaceae*, it is exceptionally long.

In the families Orchidaceae, Podostemaceae and Trapaceae, where the endosperm is absent, the embryo possesses extensively developed suspensor haustoria. A large variety of organizations is met within the Orchidaceae. It may be (a) single-celled; enlarged to become sac-like, conical, tubular or cyst like (Fig. 11.10A) *e.g., Dendrobium, Cyprepedium;* (b) uniseriate filament of 5-10 cells which grows beyond the micropyle, and upon reaching the placenta penetrates it by issuing haustorial branches (Fig. 11.10B), *e.g., Satyrium, Habenaria, Ophrys,* (c) looking like a bunch of grapes (Fig. 11.10C) *e.g., Sobralia, Epidendrum*, (d) the suspensor initial divides by three vertical divisions and the eight cells thus formed elongate downward enveloping more than half of the embryo (Fig. 11.10D), *e.g., Vanda, Luisia, Cottonia,* (e) the zygote divides to produce an irregular mass of 6-10 cells, some of which situated towards the micropylar end, elongate and form tubular structures (Fig. 11.10E) *e.g., Eulophia, Cymbidium.*

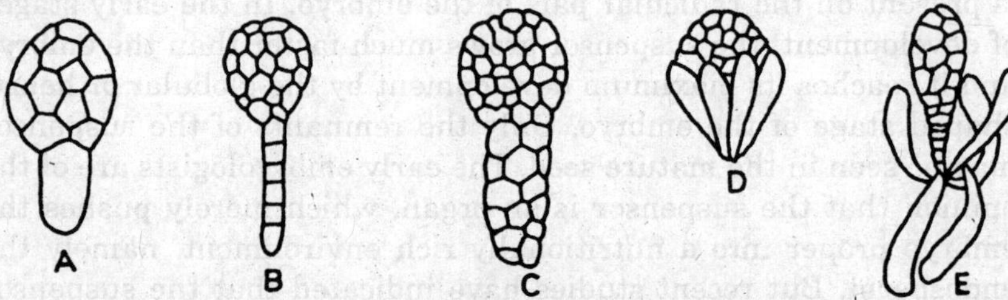

Fig. 11.10. Some suspensor types found in orchids (after Swamy, 1949)

In *Dicraea stylosa* of family Podostemaceae, the basal cell (*cb*) enlarges and contains two hypertrophied nuclei. During further development it gives out several thin-walled haustorial branches which grow in between the two integuments.

The cells of the suspensor, stain very highly when compared to the embryo proper. In spite of its great diversity, the suspensor has not attracted the attention of the investigators till recently. During recent times, ultrastructural, cytochemical, biochemical, physiological studies were undertaken in the suspensor. These studies have indicated that the suspensor play a more active and dynamic role. The suspensor cells are similar in structure to transfer cells which are involved in the short distance transport of the materials. It indicates that the suspensor is actively involved, in the absorption of the nutrients from various ovular and extra-ovular tissues and translocate the same to the embryo proper.

Ultrastructural studies have shown that the suspensor cells are more vacuolated and consists of fewer ribosomes than the embryonal cells. Numerous plasmodesmatal connections have been observed in the suspensor cells. The suspensor cells contain rough endoplasmic reticulum, mitochondira, dictyosomes, plastids and ribosomes. In *Phaseolus*, suspensor cells possess only a few small vacuoles. The giant basal cells possess leucoplasts of varying shape and structure. The cells with leucoplasts are more active in early stages of embryogeny and concerned with the synthesis of lipids.

Filamentous, cup-shaped and irregular shaped plastids are common in the suspensor cells. These plastids also vary in their internal structure like fibrils and regular or irregular thylakoid layer. The function of these suspensor plastids in providing nutrition to the developing embryo is well established in several genera like *Pisum, Ipomoea, Stellaria* etc. Another feature of interest in the suspensor cells is endopolyploidy. In *Phaseolus vulgaris* the ploidy level varies in different varieties between 8n and 2048 n.

Role of Suspensor

1. Absorption of Nutritive Substances

The suspensor plays a very important role in the nutrition of the embryo. The suspensor cells are metabolically very active and consist of abundant cell organelles. The basal cells of the suspensor show wall labyrinths, which indicate that its main function is the absorption of nutrients from the maternal tissues.

2. **Synthesis of Phytohormones**

Several growth regulators have been observed in suspensor cells. Alpi *et al.* (1975) extracted Gibberellic acid like substance from the cells of the suspensor and embryo. Gibberellic acid concentration was found to be more in the suspensor at the heart-shaped stage of the embryo. In *Phaseolus*, biologically active cytokinins are found in suspensor cells.

3. **Synthesis of Nutritive substances**

Satina and Rietsema (1959) have noted the presence of oil globules in the basal cell of the proembryo. In *Stellaria*, proteinaceous materials were reported by Pritchard (1964). Several hydrolytic and oxidative enzymes are also found in the suspensor cells indicating high metabolically active state.

4. **Transport of Nutrients**

The presence of plasmodesmatal connections in the cells of the suspensor has indicated its role in the transport of nutrients to the embryo.

Thus the suspensor plays a very dynamic role not only in nourishing the embryo proper at specific developmental stage, but also in exercising control on its growth by supplying several important phytohormones.

NUTRITION OF THE EMBRYO

During the earlier stages of development, the embryo receives nourishment from the surrounding cells of the embryo sac, namely nucellus and integuments. The disappearance of plasmodesmatal connections during fertilization causes breaking up of the nutrient flow from the maternal tissues to the embryo. Sometimes, the plasmodesmata reappear at the heart-shaped stages of the embryo and assist in the absorption of food from the cells of endosperm. After fertilization, the suspensor also pushes the embryo deep into the endosperm in order to enable the embryo in a nutritionally favourable condition.

The nutrition of the embryo can be understood in two different ways: (i) By examining the different structures that are associated with the nutrition of the embryo in nature and their chemical analysis (*in vivo* studies), (ii) by culturing the excised embryos at various stages of its development so as to understand its nutritional requirements (*in vitro* studies).

In vivo studies

Endosperm

The chief source of nutrition to the embryo in nature is the endosperm. The nutritive role of the endosperm can also be evidenced by its large pool of reserve food materials and growth regulating substances in it.

Coconut milk is the liquid endosperm of *Cocos nucifera*. Due to its easy availability in required time and quantity, the chemical analysis of coconut milk has been undertaken. The studies have revealed that it is rich in inorganic ions, amino acids, sugars and growth regulators. The other endosperm, which can readily be obtained, is the corn endosperm at milk stage. It also contains large quantities of carbohydrates, amino acids and protein complexes. The concentration of amino acids decreases with the increase in the age of the endosperm. IAA and several purine compounds are also found in the young corn endosperm. A cytokinin by name zeatin is also extracted recently from the corn endosperm. Not only the maize endosperm, but also the endosperm of other cereals like wheat, rye, barley, rice also consists of auxin. Presence of Gibberellins has been reported by Phinney *et al.* (1957) in taxa like *Pyrus, Persea, Echinocystis* etc.

Coconut milk is known to support the growth of isolated young embryos of many plants hence it is widely used in tissue culture studies. The endosperm extract of the plants can also induce divisions in mature and differentiating tissues. When the endosperm is absent as in certain families like Orchidaceae, Podostemaceae, Orobanchaceae, the embryo depends for its nutrition on the other structures like suspensor haustoria. The suspensor haustoria are very well developed in those families and act as substitute for the endosperm. In certain plants, the endosperm also develops haustoria from micropylar, chalazal and lateral positions which penetrate into the surrounding ovular tissues to absorb nutrients from them.

Pseudo Embryo sac

This is characteristic feature of the family Podostemaceae in which the endosperm is absent. The loss of endosperm is compensated by the presence of suspensor haustoria and pseudo embryo sac in Podostemaceae. Some of the nucellar cells situated below the MMC enlarge longitudinally and their walls become thin and fragile. At the quadrant stage of the embryo, the thin nucellar cells completely break down and produce a large cavity called the pseudo-embryo sac. The pseudo-embryo sac consists of several free nuclei and dense cytoplasm. The contents of the pseudo-embryo sac are gradually consumed by the developing embryo. In some genera of the family Podostemaceae, the pseudo embryo sac is formed even during the meiotic divisions of the megaspore mother cell.

Perisperm

In majority of the angiosperms, the nucellus is generally consumed by the developing embryo sac and hence no traces of it can be seen in the mature stages. But there are certain families like Amaranthaceae, Aizoaceae, Nyctaginaceae, Cannaceae, Capparidaceae, Piperaceae, Portulacaceae, Zingiberaceae in which the nucellus not only persists, but also stores lot of food materials. Such persistent nucellus is known as the perisperm. In these taxa where perisperm is present, the endosperm is much reduced and acts as an intermediary tissue between the perisperm and the embryo. It absorbs the food present in the perisperm and transports the same to the embryo.

Piper nigrum (Piperaceae) is a well known example of the taxa in which the perisperm is present. 95 percent of the mature seed is occupied by the perisperm. The cells of the perisperm are large, densely cytoplasmic and store oil globules. Starch is also present in the nucellar cells surrounding the embryo sac and in the cells of the nucellar epidermis.

Chalazopserm

Rarely, in certain taxa like *Cyanastrum*, the endosperm and major portion of the nucellus degenerate and disappear during seed development.

But only a few cells of the nucellus, just below the embryo sac and above the vascular bundle persist, divide actively and form a prominent structure called the **chalazosperm**. The cells of the chalazosperm are large, loosely arranged, filled with fat and starch, and function as a substitute for the endosperm.

In vitro Studies

Raghavan (1966) recognised two phases in the development of the embryo in relation to its nutrition. In the first phase which lasts up to the globular stage, the embryo is dependent for its nutrition on the surrounding tissues and hence he termed this phase as the **heterotrophic phase**. In the second phase, which starts from the heart-shaped stage onwards, the embryo is independent for its nutrition and hence this stage can be regarded as the **autotrophic phase**.

Basing on the above, it is relatively much easier to culture the isolated older embryo in a simple nutrient medium consisting of mineral salts and sucrose. But, the culture of the younger embryo poses many problems, because they require an elaborate medium. Van Overbeek *et al.* (1941, 1942) established the same through their studies on the *Datura* embryo. Similarly, Raghavan (1966) also recorded an increasing autonomy of the embryo with age in *Capsella bursa-pastoris*.

Another important factor with regard to the nutrition of the embryo is the osmotic value of the nutrient medium. Ryczkowski (1960-1972) have noted high osmotic value in the younger ovules. The osmotic pressure, viscosity and specific gravity are decreasing in the embryo sac with the increase in their age. The same thing can be seen in the *in vivo* studies also i.e., mature embryos grew well in the cultures even with 2 percent sucrose in the medium. However, the younger embryos require higher levels of sucrose for their normal development. In the culture of excised globular embryo of *Capsella bursa-pastoris*, Raghavan and Torrey (1969) noted a stimulatory affect with the addition of 10-12 percent sucrose to the medium. However, if IAA, kinetin and adenine sulphate are added to the nutrient medium, 2 percent sucrose is enough to cause satisfactory growth of the embryo. The addition of abscissic acid in the nutrient medium also increases its osmotic value and prevents the precocious germination of excised immature embryo of cotton (Dure, 1975).

The nutritional requirement of the embryos varies with their stage of development. For a successful development of an embryo, starting from the young stage, it has to be changed to different culture plates with different compositions. But Monnier (1976, 1978) developed a culture method which allows the complete development of the embryo of *Capsella* from very early stage up to the germination in the same culture plate without changing the embryo from its original position.

References

Batygina, T.B. 1969. On the possibility of a new type of embryogenesis in angiosperms. *Rev. Cytol Biol. Veg.* **32**: 335-341.

Bhojwani, S.S., Bhatnagar, S.P. and Dantu, P.K. 2015. *The embryology of angiosperms*. Vikas Publishing House, New Delhi.

Crete, P. 1963. Embryo. In Maheswari, P. (ed) *Recent advances in the Embryology of Angiosperms*. Int. Soc. Plant Morphologists, Delhi, pp. 171-220.

Murgai, P. 1962. Embryology of *Paeonia* with a discussion on its systematic position. In: *Plant Embryology - A Symposium*. CSIR, New Delhi, pp. 215-22.

Natesh, S. and Rau, M.A. 1984. The Embryo. Johri, B.M. (ed.) *Embryology of Angiosperms*. Springer Verlag, Berlin, pp. 377-443.

Periasamy, K. 1977. A new approach to the classification of angiosperm embryos. *Proc. Indian Acad. Sci.* Sect. B. **86**: 1-13.

Raghavan, V. 1976. *Experimental embryogensis in vascular plants*. Academic Press, New York.

Schnarf, K. 1929. *Embryologie der Angiospermen*. Borntraeger, Berlin.

Suárez, F.M. and Bozhkov, V.P. 2010 . *Plant Embryogenesis* Methods in Molecular Biology 472

Yakovlev, M.S. and Yoffe, M.D. 1957. On some peculiar features of embryogeny in *Paeonia*. *Phytomorph.* **7**: 74-87.

Yamazaki, T. 1974. A system of Gamopetalae based on embryology. *J. Fac. Sci. Univ.*, Tokyo Sect. III. *Bot.* **11**: 263-281.

CHAPTER – XII

POLYEMBRYONY

Polyembryony can be defined as the occurrence of more than one embryo in a seed (Maheshwari, 1950). This phenomenon was first discovered by Leeuwenhoek (1719) in *Citrus aurantium.* But the term polyembryony was coined by Braun (1859). Since then, this phenomenon was reported in a wide variety of plants including several gymnosperms. In fact, polyembryony was most predominant in gymnosperms than angiosperms.

There are several types of polyembryony and they have been variously classified by different workers. Ernst (1918) and Schnarf (1929) classified the polyembryony into two types namely: **True Polyembryony** and **False Polyembryony**. According to them in the True Polyembryony, the plural embryos develop within a single embryo sac and in False Polyembryony, they arise from different embryo sacs. Maheshwari (1950, 1963) opines that the above classification is the most convenient and he also adopted the same.

On genetic basis, Yokovlev (1967) classified the polyembryony into two types namely: (1) Gametophytic polyembryony and (2) Sporophytic polyembryony. In the first type, the embryos arise from a cell of the female gametophyte either with or without fertilization. But in the second type, the embryos arise from the zygote or proembryo or some other sporophytic cells of the ovule like the nucellus or the integument.

Bouman and Boesewinkle (1969) classified the polyembryony into four categories on the basis of the origin of additional embryos (a) from sporophytic cells of the parental generation *i.e.,* nucellus or integuments. (b) from the new sporophyte (proembryo or fertilized egg); (c) from the gametophytic cells (by one or more embryo sacs in the same ovule) and (d) from the male gametophyte.

Lakshmanan and Ambegaokar (1984) classified the polyembryony broadly into simple and multiple, depending upon the presence of one or more embryo sacs in the same ovule. In simple polyembryony, the embryos develop within a single embryo sac, but in multiple polyembryony, they are produced from two or more embryo sacs, in the same ovule. Simple polyembryony may be sexual or asexual.

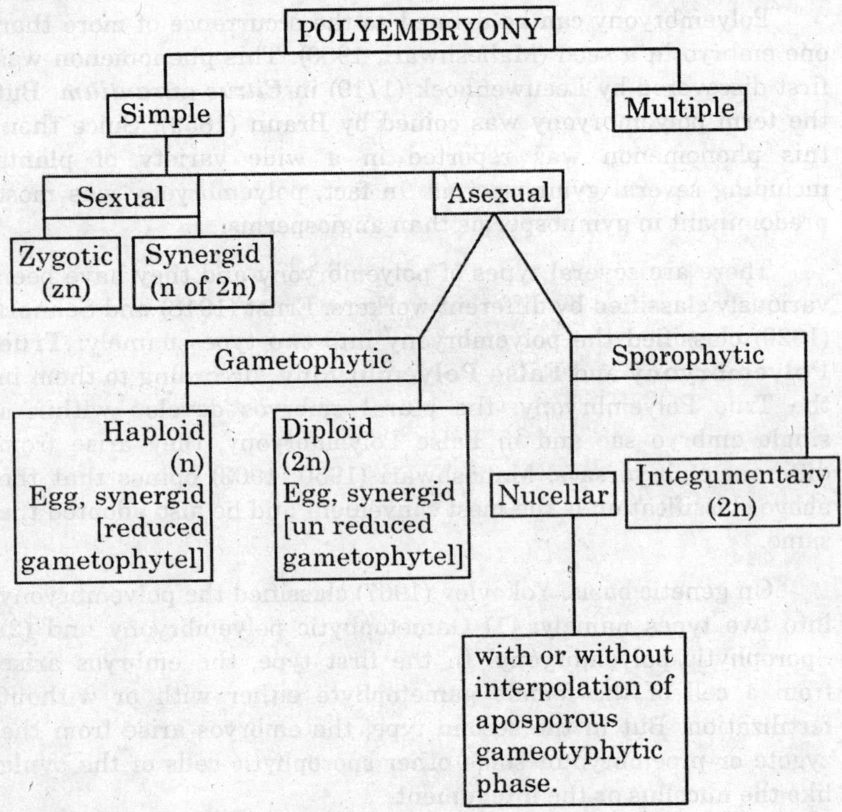

Nucellar Polyembryony

Nucellar polyembryony is also known as adventive polyembryony. The nucellar cells which are destined to develop into embryos grew at the expense of others, accumulate dense cytoplasm and starch. These cells repeatedly divide and form the embryonic masses of which some of them develop further into embryos after projecting into the embryo sac cavity. These adventive embryos can be distinguished from the sexual embryos by their lateral position, irregular shape and lack of suspensor. But in *Citrus microcarpa* a well defined suspensor has been observed (Rangaswamy, 1959). The cotyledons are unequal in the nucellar embryos. Rarely tricotyly was also seen. The origin of nucellar embryos may be autonomous or stimulated.

Nucellar embryos have been observed in the families Rutaceae, Anacardiaceae, Cactaceae, Combretaceae, Poaceae, Malpighiaceae, Urticaceae, Myrtaceae and Orchidaceae. Nucellar polyembryony is reported in 16 out of the 172 Crassinucellate families of Angiosperms (Rangaswamy, 1981).

Some species of *Citrus* are polyembryonate (*C. microcarpa, C. reticulata, C. unshiu*), while others are monoembryonate (*C. grandis, C. limon*). The number of seedlings per species varies from species to species; *Citrus microcarpa* produces 21 seedlings and *C. unshiu* as many as 40. The nucellar embryos show a well defined suspensor and therefore it is rather difficult to distinguish them from zygotic embryos. In the family Rutaceae adventive polyembryony has also been recorded in *Poncirus trifoliatus, Aegle marmelos* and *Toddalia asiatica.*

Adventive polyembryony has also been recorded in mango (Fig. 12.1 A,B). In some varieties of *Mangifera indica,* the adventive embryos originate from the nucellar tissue (Fig. 12.1 B). The zygotic embryo is also present, but in some varieties it degenerates early. In one instance as many as 50 embryos are seen in a single seed of *Mangifera indica.*

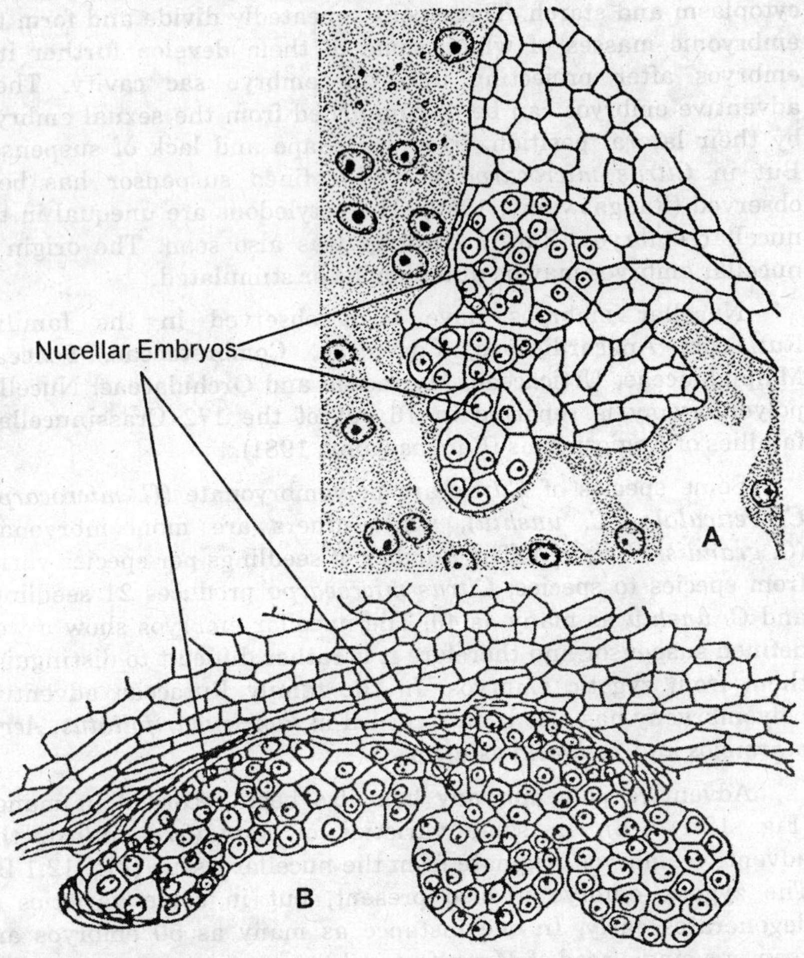

Fig. 12.1. Adventive embryony in *Mangifera*. **A.** *M. odorata*; **B.** *M. indica* (after Sachar and Chopra)

Adventive polyembryony has also been observed in the family Cactaceae. In *Opuntia dillenii* (Maheshwari and Chopra, 1955) the egg degenerates without undergoing any division and several adventive embryos develop from the nucellar cells in the upper half of the embryo sac. But only one attains maturity. In *Mammillaria tenuis* (Tiagi, 1957) while several nucellar embryos develop after fertilization, only one reaches maturity. The zygotic as well as nucellar embryos show a well-defined suspensor.

Orchidaceae is another family which shows nucellar polyembryony. Nucellar polyembryony may occur before or after pollination and fertilization of the egg cell. In *Zeuxine sulcata* (Swamy, 1946), two or three nucellar embryos oringinate from the nucellar epidermis in the micropylar region. Nucellar embryos have also been recorded in *Spiranthes cernua.*

In the family Poaceae Nucellar polyembryony has been reported in *Eragrostiella bifaria* (Venkateswarlu and Devi, 1964; Febulaus and Pullaiah, 1991) (Fig. 12.2 A-E), *Cenchrus ciliaris* (Shanthamma and Narayan, 1977; Febulaus and Pullaiah, 1991) (Fig. 12.2 F), *Pennisetum mezianum* (Shanthamma and Narayan, 1977) etc. These nucellar embryos normally arise from the cells of the nucellus and enter the embryo sac cavity and there they complete their further development as in *Eragrostiella* and *Perotis* or they do not enter the embryo sac as in *Pennisetum* and *Cenchrus* (see Pullaiah and Febulaus, 2000). Das and Islam (1977) reported that nucellar embryos arise from nucellar epidermis in *Cenchrus pennisetiformis* and *C. setigerus.*

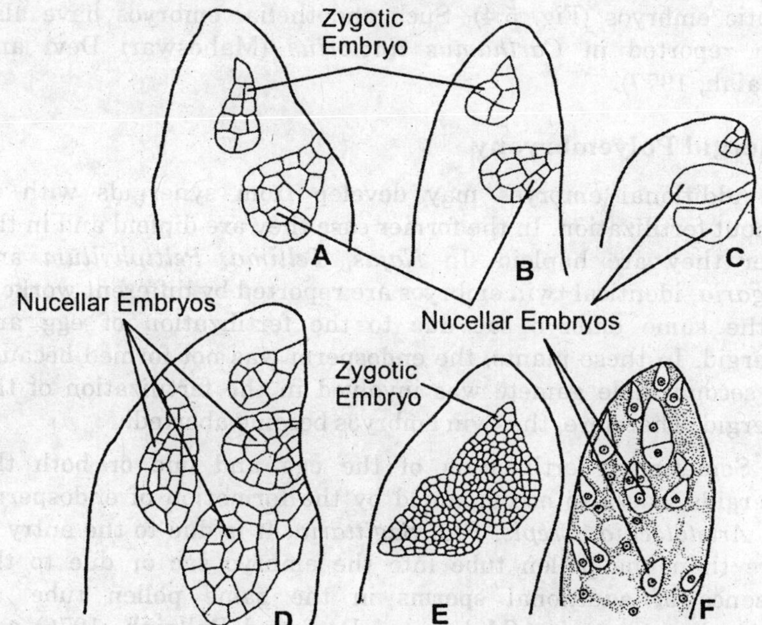

Fig. 12.2. Nucellar polyembryony. **A-E.** *Eragrsotiella bifaria* (Febulaus and Pullaiah 1992); **F.** *Cenchrus ciliaris* (Febulaus and Pullaiah, 1995).

Integumentary Polyembryony

Sometimes embryos develop from cells of the integument. This type of embryo development from integumentary cells is not dependent on fertilization. The epidermal and subepidermal cells of the inner integument develop into adventive embryos in *Euonymus.*

In the asexual individuals of *Spirathes cernua* the terminal cell and the lower adjoining cells enlarge and become richly cytoplasmic. They divide and produce two to six proembryos in the micropylar region of the ovule. These proembryos, however, do not differentiate further.

Maheswari Devi and Pullaiah (1976, 1977) reported interesting case of endothelial embryos in *Melampodium divaricatum*. The embryo sac in this species usually degenerates. Some of the cells of the integumentary tapetum become richly cytoplasmic and divide repeatedly to form several proembryos. Such embryos lack suspensor and in this respect differ from zygotic embryos (Fig. 5.4). Such endothelial embryos have also been reported in *Carthamus tinctorius* (Maheswari Devi and Pullaiah, 1977).

Synergid Polyembryony

Additional embryos may develop from synergids with or without fertilization. In the former case they are diploid and in the latter they are haploid. In *Najas, Tellima, Peltiphyllum* and *Fragaria*, identical twin embryos are reported by different workers in the same embryo sac due to the fertilization of egg and synergid. In these plants, the endosperm was not formed because the second male gamete was involved in the fertilization of the synergid. Therefore, the twin embryos become aborted.

Sometimes, fertilization of the egg and one or both the synergids was also accompanied by the formation of endosperm, *e.g., Aristolochia, Crepis, Poa, Sagittaria.* It is due to the entry of more than one pollen tube into the embryo sac or due to the presence of additional sperms in the same pollen tube. In *Carthamus tinctorius* (Maheswari Devi and Pullaiah, 1976) and *Azadirachta indica* (Fig. 12.3G) triplets develop due to fertilization of the egg and both the synergids (Fig. 12.3A,G).

Synergids may also develop into embryos without fertilization and in which case there will be formation of haploid-diploid twins, *e.g. Arabis lyalli* (Fig. 12.3B), *Dioscorea composita* (Fig. 12.3C), *Lilium* (Fig. 12.3F), *Argemone, Orchis.* In *Pennisetum dubium* the synergid autonomously develops in to an embryo (Gildenhuys and Brix, 1959). Shobha and Sindhe (1987) observed sporadic occurrence of twin embryos in *Pennisetum pedicellatum.* The location and shape of the extra embryo in the embryo sac reveal that it is of synergid origin.

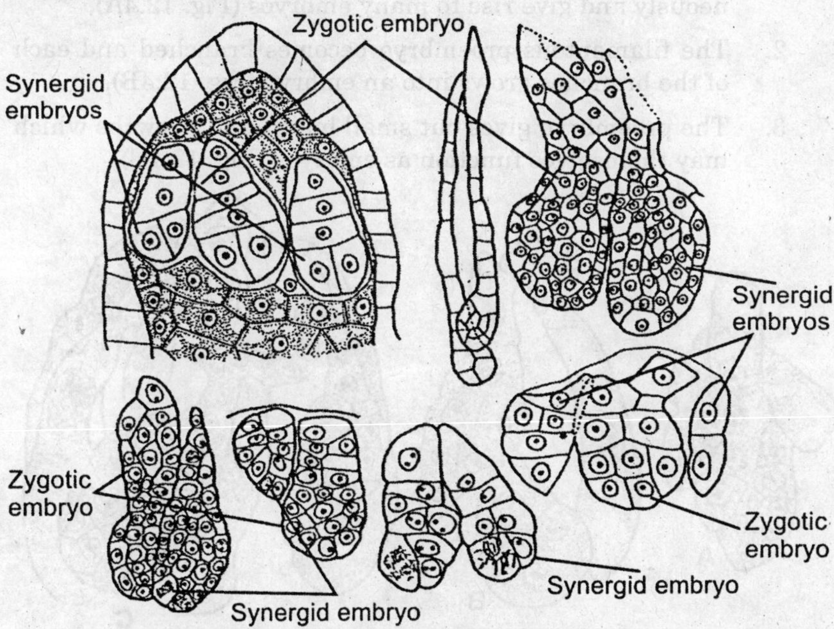

Fig. 12.3. Synergid polyembryony. **A.** *Carthamus tinctorius* (after Maheswari Devi and Pullaiah, 1976); **B.** *Arabis lyalli* (after Lebegue, 1948); **C.** *Dioscorea composita* (after Thirumaran and Lakshmanan, 1982); **D.** *Pennisetum squamulatum* (after Shinde et al., 1980); E. *Argemone mexicana* (after Sachar, 1955); **F.** *Lilium martagon* (after Cooper , 1943); **G.** *Azadirachta indica* (after Nair and Kanta, 1961).

Zygotic and Suspensor Polyembryony

The clevage polyembryony is quite common among Gymnosperms and is much less frequent in angiosperms. Among angiosperms clevage polyembryony is quite common in Orchids.

Jeffrey (1895) reported in *Erythronium americanum* that the zygote undergoes few divisions and forms an embryonic mass. Two or three embryos develop as outgrowths from the lower as independent embryos.

Swamy (1943) recorded the following variations in *Eulopia epidendraea.*

1. The zygote divides irregularly to form a mass of cells of which those lying towards the chalazal end grow simultaneously and give rise to many embryos (Fig. 12.4A),
2. The filamentous proembryo becomes branched and each of the branches grows into an embryo (Fig. 12.4B),
3. The proembryo gives out small buds or outgrowths which may themselves function as embryos (Fig. 12.4C).

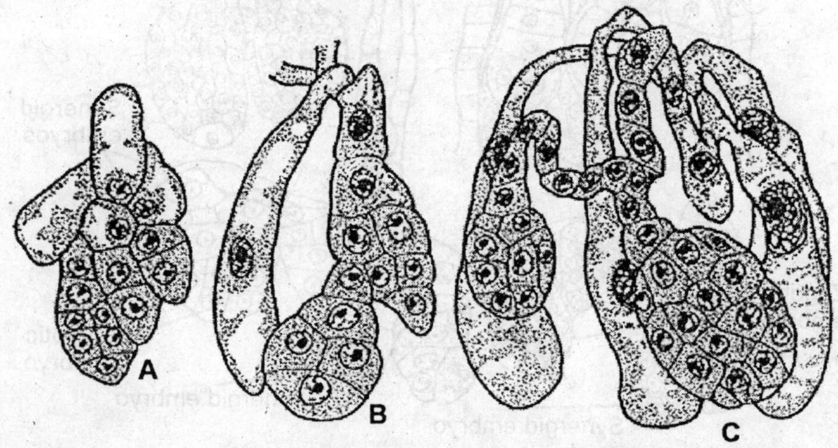

Fig. 12.4. Zygotic and suspensor polyembryony in *Eulophia epidendraea* (after Swamy, 1943)

Like *Erythronium* and *Eulopia*, in *Habenaria platyphylla* and *Geodorum densiflorum* also the zygote produces an irregular mass of cells, and some cells at the apical end develop simultaneously multiple proembryos.

Clevage polyembryony has also been reported in other families in *Tulipa gesneriana, Cocos nucifera, Vincetoxicum nigrum.* In the family Poaceae clevage polyembryony has been reported in *Poa pratensis* (Nishimura, 1922), *Zea mays* (Randolph,

1936), *Dicanthium annulatum* (Reddy and D'Cruz, 1969) and rarely in *Cenchrus ciliaris* (Gupta and Yashvir, 1971).

Suspensor polyembryony is a common feature in *Exocarpus*, a member of the Santalaceae. As many as six embryos may develop simultaneously in an ovule by the proliferation of suspensor cells. Eventually only one of them takes the lead and reaches maturity. In *Zygophyllum fabago* the cells of the uniseriate suspensor develop buds. Suspensor polyembryony is of common occurrence in Acanthaceae and is seen in *Dipteracanthus patulus, Acanthus ilicifolius* etc.

Multiple Polyembryony

In multiple polyembryony additional embryos are produced from two or more embryo sacs in the same ovule. Multiple embryo sacs may arise from (a) derivatives of the same megaspore mother cell, (b) derivatives of two or more megaspore mother cells, and (c) from nucellar cells by apospory. Twin embryos originating from different embryo sacs are reported in *Juglans nigra* (Robyns, 1938), *Citrus* (Bacchi, 1943), *Poa pratensis* (Nielsen, 1946), *Pennisetum ciliare* (Fischer *et al.*, 1954), *Paspalum dilatatum* (Smith, 1948), *Dicanthium annulatum* (Reddy and D'Cruz, 1969), *Spinifex littoreus* (Lakshmanan and Jayalakshmi, 1980), *Capillipedium huegelli* (Choda and Bhanwra, 1980) and *Casuarina equisetifolia* (Swamy, 1948). In *Trifolium* and *Atraphaxis,* the nucellar cells from the chalaza produce additional embryo sacs and additional embryos. Fusion of two or more ovules and the development of additional embryos from the fused structure are observed in *Rhizophora* and *Chaenostoma.*

TWINS AND TRIPLETS

The frequency of twins and triplets in natural populations is generally rare. Because of the low frequency, the developmental studies to trace their exact origin are also limited. Basing on cytological and genetical studies of the seedlings, one can guess the possible source of their origin. Twins generally occur in the following combinations Haploid-Haploid; Diploid-Diploid; Haploid-Diploid; Haploid-Triploid and Diploid-Triploid.

(i) Haploid-Haploid

In mature seeds the occurrence of haploid-haploid twins is rare. An examination of 49903 dry seeds of maize, by Sarkar and Coe (1966) showed 49 twins. Of these, 15 pairs were haploid-haploid. Due to complete similarity, the authors concluded that each twin originated from a single embryo sac. Whether the twin developed by cleavage of the haploid embryo or from two cells of the embryo sac was not determined. They presumed that cleavage of the haploid embryos resulted in twins.

(ii) Diploid-Diploid

In some kernels of *Zea mays*, Randolph (1936) found two embryos arranged in a parallel manner and such kernals produced genetically identical seedlings. Hence Radolph, opines that these dipoid-diploid twins might have been originated due to the cleavage of the zygotic embryo. Twins arising by cleavage of the embryo has been reported in *Dicanthium annulatum* (Reddy and D'Cruz, 1969) and *Cenchrus ciliaris* (Gupta and Yashvir, 1971). Diploid-diploid twins can also be produced by the fertilization of more than one cell of the embryo sac. All the nuclei of a monosporic embryo sac are genetically alike and produce identical twins if they are fertilized by the male gametes released by the same pollen tube. Nucellar budding can also give rise to identical twins and triplets. Sindhe *et al.* (1980) reported that diploid-diploid twins arise by fertilization of a synergid and egg in *Pennisetum squamulatum* (Fig. 12.2D).

(iii) Haploid-Diploid

Haploid-diploid twins were reported in *Linum* (Kappert, 1933), *Oryza* (Ramiah *et al.*, 1935), *Solanum tuberosum* (Muntzing, 1937), *Secale* (Kostoff, 1939) and *Cocos nucifera* (Whitehead and Chapman, 1962). In all these, the diploid member is developed from the fertilized egg and the haploid one from the unfertilized cell of the same embryo sac.

(iv) Diploid-Triploid

Diploid-triploid twins are reported by Yamamoto (1936) and Kostoff (1939) in *Triticum vulgare* and *Secale cereale*, respectively.

They believe that the triploid embryos arose from the endosperm, but failed to produce any positive evidence. It was more like that the triploid embryo originated by the fertilization of aposporous embryo sac or by the fusion of haploid embryo sac with two male gametes or unreduced male gamete. The diploid embryo arose from a fertilized cell of the embryo sac.

(v) Haploid-Triploid

Haploid-triploids are relative very rare. But one such case was reported by Nissen (1937) in *Phleum*. The haploid member might have developed from an unfertilized cell and the triploid one by the fertilization of an unreduced egg cell.

Randell and Rick (1945) studied 405 multiple seedlings in *Asparagus officinalis*. In these, 97 percent were twins, 11 were triplets and one was quadruplet. In the reciprocal crosses, in two varieties of *Capsicum,* Morgan and Rappleye (1954) isolated multiple seedlings which are mostly unattached twins, conjoined twins and triplets.

CAUSES OF POLYEMBRYONY

Several theories have been put forth by different workers to explain the causes of polyembryony. But still, this question is debatable and the latest techniques of the advanced microscopy, histochemistry, genetics, physiology and biochemistry may further our knowledge in near future. The different view points are:

1. Ernst (1918), Baker (1960) and Cesca (1961) opined that the hybridization might be an important factor which, leads to polyembryony.
2. Haberlandt (1921, 1922) put forth the necrohormone theory according to which, some of the degenerating nucellar cells of the ovule secrete the necrohormones which in turn promotes the development of other nucellar cells to develop into embryos. Haberlandt states that he induced nucellar polyembryony in *Oenothera* by damaging some of the nucellar cells of the ovule by either pricking them with fine needle or by gentle squeezing of the ovary. Subsequent workers like Beth (1938) and Van Overbeek *et al.* (1941) failed to induce nucellar

polyembryony in *Oenothera* and *Datura* by Haberlandt method. Hence they seriously questioned the Haberlandt necrohormone theory.

3. Kappert (1933) considered that the polyembryony in *Linum* is a recessive hereditary character governed by a group of multiple genes.
4. Leroy (1947) believes that adventive embryony in *Mangifera indica* was probably due to the effect of one or more recessive genes.
5. Atabekova (1957) noted that the production of polyembryonic seeds in certain races of *Lupinus* is hereditary.
6. Furusato *et al.* (1957) showed that the number of embryos in *Citrus* seeds was influenced by various factors like the age of tree, nutritional status of the plant and the orientation of the branch of the tree.
7. The studies on mono and polyembryonate species of *Citrus* have indicated that monoembryonic condition is due to the synthesis and release of certain volatile (ethanol) and non-volatile (IAA, IBA, GA_3) embryogenic inhibitors which were not found in the ovules of polyembryonate species.

Significance of Polyembryony

Polyembryony plays an important role in plant breeding and horticulture. The seedlings developed through nucellar polyembryony are proved to be far more superior than the clones multiplied through vegetative propagation. In *Citrus*, the seedlings developed by nucellar polyembryony are free from viral diseases, and show great vigour when compared with the seedlings of vegetative propagation. In *Mangifera, Citrus* and several other fruit trees, polyembryony helps in the large scale propagation of selected genotypes. In elite tree (plus trees), uniform seedlings of the parental types can be produced in large numbers through polyembryony.

Twin seedlings provide an excellent source of haploids. Though the haploids by themselves are weak and valueless, true breeding, fertile, homozygous diploids can easily be produced from them by colchicine treatment. Recognizing the importance of haploids in plant breeding, several attempts were made by earlier investigators towards the artificial induction of haploids. Some of the methods, they employed are (a) by subjecting of the flowers to alternate high and low temperatures, (b) application of X-rayed or foreign pollen, (c) delayed pollination, (d) treating the ovules with various chemicals and hormones.

In recent times, almost all the portions of the plant body can be successfully cultured and somatic embryos can be generated from them. Somatic embryos so produced can be capsulated in a suitable hydrogel and 'artificial seeds' are produced in several crop plants. These artificial seeds can be sown in the field by using suitable mechanical devices similar to those used for normal seeds.

References

Kishore, K. 2014. Polyembryony. In: K.G.Ramawat *et al.* (eds.). *Reproductive Biology of Plants.* CRC press.

Lakshmanan, K.K. and K.B. Ambegaokar. 1984. Polyembryony. In: *Embryology of Angiosperms* (ed. B.M. Johri) Springer-Verlag, Berlin, pp. 445-474.

Pullaiah, T. and G.N.V. Febulaus. 2000. *Embryology and apomixis in Grasses.* Regency publications, New Delhi.

CHAPTER – XIII

APOMIXIS

The term apomixis was first coined by Hacke (1893). Apomixis as defined by Winkler (1908), is a process, in contrast to amphimixis (sexual reproduction), the sporophyte is produced without any gametic fusion. Winkler included vivipary and other forms of vegetative propagation such as bulbils, runners, rhizomes etc. under apomixis. But Gustafson (1946, 1947), Maheshwari (1950) and Battaglia (1963) have not included vegetative reproduction and adventive embryony in apomixis. According to Battaglia (1963) apomixis may be defined as the production of sporophyte from gametophyte without sexual fusion.

Maheshwari (1950) and Battaglia (1963) classified apomixis into two types: **Recurrent** (apomixis in unreduced embryo sacs) and **Non-recurrent** (apomixis in reduced embryo sacs). Heslop-Harrison (1972) does not recognize the non-recurrent type as a part of apomixis as it includes the process of meiosis in the formation of reduced embryo sacs.

Some authors (Nogler, 1984) used the term Gametophytic apomixis. According to Battaglia (1992) the term apomixis means loss of mixis and such a loss must be unequivocally referred to the gametophyte. Consequently, it follows that "sporophytic apomixis" has no logical meaning and cannot be established. Hence its equivalent term gametophytic apomixis also is rejected.

Approximately 450 species belonging to 40 families of plants show apomixis, of which more than 250 species are from Poaceae (Gramineae). Apomixis is especially prevalent in polyploids, auto as well as allopolyploids, wherein meiosis is very frequently disturbed because of difficulties in pairing and segregation of chromosomes – apomixis bypasses sexual reproduction and enables those plants to reproduce. Many wild species belonging to

Poaceae, Asteraceae and Rosaceae show apomixis. Some species of Rutaceae, Liliaceae, Urticaceae and some other families where diploidy is observed also show this trait.

While studying apomixis two aspects have to be known. These are development of embryo sac and development of embryo in apomictic plants.

DEVELOPMENT OF EMBRYO SAC

In apomictic plants embryo sac may develop in aposporous or diplosporous manner.

Apospory

Apospory was discovered by Rosenberg (1906, 1907) in the species of *Hieracium* subgenus *Pilosella.*

In this type, somatic cells, usually nucellar cells enlarge and function as initials of embryo sac. These initials enlarge and undergo mitotic divisions and develop into embryo sacs.

Apospory is broadly divided into Hieracium type and Pennisetum type.

(i) Hieracium type

One or more somatic cells, usually nucellar cells enlarge and function as initials of embryo sac (Fig. 13.1.A-C, F,G). These aposporous initials enlarge and develop vacuoles. The nucleus undergoes mitotic division resulting in two nuclei (Fig. 13.1 H) which are diploid. The spindle of the first mitosis is usually parallel to the nucellus. The two nuclei move to the opposite poles and they are separated by a vacuole. These two nuclei undergo two more mitotic divisions resulting in eight nuclei, four at each pole. They organize into 3-celled egg apparatus and three antipodals while two nuclei fuse in the centre forming tetraploid secondary nucleus (Fig. 13.2).

This type was first reported by Rosenberg (1906,1907) in the species of *Hieracium* subgenus *Pilosella.*

Hieracium type also occurs in *Artemisia nitida* (Chiarugi, 1926), *Coreopsis bicolor* (Gelin, 1934), *Crepis* sp. (Stebbins and Jenkins, 1939), *Leontodon hispidus* (Bergman, 1935), all belonging to the family Asteraceae. In other families it has been reported in *Poa compressa* (Armstrong, 1937), *P. pratensis* (Akerberg, 1939, Tinney, 1940), *P. granitica* (Skalinska, 1959), *Hierochloe* (Weimarck, 1967, 1970), *Hypericum perforatum* (Noack, 1939), *Ochna serrulata* (Chiarugi and Francini, 1930), *Ranunculus cassubicus* (Izmailow, 1965), *R. auricomus* (Nogler, 1971), *Beta* section *Corollinae* (Jasem, 1976), *Alchemilla* subg. *Aphanes* (Hjelmquist, 1959), *Malus hupahensis* (Dermen, 1936), *Potentilla canescens*, *P. argentea*, *P. praecox* (Rutishauser, 1943, 1948) *Elatostema acuminatum* and *E. penninerve* (Fagerling, 1944).

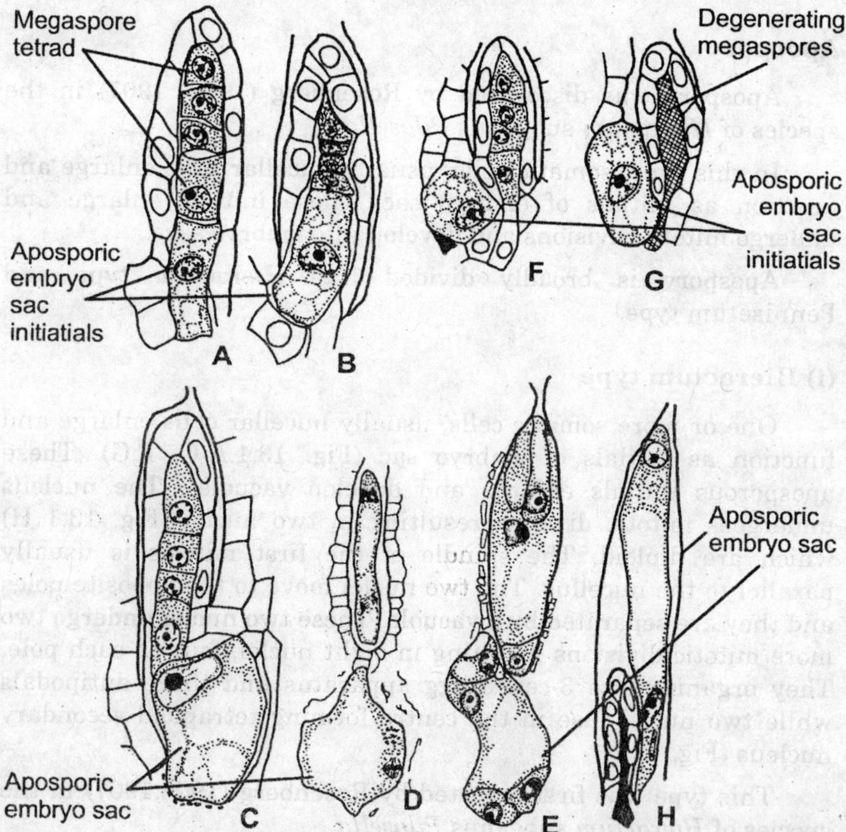

Fig. 13.1. Development of aposporic embryo sacs in *Hieracium excellans* (A-E) and *H. flagellare*; (F-H) (after Rosenberg, 1907)

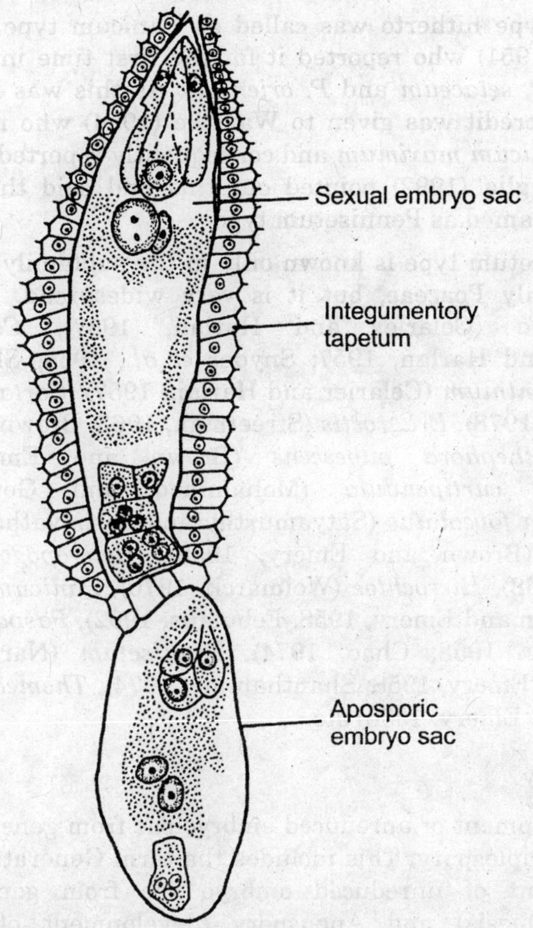

Fig. 13.2. Sexual and aposporic embryo sacs in *Melampodium divaricatum* (Maheswari Devi and Pullaiah, 1976)

(ii) Pennisetum (Panicum) Type

In this type one or more somatic cells enlarge and function as embryo sac initials. The embryo sac initial enlarges and develops vacuoles. Its nucleus moves to the micropylar end. It undergoes mitotic division and the two nuclei remain in the micropylar end. These two nuclei undergo one more mitotic division resulting in four nuclei. These organize into 3-celled egg apparatus, while the fourth one functions as polar nucleus. Antipodals are absent.

This type hitherto was called as Panicum type. But it was Narayan (1951) who reported it for the first time in *Pennisetum villosum*, *P. setaceum* and *P. orientale*. As this was in the Ph.D. thesis the credit was given to Warmke (1954) who reported this type in *Panicum maximum* and consequently reported as Panicum type. Battaglia (1992) pointed out this and said that this type should be named as Pennisetum type.

Pennisetum type is known only in the subfamily Panicoideae of the family Poaceae, but it is very widespread. It occurs in *Bothriochloa* (Celarier and Harlan, 1957), *Capillipedium* (Celarier and Harlan, 1957; Snyder *et al.*, 1955, Shanthamma, 1982), *Dicanthium* (Celarier and Harlan, 1957), *Cortaderia jubata* (Philipson, 1978), *Eragrostis* (Streetman, 1963; Brown and Emery, 1958), *Anthephora pubescens* (Brown and Emery, 1958), *Boutelouva curtipendula* (Mohammad and Gould, 1966), *Eremopogon foveolatus* (Satyamurthy and Seshavatharam, 1983), *Eriochloa* (Brown and Emery, 1958), *Heteropogon contortus* (Gupta, 1968), *Hierochloe* (Weimarck, 1970), *Panicum* (Warmke, 1954; Brown and Emery, 1958, Febulaus, 1992), *Paspalum* (Brown and Emery, 1958; Chao, 1974), *Pennisetum* (Narayan, 1951; Brown and Emery, 1958; Shanthamma, 1974), *Themeda*, *Urochloa* (Brown and Emery, 1958) etc.

Diplospory

Development of unreduced embryo sac from generative cell is known as diplospory. This includes the term Generative apospory (development of unreduced embryo sac from generative cell without meiosis) and Aneuspory (development of unreduced embryo sac from generative cell with irregular meiosis) of Battaglia (1963). Diplospory was first described in *Antennaria alpina* by Juel (1900).

Diplospory is broadly divided into three types – Antennaria type, Taraxacum type and Ixeris type.

(i) Antennaria type

In Antennaria type, meiosis in megaspore mother cell is completely omitted. Megaspore mother cell, after a long interphase enlarges and undergoes mitotic division (Fig. 13.3 A-D, 13.4 A,B). The resulting two nuclei move to the opposite poles separated by a

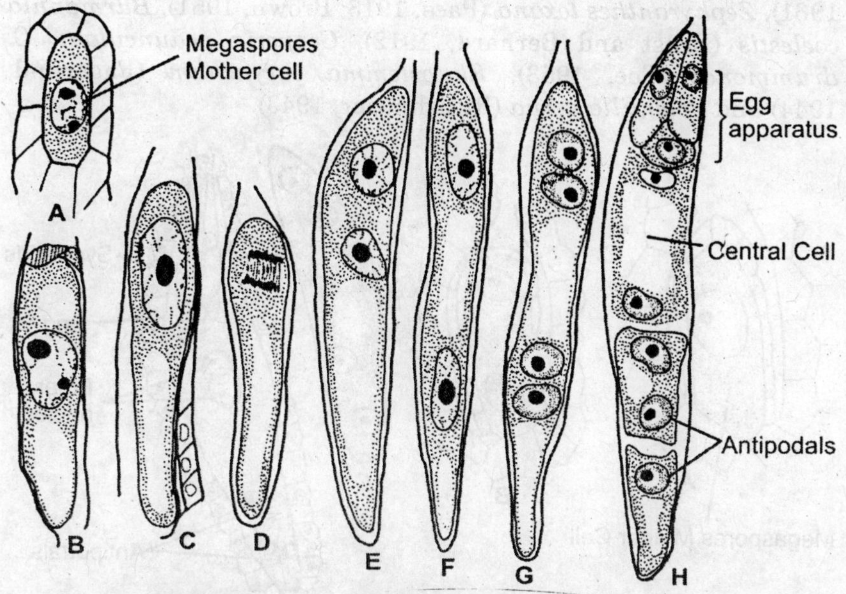

Fig. 13.3. Development of Diplosporous embryo sac in *Eupatorium glandulosum* (after Holmgren, 1919). **A.** Megaspore mother cell; **B-D.** Mitotic division; **E-F.** Two-nucleate embryo sac; **G.** Four-nucleate embryo sac; **H.** Organised embryo sac.

vacuole (Fig. 13.3 F). These two nuclei undergo two more mitotic divisions resulting in 8 nuclei, 4 at each pole. Of these, three organize into 3-celled egg apparatus, 3 at the chalazal end organize into 3 antipodals, while the other two function as polar nuclei (Fig. 13.3 G,H, 13.4 C,D). This type was first reported in *Antennaria alpina* by Juel (1900). This type occurs in *Antennaria alpina* and other species of this genus (Stebbins, 1932; Urbanska-Worytkiewicz, 1974), *Arnica* (Afzelius, 1936; Flovik, 1940; Engell, 1970), *Hieracium* subg. *Archieracium* (Gustaffson, 1935, 1946, 1947; Bergman, 1935, 1941; Gentchef, 1937), *Parthenium argentatum*, *P. incanum* (Esau, 1944, 1946), *Leontopodium alpinum* (Sokolowska- Kulczycka, 1959), *Brachycome ciliaris* (Davis, 1964), *Erigeron Eupatorium* (Holmgren, 1919; Pullaiah, 1981), all belonging to the family Asteraceae. In other families it has been recorded in *Poa alpina* (Hakanson, 1943), *P. nervosa* (Grun, 1935), *P. nemoralis*, *P. glauca*, *P. palustris*, *Calam-ogrostis stricta* (Nygren, 1946; Greene, 1984), *Nardus stricta* (Rychlewski,

1961), *Zephyranthes texana* (Pace, 1913; Brown, 1951), *Burmannia coelestis* (Ernst and Bernard, 1912), *Cooperia pedunculata, C. drummondi* (Coe, 1953), *Elatostemma eurynchum* (Fagerlind, 1944) and *Potentilla verna* (Rutishauser, 1943).

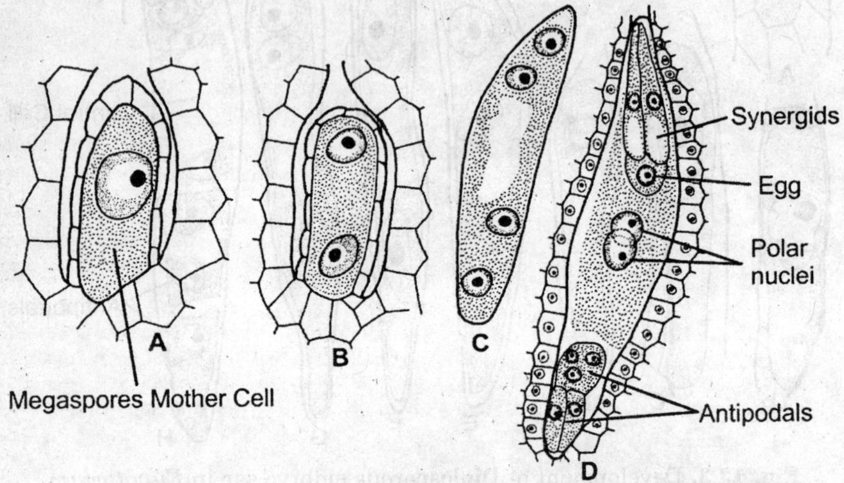

Fig. 13.4. Development of Diplosoporous embryo sac in *Eupatorium odoratum* (after Maheswari Devi and Pullaiah, 1977). **A.** Megaspore mother cell; **B.** Two-nucleate embryo sac; **C.** Four-nucleate embryo sac; **D.** Organised embryo sac.

(ii) Taraxacum type

In this type the megaspore mother cell enters meiosis, but due to asynapsis, there is mostly no pairing. The univalents remain scattered over the whole spindle. Meiosis I ends in a restitution nucleus (Fig. 13.5 A). Hence Meiosis II continues with unreduced chromosome numbers. Meiosis II is regular and produces a dyad of unreduced cells of which the lower dyad functions, while the upper dyad cell degenerates (Fig. 13.5 B-D). The nucleus of the functional dyad cell undergoes three mitotic divisions resulting in an 8-nucleate, 7-celled embryo sac (Fig. 13.5 E,F). This type was reported in *Taraxacum officinale* by Juel (1904, 1905). It has also been reported in *Taraxacum* (Osawa, 1913; Fagerlind, 1947; Battaglia, 1948), *Chondrilla juncea* (Rosenberg, 1912; Poddubnaja -Arnoldi, 1933; Bergman, 1950), *Antennaria carpatica* (Bergman, 1951), *Erigeron* (Holmgren, 1919;

Fagerlind, 1947; Battaglia, 1950) (all belonging to the family Asteraceae), *Balanophora globosa, B. japonica* (Kuwada, 1928; Fagerlind, 1945), *Arabis holboelli* (Bocher, 1947, 1951), *Wickstroemia viridiflora* (Fagerlind, 1940), *Agropyron scabrum* (Hair, 1956) and species of *Paspalum* (Chao, 1974, 1980).

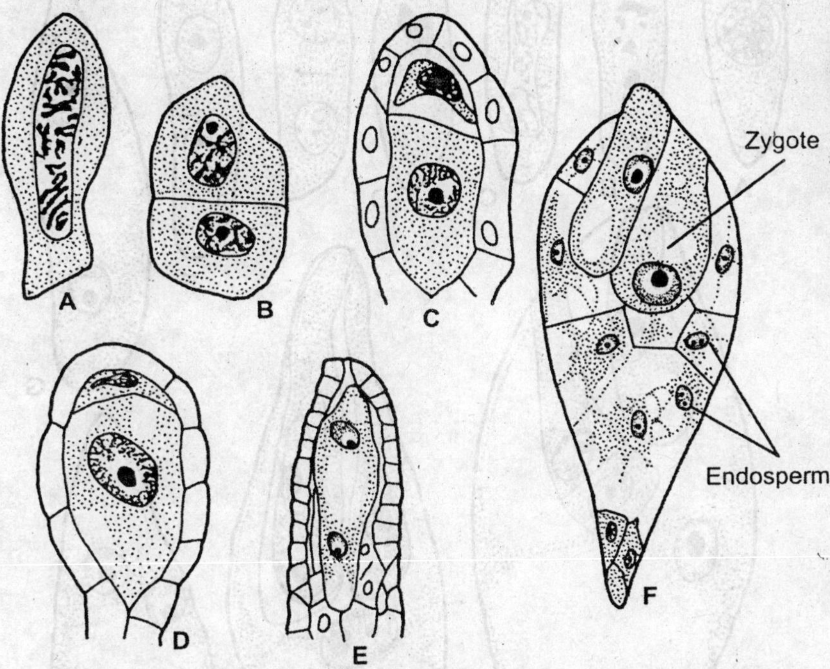

Fig. 13.5. Development of Diplosporous embryo sac in *Taraxacum albidum* (after Osawa, 1913). **A.** Megaspore mother cell in meiosis; **B.** Dyad cells; **C-D.** Upper dyad cell degeneration; **E.** Two nucleate Embryo sac; **F.** Zygote and endosperm.

(iii) Ixeris type

As in the Taraxacum type, here meiosis I gives rise to a restitution nucleus (Fig. 13.6 A-C). Meiosis II is regular, but wall formation is omitted and thus a single binucleate spore is formed (Fig. 13.6 D-G). Two more nuclear divisions lead to the development of 8-nucleate unreduced embryo sac (Fig. 13.6 H, I). The Ixeris type was first described by Okabe (1932) in *Ixeris dentata.* Besides this, this type has also been reported in *Erigeron annuus* (Holmgren, 1919), *E. karwinskianus* var. *mucronatus*

(Carano, 1921, 1924), *Rudbeckia laciniata*, *R. speciosa* (Battaglia, 1945, 1946) and *Statice oleaefolia* (D'Amato, 1940).

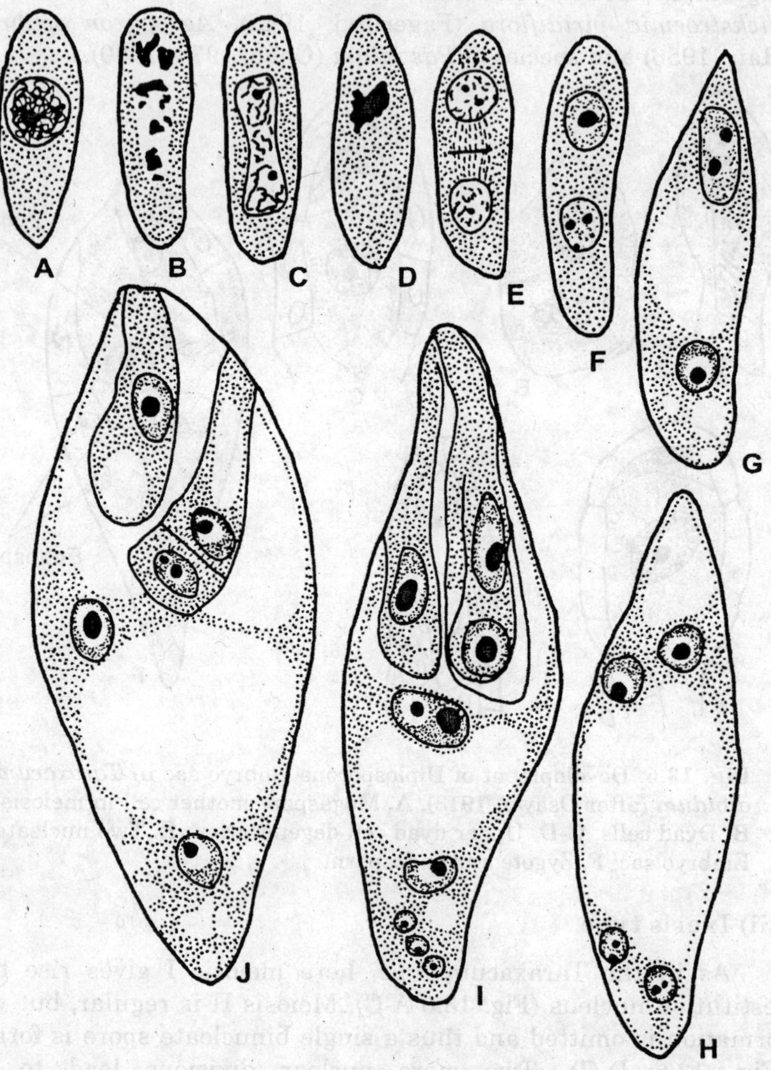

Fig. 13.6. Development of diplosporous embryo sac in *Ixeris dentata* (after Okabe, 1932). **A-B.** Irregular Meiosis in Megaspore Mother Cell; **C.** Restitution nucleus; **D-E.** Meiotic Division II; **F-G.** Two-nucleate embryo sac; **H.** Four-nucleate embryo sac. **I.** Mature embryo sac; **J.** Embryo and endosperm.

DEVELOPMENT OF EMBRYO

Diplospory and apospory result in the formation of unreduced embryo sacs. The sporophyte develops from this unreduced gamete without involvement of sexual fusion. Embryo may develop in pseudogamous, semigamous manner or through parthenogenesis.

Pseudogamy

Here pollination takes place, but the male nucleus degenerates inside the egg. Embryo develops from the egg. Focke (1881) discovered Pseudogamy in *Rubus*.

In many pseudogamous apomicts the embryo starts to develop only after initiation of endosperm.

Pseudogamy has been reported in *Hypericum perforatum* (Noack, 1939), *Ranunculus auricomus* (Rutishauser, 1954), *R. cassubicus*, *Hierochloe* (Weimarck, 1967), *Panicum maximum* (Warmke, 1954), *Pennisetum ciliare* (Snyder *et al.,* 1955), *P. dubium* (Gildenhuys and Brix, 1959), *Zephyranthes texana* (Pace, 1913), *Parthenium argentatum*, *P. incanum* (Esau, 1946), *Cooperia pedunculata* (Coe, 1953), *Arabis holboelli* (Bocher, 1951), *Malus sieboldi* (Olden, 1953) and *Paspalum secans* (Snyder, 1957).

Semigamy (Hemigamy)

The nucleus of the second male gamete enters the cytoplasm of the egg cell, but does not fuse with its nucleus, and divides independently (Fig. 13.7). No fusion occurs and the embryo shows coexistence of male and female nuclei (Fig. 13.7). According to the position of the sperm nucleus inside the egg, a variable part of the embryo possesses nuclei of purely male origin. Frequently the sperm derivatives constitute the suspensor of the embryo. Semigamy was first observed by Nawashin (1898) in *Rudbeckia speciosa* and later by Pace (1913) in *Zephyranthes texana*. This phenomenon was first described by Battaglia (1945a) and he coined the term Semigamia.

Solntseva (1974) calls this as Hemigamy. This phenomenon has also been observed in *Rudbeckia sullivantii, R. laciniata* (Battaglia, 1947), *Cooperia pedunculata, C. drummondii* (Coe, 1953). Rao and Narayana (1980) reported semigamy in *Coix aquatica* of the family Poaceae. This report of semigamy in *Coix aquatica* is the fourteenth one among angiosperms, the sixth in monocotyledons and the first in the family Poaceae.

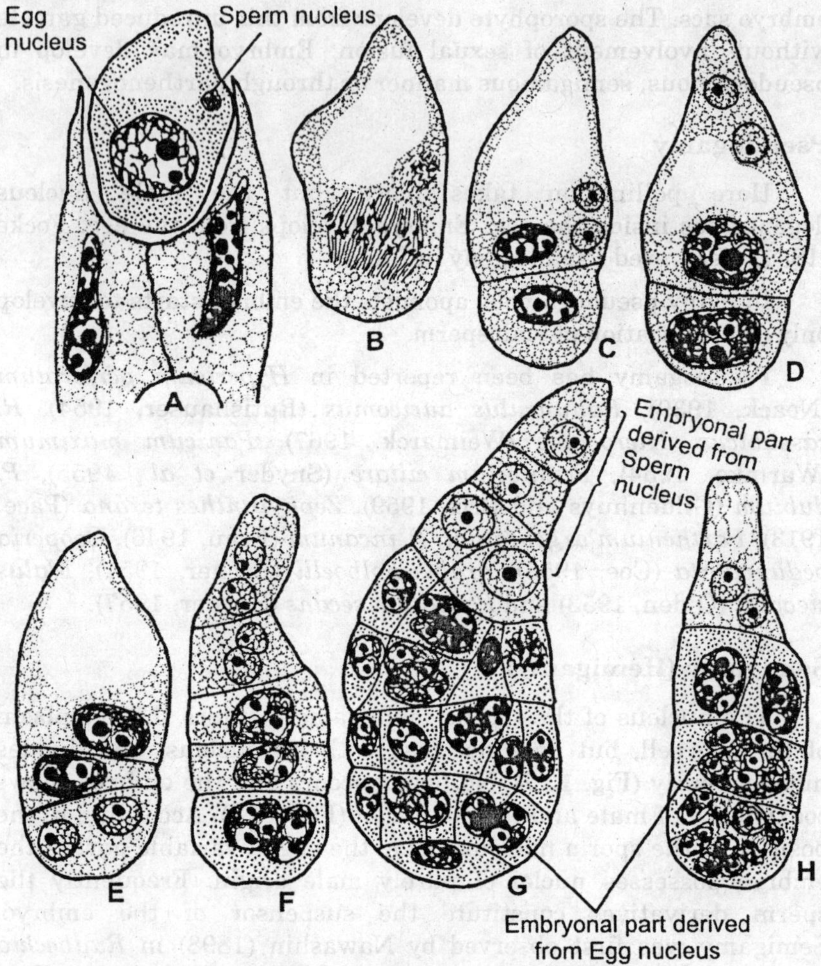

Fig. 13.7. Semigamy in *Rudbeckia* (after Battaglia, 1947)

Parthenogenesis

Development of embryo from unfertilized egg is called Parthenogenesis. In this process even pollination stimulus is absent. If pollination stimulus is there then it is called as **Pseudogamy**. There are quite a large number of examples for parthenogenesis. These include *Alnus rugosa* (Woodworth, 1930), *Burmannia coelestis* (Ernst and Bernard, 1912), *Antennaria* (Afzelius, 1936), *Chondrilla* (Poddubnaja-Arnoldi, 1933), *Crepis* (Stebbins and Tenkins, 1939), *Erigeron* (Holm-gren, 1919; Gustaffson, 1946-47), *Eupatorium* (Holmgren, 1919; Pullaiah, 1981), *Hieracium* subg. *Pilosella* (Bergman, 1935), *Ixeris dentata* (Okabe, 1932), *Taraxacum* (Murbeck, 1904; Gustaffson, 1946-47; Battaglia, 1948), *Pennisetum* (Choda *et al.*, 1981), *Poa* (Tinney, 1940; Akerberg, 1941, 1943), *Alchemilla* (Murbeck, 1901; Boos, 1917), *Potentilla* (Gentcheff and Gustaffson, 1940), and *Elatostemma* (Fagerlind, 1944).

Androgenesis

The development of haploid embryo from the sperm nucleus is called androgenesis. Here the egg nucleus degenerates and the sperm nucleus functions in the cytoplasm of the egg cell. This has been reported in *Crepis tectorum* by Gerassimova (1963).

Polyploidy and Apomixis

The close connection that exists between polyploidy and apomixis prompted some students of apomixis to advance the suggestion that polyploidy is a possible cause for the origin of apomixis.

In some grass genera like *Bothriochloa*, *Capillipedium* and *Dichanthium* (Borgaonkar, 1963) diploids are sexual, tetraploids exhibit various degrees of sexuality and higher polyploids are either completely sexual or essentially obligate apomicts. Norman *et al.* (1989) reported that *Paspalum intermedium*, *P. quadrifarium*, *P. haumanii*, *P. brunneum*, *P. rufum* and *P. maculosum* have both diploids (2n=20) and tetraploids (2n=40) within the same species. The diploid plants are sexual, while tetraploids are facultative apomicts. In the genus *Panicum* accessions from Kenya and Tanzania studied by Combes and Pernes (1970) were shown to be either tetraploid (2n = 32) and apomicts or diploid (2n = 16) and fully sexual. Similarly, in *Themeda triandra* (Liebenberg, 1990) diploids and two of the tetraploids are sexual, while penta and hexaploids are obligate apomicts. In *Dichanthium* and *Bothriochloa* (Harlan and deWet, 1975) crossing experiments have revealed that hexaploids are facultative apomicts.

In *Calamogrostis* (Nygren, 1948, 1951) chromosome doubling by colchicine induces change from sexuality to apomixis. A very similar situation exists in *Panicum maximum*, an important forage grass in certain tropical regions (Savidan *et al.*, 1978).

However, there is no regular association between polyploidy and apomixis as seen in a large number of sexual polyploids and the presence of a few apomictic diploids as in *Nardus stricta* 2n = 26 (Rychlewski, 1961), *Panicum antidotale* 2n=18 (Shanthakumari, 1961), *Pennisetum ramosum* (Narayan, 1962), *P. hohenackeri* 2n = 18 (Kurup, 1973). In the genus *Pennisetum*, while forms with higher chromosome numbers are sexual as in *Pennisetum macrostachyum* 2n = 68 and *P. latifolium* 2n = 36, while *P. clandestinum*, *P. setaceum* (Narayan, 1951), *P. hohenackeri* (Kurup, 1973) and *P. mezianum* (Shanthamma, 1973) with lower chromosome numbers are apomictic. All these lead to the conclusion that in the genus *Pennisetum* polyploidy could not be a factor in the origin of apomixis. It is the genotype that determines the presence or absence of apospory and not the chromosome number as such. Stebbins (1950) is of the opinion that polyploidy may "reinforce the action of genes favouring apomixis'.

Causes of Apomixis

According to Ernst (1918) hybridization is responsible for apomixis. This was criticized by Holmgren (1919) who stated that hybridity is only one of the prerequisites for apomixis. Hybridity and heterozygosity are now understood as a consequence rather than a cause of apomixis.

In facultative apomicts, environmental conditions seem to play a key role in the shift from sexual mode of reproduction to apomixis. *Deschampsia cespitosa* reproduces sexually in Sweden and apomictically in California.

It has been shown in *Dichanthium aristatum* (Knox, 1967; Knox and Heslop-Harrison, 1963) that there is a tendency for more apomictic embryo sacs in short days, while long days promoted sexual reproduction. However, McWilliam *et al.* (1970) could demonstrate no relationship between photoperiod and apomixis in aposporous *Hyparrhenia hirta*.

Genetics of Apomixis

Apomixis has been shown to be genetically controlled (Asker, 1979; Bashaw, 1980; Nogler, 1984). It is usually controlled by one or a few genes that are dominant in some species and recessive in others. More genetic information is available on apospory than on other apomictic mechanisms.

Taliaferro and Bashaw (1966) hypothesized apospory to be controlled by two genes exhibiting epistasis in *Cenchrus ciliaris*. In *Panicum maximum*, apomixis was found to be recessive and controlled by two genes with modifiers (Hanna *et al.*, 1973). Savidan (1981) reported apomixis to be dominant and independent of sexuality and probably controlled by one gene per genome in the *Bothriochloa-Dichanthium* complex. Apomixis in *Paspalum notatum* has been shown to be recessive to sexuality and controlled by a few genes (Burton and Forbes, 1969). In research to transfer obligate apomixis from *Pennisetum squammulatum* to cultivated pearl millet, Dujardin and Hanna (1984) found that the apospory was controlled by a major dominant gene. Nogler (1984) summarized inheritance studies on apospory in *Ranunculus* that show it to be controlled by a single dominant factor, whereas Asker (1970) reported apospory in *Potentilla* to be recessive.

Apomixis and Plant Breeding

Asexual seeds are formed without pollination and/or fertilization. Since the apomictic offsprings are genetically identical to the mother plants it offers an inexpensive way to clonally multiply selected lines/varieties of large volume agronomic crops. During the past three decades there has been a reviewed interest to understand the genetics of apomixes employing modern methods of molecular biology, so that the crop plants can be genetically modified for the apomictic trait. Although apomixis has been known for many years, it is only recently that its potential to transform the agricultural scene is being widely realized. Indeed, according to one view (Jefferson, 1994), exploitation of this single trait in plant breeding can usher in the greatest change in agricultural practice since the dawn of cultivation. As to the significance of apomixis, plant breeders, employing hybridization, have combined genes from different plants, including wild species. Nevertheless it is necessary that a prized hybrid be multiplied in a way that the progeny is identical to the parent *i.e.,* selected hybrid. In crops, hybrids are highly prized on account of their 'hybrid' vigour, but their multiplication through seeds poses a serious problem, and it is at present almost impossible, since normal sexual reproduction involves segregation and recombination of chromosomes, leading to mixed progeny. If, however, some way could be found for stopping reduction division in the megaspore mother cell and obtain embryos, say parthenogenetically without fertilization (or alternatively from nucellar cells), a particular hybrid can be 'fixed' with seeds which all form plants true to the parental type. This can be great boon to agriculture.

Since apomictic plants are derived from embryos that have genetic constitution of the female parent, they can greatly facilitate hybrid seed production and be valuable because heterosis will be perpetuated through successive generations via seed. Thus, a prized hybrid, once produced, can be multiplied vegetatively via apomictic seeds at no additional cost to the farmer. The hybrid will not lose its genetic makeup unavoidable in sexual crossing and will breed true. The farmer can also be spared incurring expenditure on purchasing expensive hybrid seeds which normally he must procure every year from seed companies (specializing in maintaining male sterile and restorer lines and related techniques) and which on average at least 10 times costlier. It is this attractive prospect that has awakened world-wide interest in introducing of apomixis in crops (Bashaw, 1980; Hanna and Bashaw, 1987; den Nijs and Van Dijk, 1993; Khush, 1994; Hanna, 1995; Vielle *et al.*, 1996).

Crossing experiments between sexual crops and related wild species have been performed in recent years. Such experiments have given valuable theoritical information and proved the possibility to transfer resistance from the wild species. Moreover they have shown that the introduction and manipulation of apomictic reproduction in sexuals is no impossibility.

(i) Wheat

Wheat (*Triticum aestivum*) can be crossed with the apomict *Agropyron scabrum*. It is uncertain, if this possibility can be practically exploited. It may be mentioned that pseudogamy has been reported in intergeneric hybrids between *Triticum* and *Avena* (Kruse, 1969). According to the author, the mode of reproduction in this case depended either on the intrageneric hybridization or on the UV-irradiation preceding pollination. Efforts are also underway to transfer apomictic trait in wheat from *Elymus* (Liu *et al.*, 1995) as well as to isolate the gene for apomixis.

(ii) Sugar Beet

In the genus *Beta* apomixis occurs in tetraploid *Beta lomatogona* (sexual diploids also occur within this species), tetraploid *B. intermedia* (of hybrid origin) and hexaploid *B. trigyna.* When diploid and tetraploid sugar beets are pollinated by tetraploid aposporous *B. lomatogona*, triploid and tetraploid hybrids are produced, respectively, with predominant apomictic reproduction. An exploitation of resistance from wild species, thus, now seems possible.

(iii) Maize

In maize, several genes are known to affect the reproductive behaviour, for instance by influencing meiosis. It would not be unbelievable to produce a functioning apomictic system by the help of induced mutations. So far, more interest has been attracted to crosses with apomictic forms of *Tripsacum dactyloides*. The chromosome number of diploid maize 2n = 20 and of apomictic *Tripsacum dactyloides* 2n = 72 (sexual forms of the later species have 2n = 36). By crossing diploid or tetraploid maize with *Tripsacum* and by back-crossing, hybrids have been obtained which represent a variety of combinations between the genomes of parental species. Plants with 36 *Tripsacum* and 20 *Zea* chromosomes give a very variable off-spring after pollination with maize. Among others, partially apomictic plants with the chromosome numbers 2n = 28 (dihaploid), 38 and 56 occurred (Engle *et al.*, 1974).

The occurrence of "Pseudi apomixis" in certain hybrids with 2n = 46 should be mentioned (Harlan *et al.*, 1978). In the meiosis, often 18 bivalents and 10 univalents are formed. The latter (=the maize chromosomes) are eliminated in the first meiotic division. After the second division, no cell wall is formed, and the egg cell retains the 36 *Tripsacum* chromosomes. A back cross with tetraploid maize restores the chromosome number 2n = 46. This process has been reported for several generations.

Resistance genes have already been transferred from *Tripsacum* to *Zea*. To obtain high yielding apomicts by *Zea-Tripsacum* hybridization remains a difficult task. Petrov *et al.* (1973) claim to have transferred "genes, controlling apomictic elements, from the chromosomes of *Tripsacum*, to those of maize". This could be an important step towards the introduction of apomixis into cultivated maize.

Kindiger *et al.* (1996) reported that an apomictic form of hybrid maize would provide an immortalized line which would be stabilized against genetic change. The development of agronomically superior, apomictic maize hybrids could provide a superior level of food security in developing nations as well as altering commercial and public maize breeding programmes-systems in developed nations. Back cross selections obtained from an apomictic 38-chromosome (20 maize + 18 *Tripsacum*) maize-*Tripsacum* hybrid have resulted in the development of apomictic, 39 chromosome individuals with 30 maize + 9 *Tripsacum* chromosomes. The identification of these materials advances two major objectives (i) the elimination of nine *Tripsacum* chromosomes which do not possess gene(s) controlling apomictic reproduction and (ii) the continued refinement of an apomictic maize line. This study was conducted to determine whether the 30 chromosome materials reproduce by apomixis. The generation of these materials could result in the eventual development of a hybrid maize with an apomictic reproductive system (Kindiger *et al.*, 1996).

(iv) Forage Grasses

An interesting project aiming at the development of new apomictic forage grasses is carried out at Gatersleben by Grober *et al.* (1974, 1976). According to the authors, spontaneous hybridi-zation between species or genera is a pre-requisite for the origin of apomixis within a plant group, vigorous, but more or less sterile hybrids may be propagated vegetatively, until by several steps of evolution and natural (or human) selection a functioning apomict may arise.

Grober *et al.* (1974) performed generic and specific crosses between species belonging to the genera *Bromus, Festuca* and

Lolium. More than 100 crossing combinations were realized and maternal plants of different origin were used to enhance the formation of hybrid off-spring. In many cases, embryo culture was necessary to get the adult hybrids. The hybrids obtained were in many cases sterile, but often very vigorous.

To date only one apomictic forage grass species has been improved through breeding. Bashaw (1980) developed three improved cultivars of apomictic bufel grass through hybridization between sexual and apomictic clones.

Introgression of Apomixis and Mapping the Apomictic Genes

S.C. Maheshwari *et al.* (1998) in an excellent review have given various aspects of mapping the apomictic gene. Since only a few genes (probably a single key gene) seem to be involved in the control of apomixis, many breeders have been encouraged to introgress the apomictic trait in crops from wild relatives. Although almost all the studies to date have relied on classical and empirical breeding methods, such as repeated back-crossing by a donor of the gene, some notable progress has been made using these methods. Example wise, the aposporous trait found naturally in *Pennisetum squammulatum* is being introduced into *Pennisetum glaucum*, a cultivated species by Ozias-Akins and co-workers (1993). One resulting line has the apomictic gene located on a supernumerary *squammulatum* chromosome. Fortunately, RFLP markers are now available for several important crop plants and efforts are being made also to develop RFLP maps. Two molecular markers co-segregate with apomixis, thereby giving a valuable opportunity to locate the apomictic gene and work on this is being actively pursued by this group in USA. Similar efforts are underway to transfer the apomictic trait also across genera, for example in Maize (Leblanc *et al.*, 1995) from *Tripsacum*, a wild relative, and in wheat (Liu *et al.*, 1995) from *Elymus* as well as to isolate gene for apomixis. According to S.C. Maheshwari *et al.* (1998) there is some progress towards transferring apomixis from wild to cultivated species and also identifying gene loci responsible for apomixis. Nonetheless, so far, the identity of the genes or of the products and their function(s) are unknown and this is clearly a task for the future.

Identifying Apomixis

Apomixis of various types and at various frequencies would probably be found in more plant species if research efforts were concentrated in this area (Hanna and Bashaw, 1987). Unfortunately, it is easy for plant breeders to overlook possible sources of apomixis due to a lack of understanding of the apomictic mechanism. Failure to obtain F_1 plants from a cross or production of uniform progeny in an F_2 population may be blamed on poor emasculating or crossing technique. Although absence of F_1 plants in crosses and plant uniformity in F_2 population do not guarantee apomictic reproduction, they are indicators of apomixis that should be evaluated.

There is no single method or technique for identifying and confirming apomixis. A combination of cytological, genetic and progeny tests can be used to detect apomixis, identifying apomictic mechanism, and determine the frequency of apomictic reproduction. The following discussion will concentrate on methods for detecting apomixis in sexual species because the greatest impact of apomictic reproduction would be to use it in the breeding of economically important cereals, fruits and vegetables. The same general principle would apply for detecting sexuality in obligate apomictic species (Bashaw, 1962; Hanna *et al.*, 1973).

Breeding behaviour

a. Uniform progeny from a heterozygous plant or an open-pollinated species is one of the best indicators of obligate apomixis. Obligate apomixis will result in uniform progeny identical to the maternal parent on which the seed was produced. Progeny testing of open-pollinated seed from cross-pollinated species is a quick and easy way to screen a large number of introductions.

b. Production of maternal phenotypes in crosses is another indicator of apomixis, but further studies are needed to eliminate the possibility that the crossing technique might have allowed selfing. This would be especially important to check in self-pollinated species or in homozygous lines.

c. Use of a dominant marker gene in a male pollinator carefully crossed onto various accessions of plants suspected of reproducing by apomixis. Production of maternal progeny with the recessive gene indicates apomixis. This technique is more efficient if a marker gene identifiable as early as the seedling stage is used.

d. Unusually high seed set and/or uniform maternal progeny in plant with irregular chromosome numbers or behaviour such as in wide crosses, aneuploids and odd polyploids may indicate apomictic reproduction. In genotypes that are similar except for reproduction behaviour, apomicts will usually set more seed than sexual plants, provided viable pollen is available.

Apomictic reproduction eliminates the meiotic process which can produce unbalanced and lethal female gametes.

Seed and Seedling Characteristics of Apomicts

Multiple seedlings per seed, seed with two or more fused ovaries, multiple stigmas and multiple ovules per floret (Hanna *et al.*, 1970, 1973) may indicate apomixis. Ovaries/pistils with multiple stigmas and multiple ovules per floret often result in seed with fused ovaries. Multiple seedlings per seed can be due to: (1) the development of multiple aposporous embryo sacs, (2) embryos develop in both sexual and apomictic embryo sacs in the same ovule (one or more seedlings may be morphologically different than the rest), or (3) development of embryos in two separate sexual embryo sacs in an ovule (plants may be morphologically different). Cytological studies are needed to confirm these characteristics.

Cytological Tests

Cytological studies are useful for detecting apomixis and are essential for determining the apomictic mechanism.

Apospory

Multiple embryo sacs clustered together and or randomly distributed through out the ovule, usually four or fewer nuclei per embryo sac, and lack of antipodals in the embryo sac at anthesis indicate apospory. Single aposporous embryo sacs also are possible and can be detected by their lack of antipodals. The lack of antipodals in a mature embryo sac is one of the most reliable distinguishing characteristic of apospory.

Diplospory

Cytological studies to detect diplospory must be conducted between the stage when the megaspore mother cell differentiates and early development of the embryo sac. This stage is synchronized with microsporogenesis in many species. The lack of meiosis and chromosome reduction (may be difficult to determine) and the lack of a linear tetrad of spores resulting from meiosis (more easily determined) are characteristic of diplospory. In one type of diplospory the megaspore mother cell goes through one mitotic division to produce a dyad of megaspores of which one cell functions and the other degenerates (Gustaffson, 1946; Nogler, 1984). This degenerating disomic megaspore should not be confused with the three degenerating micropylar spores in sexual reproduction. A mature diplosporous embryo sac may look identical to a mature sexual embryo sac at anthesis.

Adventive Embryony

Adventive embryony can be identified by the presence of proembryos that develop directly into embryos from somatic cells in the ovule (Bashaw, 1980a). No apomictic embryo sacs are formed in adventive embryony; however, sexual embryo sac may be formed in the same ovule. The sexual embryo sacs are needed for the formation of endosperm.

Cytological Technique

Systematic examination of anatomical features throughout development of an ovule is needed to thoroughly understand the cytological basis for the various reproductive mechanisms (Bashaw, 1980a). Embedding ovules in paraffin or a resin, sectioning and staining with saffranin-fast green as outlined in Gurr (1965) as well as other plant microtechnique books is satisfactory for studying general morphological detail. Chromatin stains are needed for observing chromosome detail. In histological studies, care should be taken to properly align the ovules so that longitudinal sections through the gametophytic tissue are obtained (Bashaw, 1980a).

Clearing methods using aromatic esters that greatly reduce the time needed to prepare samples before observations have been reported recently (Young *et al.*, 1979; Crane and Carman, 1987). These techniques allow one to use interference contrast microscopy to observe the contents of the entire ovule by changing the focal level. These methods make possible the determination of the reproductive mechanism in 2 or 3 days after ovule collection

References

Asker, S.E. and L. Jerling 1992. *Apomixis in plants.* C.R.C. Press, Boca Raton, USA.

Maheshwari, S.C., N. Maheshwari, J.P. Khurana and S.K. Sopory. 1998. Engineering apomixis in crops: A challenge to plant molecular biologists in the next century. *Curr. Sci.* **75**: 1141-1147.

Pullaiah, T. 1984. *Embryology of Compositae.* Today & Tomorrows. New Delhi.

Pullaiah, T. and G.N.V. Febulaus. 2000. *Embryology and apomixis in Grasses.* Regency Publications, New Delhi.

CHAPTER – XIV

EMBRYOLOGY IN RELATION TO TAXONOMY

The role of embryology in taxonomy was first brought to prominence by the German embryologist Schnarf in 1931. Subsequently, enormous work has been done on this aspect and the data collected has proved to be of considerable taxonomic significance. Often a single embryological character can mark out the family, *e.g.*, pseudoembryo sac in the Podostemaceae, composite endosperm in the Loranthaceae and formation of a single pollen grain from a pollen mother cell in the Cyperaceae. The importance of embryology in taxonomy has been summarized by many embryologists like Maheshwari (1950), Johri (1963), Davis (1966), Poddubnaja-Arnoldi (1976), Herr (1984), Batygina and Yakovlev (1980-1990) and Johri *et al.* (1992).

Embryological Features of Taxonomic Importance

(1) *Anther*: Number of vertical rows of cells of archesporium, type of wall development (Basic, Monocotyledonous, Dicotyledonous and Reduced), the tapetum glandular or amoeboid, number and ploidy of nuclei in tapetal cells, successive or simultaneous cytokinesis, types of tetrad.

(2) *Pollen grain*: Number and position of germ pores, number of cells at the shedding stage.

(3) *Ovule*: type of ovule (orthotropous, anatropous, campylotropous, amphitropous or circinotropous), number of integuments, micropyle formed by one (inner or outer) or two integuments, nucellus crassinucellate, pseudocrassinucellate or tenuinucellate, presence or absence of endothelium, epistase, hypostase, aril, caruncle, obturator etc.

(4) *Archesporium*: Number of archesporial cells, extent of parietal tissue.

(5) *Megasporogenesis and development of embryo sac*: Shape of the megaspore tetrad, position of functional megaspore in the tetrad; embryo sac development mono, bi- or tetrasporic, number of nuclear divisions.

(6) *Mature embryo sac*: Number and organization of cells in the embryo sac.

(7) Fertilization.

(8) *Endosperm*: Nuclear, Cellular or Helobial, presence or absence of haustoria, seeds albuminous or exalbuminous.

(9) *Embryo*: Development type (Asterad, Onagrad, Chenopodiad, Solanad and Caryophyllad), presence or absence of suspensor haustoria.

(10) *Seed coat*: Number of layers and their derivation (from one or two integuments).

(11) *Special features*: Apomixis, polyembryony.

Distribution of Embryological Characters

Among the 411 families recognized by Hutchinson (1969), 342 (83.2%) are dicotyledons so that the dicotyledon and monocotyledon ratio is 5:1. Thus any character which occurs five times more in dicotyledonous families as compared to monocotyledonous, has an equal frequency in the two groups. A comparative account of distribution of embryological characters in Dicotyledones and Monocotyledones is given in Table 14.1.

Table 14.1. Comparative distribution of embryological characters (after Kapil and Bhatnagar, 1991)

Character	State	Dicot (%)	Monocot (%)
Anther wall	Basic type	100	—
	Dicot type	97	3
	Monocot type	60	40
Tapetum	Secretory	87	13
	Amoeboid	38	62
Cytokinesis	Simultaneous	95	5

Character	State	Dicot (%)	Monocot (%)
	Successive	13	87
Microspore tetrads	Tetrahedral	96	4
	Isobilateral	14	86
Pollen	Binucleate	84	16
	Trinucleate	63	37
Ovule	Bitegmic	70	30
	Unitegmic	100	—
	Crassinucellate	79	21
	Tenuinucellate	89	11
Endosperm	Nuclear	83	17
	Cellular	97	3
	Helobial	17	83
Embryogeny	Asterad	79	21
	Onagrad	93	7
	Solanad	100	—
	Caryophyllad	55	45
	Chenopodiad	83	17
	Piperad	80	20

Application of Embryology to Problems of Taxonomy

Embryological characters are useful in delimitation of plant groups at all levels. At the highest level Angiosperms are characterized by double fertilization (syngamy and triple fusion) which is not seen in any other division. Similarly, angiosperms differ from gymnosperms in that the nutritive tissue for development of embryo in angiosperms is a post fertilization phenomenon, while in gymnosperms it is a pre-fertilization character. The Dicotyledons and Monocotyledons differ in the cytokinesis of pollen mother cells after meiosis. The Dicotyledons are characterized by the simultaneous cytokinesis, while Monocotyledons are characterized by the successive cytokinesis.

Within Dicotyledons and Monocotyledons the subclasses are also differentiated on embryological grounds. The Asteridae are characterized by unitegmic and tenuinucellate ovules, while the

Caryophyllidae are characterized by bitegmic and crassinucellate ovules.

Embryological data supplement other evidence in many cases. Polygonaceae has been associated by various taxonmists with the Centrospermae group. Polygonaceae differs embryologically from other Centrospermae group. Polygonaceae are characterized by the orthotropous ovules while Centrospermae have either Campylotropous or amphitropous ovules.

Some Families marked out by their Embryological features

Podostemaceae

An exclusively tropical family, Podostemaceae includes perennial aquatic herbs, which grow attached to stones in running water. The family shows interesting embryological features. A unique feature is the formation of pseudoembryo sac due to the disintegration of nucellar cells below the embryo sac (Figs. 5.7 & 14.1B). A combination of the following characters further mark out this family.

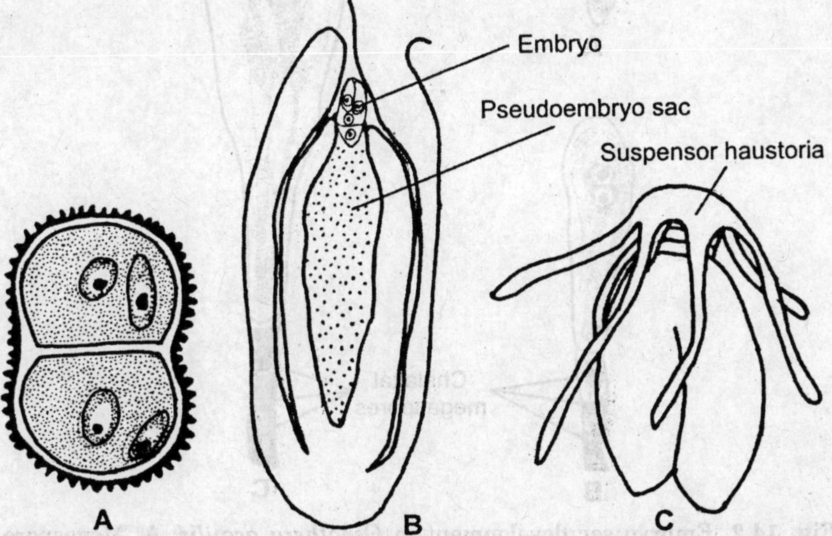

Fig. 14.1. Podostemaceae. **A.** Paired pollen grains of *Podosstemon subulatus*; **B.** Ovule of *Indotristicha ramosissima*; **C.** Mature embryo of *Dicraea elongata* (A,C after Magnus, 1913; B. after Mukkada, 1962)

1. Occurrence of pollen grains in pairs (Fig. 14.1A),
2. Anatropous, bitegmic and tenuinucellate ovule with micropyle formed by the outer integument (14.1B),
3. Bisporic embryo sac,
4. Solanad type of embryogeny,
5. Prominent suspensor haustoria (Fig. 14.1C),
6. Absence of antipodal cells (except in *Dicraea*),
7. Absence of triple fusion and consequently endosperm,
8. Presence of pseudoembryo sac.

Onagraceae

The family Onagraceae is characterized by the presence of Oenothera type of embryo sac development. Oenothera type of

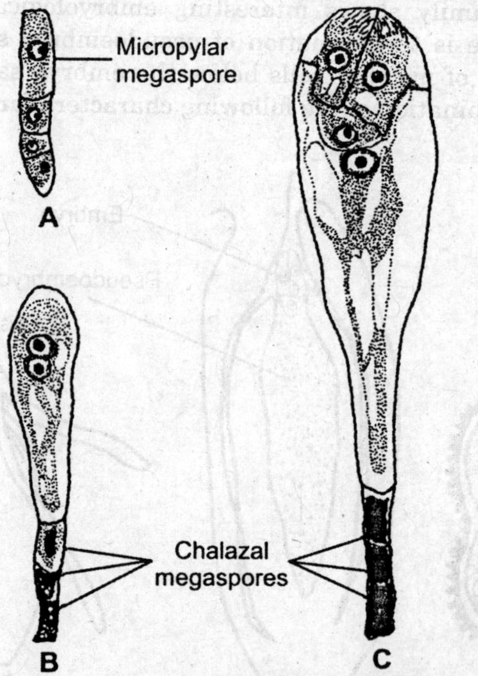

Fig. 14.2. Embryo sac development in *Oenothera acaulis*; **A.** Megaspore tetrad, the functional micropylar megaspore has enlarged; **B.** Two-nucleate stage; **C.** Mature embryo sac; the three cells below the embryo sac are persistent non-functional megaspores (after Pagni, 1958).

embryo sac development is seen in all the members of the family Onagraceae and is not found in any other family. Further of the four megaspores formed after meiosis, the chalazal three megaspores degenerate, while the micropylar megaspore develops into embryo sac (Fig. 14.2). The family is also characterised by the absence of antipodals. Onagraceae are further characterized by a diploid endosperm.

Cyperaceae

The development of microspore from the pollen mother cell in the family Cyperaceae is quite unique. In all other angiosperms the pollen mother cell divides by meiosis, resulting in four haploid nuclei, each of which forms a functional microspore. The members of the Cyperaceae exhibit a special mode of pollen development. Of the four nuclei formed after meiosis, only one functions (Fig. 3.5A-E). The functional nucleus remains in the centre and the three non-functional nuclei are cut off on one side of the cell, and they do not form pollen grains (Fig. 3.5D,E). The functional nucleus divides to form a vegetative cell and a generative cell (Fig. 3.5G). The non-functional nuclei eventually degenerate.

Loranthaceae

Members of the Loranthaceae show some embryological features, which are not seen in any other angiosperms. In this family members do not show any differentiation in ovules and placenta. Here no ovules are present as such and the placental ovular complex is known as a mamelon. The integuments have disappeared and there is a gradual reduction in the lobing and differentiation of mamelon. Finally in some genera even the mamelon is absent and the archesporial cells differentiate at the base of the ovarian cavity. The embryo sacs ascend to various heights in the style. The family is also characterized by the composite endosperm formed by several embryo sacs.

Some other examples of the value of embryology in Taxonomy

Paeonia

In most of the classical systems of classification the genus *Paeonia* is kept in the monogeneric tribe *Paeonieae* of the

Ranunculaceae. The vasuclar anatomy and certain characters of leaves and fruits led Worsdell (1908) to conclude that *Paeonia* constitutes a natural and independent group and should be assigned to a separate family Paeoniaceae which is closer to Magnoliaceae than Ranunculaceae. Kumazawa (1935) studied its wood anatomy and confirmed this conclusion. In the structure of exine Wodehouse (1936) found *Paeonia* to differ from the rest of the Ranunculaceae. Gregory (1941) noticed that the basic chromosome number of *Paeonia* is five in contrast to seven, eight or nine in the Ranunculaceae. Eames (1951) also pointed out that the floral anatomy of *Paeonia* is quite different.

Table 14.2. Embryological differences between *Paeonia* and Ranunculaceae

Feature	*Paeonia*	Ranunculaceae
Stamens	Spirally arranged, centrifugal	Spirally arranged, centripetal
Anther	Multilayered endothecium, mostly 2-layered tapetum	One-layered endothecium and one-layered tapetum
Pollen	Reticulately pitted exine, large and elongate generative cell	Granular, papillate or smooth exine, small lenticular generative cell.
Female archesporium	Multicelled, many megaspore mother cells function	Uni- or multicelled, one cell functions
Antipodal cells	Persistent, not polyploid	Persistent (ephemeral in *Adonis*), nuclei one or more than one, polyploid.
Embryogeny	Unique, Free nuclear	Onagrad or rarely Solanad type
Seed	Arillate	Non arillate

Embryological data also support the view of removing *Paeonia* from the Ranunculaceae to a separate family Paeoniaceae. The most peculiar embryological feature of this genus is its embryogeny (Yakovlev and Yoffe, 1957; Murgai, 1959, 1962). According to Yakovlev and Yoffe (1957) the zygote undergoes repeated nuclear divisions forming coenocytic structure (Fig. 11.9 A-C). Later, the nuclei become lodged in the peripheral layer of cytoplasm around a large central vacuole (Fig. 11.9D). Now, wall formation occurs, and the coenocyte becomes cellular, except in the central part (Fig. 11.9F) of which one matures into adult embryo.

Other embryological features which justify the separation of *Paeonia* from Ranunculaceae to a separate family are given in Table 14.2.

Trapa

Bentham and Hooker (1883) placed the genus *Trapa* in the family Onagraceae, while Raimann (1898), Engler and Diels (1936) and Hutchinson (1959) included it in the unigeneric family Hydrocaryaceae now called Trapaceae. Eames (1953) remarked that on the basis of anatomical evidence *Trapa* does not belong to Onagraceae. The embryological data presented in Table 14.3 also support the view that *Trapa* should be removed from Onagraceae to Trapaceae.

Table 14.3. Embryological differences between *Trapa* and Onagraceae

Feature	*Trapa*	Onagraceae
Pollen grain	Pyramidal with 3 much folded meridional crests	Bluntly triangular and basin shaped
Ovary	Semi-inferior, bilocular with a single pendulous ovule in each chamber	Inferior, mostly trilocular with many ovules per chamber on an axile placenta
Embryo sac	Polygonum type	Oenothera type
Endosperm	Absent	Present, nuclear
Embryo	Solanad type	Onagrad type
Suspensor	Well developed suspensor haustorium	Short and inconspicuous
Cotyledons	One cotyledon extremely reduced	Cotyledons equal
Fruit	Large, one-seeded drupe with prominent spines	Loculicidal capsule

Basellaceae

Several divergent views have been expressed about the systematic position of Basellaceae, a small family of herbaceous twining vines that occur chiefly in tropical America and Asia. Eichler (1878), Volkens (1893) and Lawrence (1951) include the family in Centrospermae close to Portulacaceae. A similar assignment is offered by Bessey (1915), Gunderson (1950), Takhtajan (1997) and Cronquist (1989), although these systematists have adopted the alternative ordinal name, Caryophyllales. Hutchinson (1973) includes the family among group of advanced Caryophyllaceous families which he elevates to a separate order, Chenopodiales. Hooker (1880) includes it in Chenopodiaceae as the subfamily Baselleae and Franz (1908) lowers it to a tribal rank, placing it in subfamily Montioideae of Portulacaceae.

Maheswari Devi and Pullaiah (1975) investigated the life history of *Basella rubra* and summarised the embryological features of the family Basellaceae. They also offer a comparison of those families included in Centrospermae in Table 14.4.

Basellaceae and Chenopodiaceae differ markedly embryologically, and the inclusion of Basellaceae as a subfamily of Chenopodiaceae is untenable. This conclusion is also supported by anatomical features summarized by Metcalfe and Chalk (1950). For example, stems in Basellaceae contain a single ring of vascular strands with intraxylary phloem in the larger bundles. Chenopodiaceae, on the other hand, are characterized by the development of polycyclic, anomalous secondary vascular bundles.

Basellaceae and Portulacaceae share many embryological features (Table 14.4) supporting a close relationship, as also suggested by Takhtajan (1997). However, to include Basellaceae as a tribe in Portulacaceae does not appear justified in view of certain differences.

Since the Basellaceae shows all the embryological features which predominate in Caryophyllales, the retention of Basellaceae as one of the independent families in Caryophyllales seems to be fully justified on embryological grounds.

Table 14.4. Embryology of Phytolaccaceae, Aizoaceae, Portulacaceae, Basellaceae, Caryophyllaceae and Chenopodicaceae (Adapted after Maheswari Devi and Pullaiah, 1975)

	Phytolaccaceae	Aizoaceae	Portulacacae	Basellaceae	Caryophyllaceae	Chenopodiaceae
Anther	Tetraspo-rangiate	Tetraspor-angiate	Tetrasporan-giate	Tetrasporan-giate	Bisporangiate	Tetraspora-ngiate
Wall	–	Basic type	Monocotyle-donous	Monocotyled-onous	Mono-or dicotyledonous	Monocotyle-donous
Endothecium	Fibrous	Fibrous	Fibrous	Fibrous multiseriate	Fibrous	Fibrou
Tapetum	Glandular, cells bi or multinucleate	Glandular, cells bi- or four-nucleate	Glandular, cells bi- or multinu-cleate	Glandular, cells multinucleate	Glandular, cells binucleate	Glandular, cells uni- or binucleate; occasionally amoeboid
Tetrads	Tetrahedral, iso-bilateral	Tetrahedral, iso- bilateral, decussate	Tetrahedral, iso- bilateral	Tetrahedral, iso- bilateral	Tetrahedral, decussate	Tetrahedral, iso-bilateral, decussate
Pollen grain	Two- (*Phyto-lacca*) or three-celled (*Rivina*)	3-celled	3-celled	3-celled	3-celled	3-celled
Ovule	Anacampy-lotropous, bitegmic, crassinucellate	Anacempylotro-pous, bitegmic, crassinucellate, funicular oburator develops	Anacampylot-ropous bitegmic, crasinucellate	Ana- or campylotropous, bitegmic, crassinucellate	Anacampylotr-opous, bitegmic, crasinucellate, nucellar beak conspicuous	Anacampylotr-opous, bitegmic, crassinucellate, nucellar beak conspicuous

	Phytolaccaceae	Aizoaceae	Portulacacae	Basellaceae	Caryophyllaceae	Chenopodiaceae
Microphyle	Formed by Inner integument	Inner integument	Inner integument	Inner integument	Inner integument	Inner integument
Air- space between the ineguments at their base	Present	Present	Present	Present	Present	Present
Archesporium	2- or 3-celled	Multicelled	2- or 3-celled	2- or 3-celled	2- or 3-celled	2- or 3-celled
Embryo sac development	Polygonum type	Polygonum, Endymion, Drusa, Adoxa or Penaea type	Polygonum type	Polygonum type	Polygonum type	Polygonum type Allium type in *Suaeda maritima*
Antipodal cell	Persistent, secondary multiplication in *Rivina*	Ephemeral	Ephemeral	Ephemeral	Ephemeral supernumerary due to divisions	Ephemeral, sometimes more than 3 due to further division
Endosperm	Nuclear type	Nuclear, chala-zal haustorium in *Hereroa* hesparantha	Nuclear	Nuclear	Nuclear in some taxa superficial extensionss develop	Nuclear
Embryogony	Caryophyllad type massive suspensor	Solanad, occasional suspensor polyembryony	Oragrad or Caryophyllad, occaional polyembryony	Chenopodiad	Caryophyllad, basal cell of suspensor haustorial	Chenopodiad, suspensor massive
Seed	Small, exot-estal, peris-permous	Exotestal, and endotegmic, perispermous	Small, exotestal, albuminous perispermous	Small, exotestal, perispermous	Small, exotetsal, perispermous	Small, exotestal and endotegmic, perispermous

Gentianaceae and Menyanthaceae

Benthem and Hooker (1873-1876) and Rendle (1952) split the Gentianaceae into two subfamilies: (a) Menyanthoideae including the marshy plants, and (b) Gentianoideae including all the others. Wettstein (1935) raised Menyanthoideae to a family rank, which was accepted by Don (1938) and Lindsey (1936). This was supported by both vegetative (Perrot, 1897) and floral anatomy (Lindsay, 1938) and embryology (Soueges, 1943; Maheswari Devi, 1962).

Morphological and embryological dissimilarities of the subfamilies Menyanthoideae and Gentianoideae are tabulated in Table 14.5.

Maheswari Devi (1962) is of the opinion that these differences fully justify a family rank for the Menyanthoideae which would be better designated as Menyanthaceae. The name Gentianaceae should be retained for Gentianoideae.

Table 14.5. Differences between Menyanthoideae and Gentianoideae

Feature	Menyanthoideae	Gentianoideae
Habit	Aquatic or marshy	Tetrrestrial
Morphology	Leaves alternate, corolla valvate or imbricate	Leaves opposite and decussate, corolla twisted
Anatomy	Vascular bundles collateral, sieve tubes absent in phloem	Vascular bundles bicollateral or amphicribal, sieve tubes present in phloem
Anther tapetum	Secretory type, cells bi- or multinucleate, nuclear fusion common	Secretory (*Exacum*) or amoeboid (*Canscora*), cells remain uninucleate
Placenta	Suppressed parietal placenta	Well developed parietal placenta, some times the placenta fuse at the centre to form a bilocular ovary
Endothelium	Present	Absent
Endosperm	Cellular	Nuclear
Embryogeny	Asterad type	Solanad type

CHAPTER – XV

EXPERIMENTAL EMBRYOLOGY

As mentioned in Chapter I embryology in the second half of twentieth century became an experimental science. The main objective of experimental embryology are, (1) to understand various factors that control embryological processes and (2) to control and modify these embryological processes.

Significant progress in the field of experimental embryology was made possible with the emergence of *in vitro* culture techniques. *In vitro* cell, tissue and organ culture is now a well established discipline with a wide range of contacts with other disciplines of biology like physiology, biochemistry and genetics. The three important aspects of *in vitro* culture technique are Nutrient medium, Aseptic conditions and Aeration.

Nutrient Media

A number of media have been devised for specific tissues and organs. White's (1953) medium is one of the earliest plant tissue culture media. Other media include Murashige and Skoog's (1962) medium, B_5 medium (Gamborg *et al.*, 1968) and Nitsch's medium (Nitsch and Nitsch, 1969). Most of these media contain inorganic salts of major and minor elements, vitamins and sucrose (Table 15.1). A medium with these components is known as Basal medium (BM). Growth regulators like auxins, cytokinins and gibberellins may be added to this medium. Natural plant extracts such as coconut milk, yeast extract, casein hydrolysate and various fruit juices are also added to the medium for promoting growth in some tissues.

All the constituents of the medium after weighing properly are dissolved in double distilled water. After the thorough mixing of contents in water, sugar and other organic supplements are added. Finally the volume is made up to 1 litre. The pH of the medium is adjusted around 5.8 (slightly acidic) by adding 0.1 N NaOH or 0.1N HCl. In order to make the medium semi-solid, agar in appropriate quantities is added to the liquid medium and boiled till the agar melts totally. Now equal quantities of the medium is dispensed into culture vials (test tubes, flasks or petriplates). The culture vials, containing medium are plugged either with non-absorbent cotton wrapped in cheese cloth, aluminium foils or plastic closures and autoclaved under 15 psi at 121^{0}C for 15-40 minutes. Culture medium is allowed to cool and stored at 4^{0}C until use.

While preparing a solid medium in culture tubes, it is desirable to make slants by keeping the tubes in tilted position during cooling as such slants provide larger surface area for tissue growth.

Aseptic conditions

Inoculations are generally carried out in laminar flow chamber. The laminar flow chamber is initially sterilized by UV light. Since UV radiation is harmful to the eyes, care should be taken to avoid direct exposure to UV. This is followed by switching on the hood and wiping the working surface with 95 percent ethyl alcohol, 15 minutes before use.

The plant material to be inoculated should be surface sterilized. The material is first washed with a detergent followed by a sterilant such as sodium hypochlorite (5%), or hydrogen peroxide or mercuric chloride (0.1-1.0%) solution. The so treated material is rinsed with sterilized double distilled water several times to completely wash off the sterilant.

Aeration

The tissue or organs can be cultured either on a solid medium or liquid medium. Agar-agar (a polysaccharide obtained from sea weeds) is used at a concentration of 0.8-1 percent to solidify the medium. When liquid medium is used, aeration is provided by shaking the vials on an automatic shaker. The advantages of growing tissues in agitated liquid media, are that it facilitates gaseous exchange and reduces nutrient gradient.

Table 15.1. Nutritional components of some plant tissue culture media[1]

Components	Amount (mg l^{-1})[2]			
	White's	MS	B_5	Nitsch's
Macronutrients				
$MgSO_4$ $7H_2O$	750	370	250	185
KH_2PO_4	--	170	--	68
$NaH_2PO_4.H_2O$	19	--	150	--
KNO_3	80	1900	2500	950
NH_4NO_3	--	1650	--	720
$CaCl_2.2H_2O$	--	440	150	--
$(NH_4)_2SO_4$	--	--	134	--
Micronutrients				
H_3BO_3	1.5	6.2	3	--
$MnSO_4$. H_2O	5	22.3	--	25
$ZnSO_4.7H_2O$	3	8.6	2	10
$Na_2MoO_4.2H_2O$	--	0.25	0.25	0.25
$CuSO_4.5H_2O$	0.01	0.025	0.025	0.025
$CoCl_2.6H_2O$	--	0.025	0.025	0.025
KI	--	0.83	0.75	0.75
$FeSO_4.7H_2O$	--	27.8	--	27.8
$Na_2EDTA.2H_2O$	--	37.3	--	37.3
EDTA Na Ferric salt	--	--	43	--
Sucrose (g)	20	30	20	20
Organic supplements				
Vitamins				
Thiamine HCl	0.01	0.5	10	0.5
Pyridoxine HCl	0.01	0.5	1	0.5
Nicotinic acid	0.05	0.5	1	5
Myoinositol	--	100	100	100
Others				
Glycine	3	2	--	2
Folic acid	--	--	--	0.5
Biotin	--	--	--	0.05
pH	5.8	5.8	5.5	5.8

[1] Growth regulators and complex nutrient mixtures used by various workers are not included here as the quantity and constituents of these compounds vary for specific tissue and organ

[2] Abbreviations and references: White's. 1953. *Am. J. Bot.* **40**: 517-524
MS: Murashige and Skoog. 1962. *Physiol. Plant.* **15**: 473
B_5: Gamborg *et al.* 1968. *Exp. Cell Res.* **50** 151
Nitsch's: Nitsch and Nitsch. 1969. *Science N.Y.* **163**: 85

Terminology

In order to understand the technology, one should be familier with the following terminology.

1. *Totipotency*: The ability of a plant cell to develop into a complete plant is known as totipotency.
2. *Callus*: An unorganized mass of parenchymatous tissue formed in response to wounding of plant parts or in tissue culture is called as callus (Pl.: calli).
3. *Dedifferentiation*: The phenomenon of a mature cell reverting to its meristematic state and forming undifferentiated callus tissue is termed dedifferentiation.
4. *Redifferentiation*: The ability of a dedifferentiated cell to form a whole plant or plant organs is termed as redifferentiation.
5. *Cytodifferentiation*: Cell differentiation is the basic event of development in higher organisms and conveniently referred to as cytodifferentiation.
6. *Basal medium (BM):* A medium containing inorganic salts of major and minor elements, vitamins (but without hormones) and sucrose is known as Basal medium (see Table 15.1).
7. *Androgenesis*: Development of haploid plants from the male gametophyte following a developmental pattern resembling embryogenesis, resulting from the culture of anthers or microspores.
8. *Axenic*: A pure culture of single organism that is not contaminated by symbionts or parasites.
9. *Clonal propagation*: Asexual reproduction or vegetative propagation or plants considered to be physiologically and/or genetically uniform and which originated from a single individual or explant.
10. *Cybrid*: A cytoplasmically hybrid cell with organelles from both parental sources, *e.g.*, achieved by fusion of a cytoplast with a whole cell.
11. *Embryoid*: Mass of cells which has an external morphology resembling a proembryo. Non-zygotic embryo developed in tissue cultures.

12. *Embryo rescue*: Application of embryo culture technique to obtain interspecific hybrid plants prior to abortion of the embryos due to post-fertilization hybridization barriers.

13. *Explant*: The tissue taken from a plant or seed and transferred to culture medium to establish a tissue culture system or regenerate a plant.

14. *Friability*: The property of cultured plant cells that separate easily from one another.

15. *Gynogenesis*: Development of haploid plants from the female gametophyte by culture of unfertilized ovules or ovaries.

16. *Micropropagation*: Synonymous with *in vitro* propagation especially with reference to enhanced axillary branching or adventitious plantlet regeneration.

17. *Somaclonal variation:* Variation which occurs in cultures of cells and tissues that may be either genetic or epigenetic.

18. *Excise*: Removing the part of a plant or seed in aseptic conditions for tissue culture.

19. *Auxotroph*: A mutant organism that will not grow on a minimal medium, but requires the addition of some growth factor.

20. *Vitrification*: A physical process by which a concentrated aqueous solution cooled to low or super low temperature directly solidifies into an amorphous glassy state without crystallization; the term was earlier used to hyperhydration.

21. *Synthetic seeds*: Somatic embryos encapsulated in hydrated or desiccated coating, which protects the embryo from mechanical damage during handling and allows their germination like sexual seeds. The capsule may also serve as an artificial endosperm.

22. *Bioreactor*: A vessel used for large scale aseptic culture of cells, embryos or shoots under controlled growth conditions.

23. *Batch culture*: A type of cell suspension culture in which cell or small cell aggregates, grow in a definite volume of nutrient medium.

ANTHER CULTURE

Sipra Guha and Satish Chandra Maheshwari in 1964 cultured the anthers of *Datura innoxia* with the objective of understanding the physiology of meiosis. However, they made a chance observation of development of embryo like structures in these cultures. Later on (Guha and Maheshwari, 1966, 1967) confirmed that these embryos were haploid and derived from pollen grains. Because of the usefulness of haploids this report has revolutionized the whole plant breeding programmes. Development of embryos from male cells, *i.e.,* pollen grains, is known as *Androgenesis.*

Technique of Androgenesis

Guha and Maheshwari (1964, 1966) cultured anthers of *Datura innoxia* on a basal medium supplemented with coconut milk or kinetin.

Young flower buds at the appropriate stage of pollen development are surface sterilized and anthers removed along with their filament. The anthers, after a quick dip in absolute ethyl alcohol are excised under aseptic conditions and placed horizontally on the medium. Sufficient care should be taken that the anthers are not injured during the process of anther culture, otherwise wounding causes callusing of anther wall tissues. For most species, nutrient medium solidified with agar has been used (Guha and Maheshwari, 1964). However, in species of *Nicotiana* and *Hyoscyamus*, a better yield of pollen-plantlets is obtained by using liquid medium. When dissection of buds poses problems, intact whorl of stamens as in tobacco or whole spike as in barley may be cultured.

Develoment of Androgenic Haploids

Androgenic haploids may develop in four ways (Bhojwani and Bhatnagar, 1990) (Fig. 15.1).

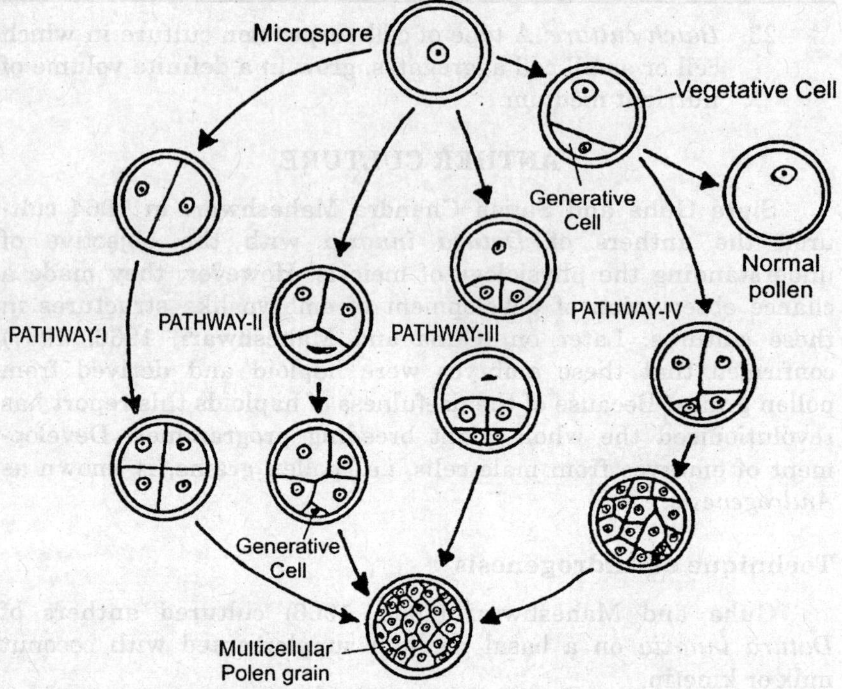

Fig. 15.1. Diagram showing development of multicellular pollen grains (after Bhojwani and Razdan, 1983)

Pathway I

Microspore divides by equal division forming two identical daughter cells. Both these cells contribute to the development of sporophyte (*e.g.*, *Datura innoxia*). In this pathway vegetative and generative cells are not formed.

Pathway II

Microspore undergoes unequal division resulting in a large vegetative cell and a small generative cell. The sporophyte develops from the vegetative cell alone (*e.g.*, *Nicotiana tabacum*, *Hordeum vulgare*, *Capsicum annuum*, *Triticale*). The generative cell may or may not divide further, but degenerates eventually.

Pathway III

Microspore undergoes unequal division resulting in a large vegetative cell and a small generative cell. The sporophyte develops from the generative cell alone (*e.g.*, *Hyoscyamus niger*).

Pathway IV

Microspore undergoes unequal division forming a large vegetative cell and a small generative cell. Both vegetative and generative cells contribute towards the formation of sporophyte (*e.g.*, *Datura innoxia*, occasionally *Datura metel*, *Atropa belladonna*) (Fig. 15.1).

Irrespective of the nature of first pollen mitosis, further development may lead to the formation of embryoids and plantlets or a mass of callus (Fig. 15.2). In *Nicotiana*, *Datura*, *Atropa*, *Capsicum*, *Petunia*, *Brassica* and *Solanum*, the pollen grains produce embryoids and plantlets directly without the intervention of a callus phase. In many of the monocotyledonous plants like *Oryza*, *Triticum*, *Hordeum* and *Lolium* no embryoid formation ever occurred, instead, the pollen grain developed into unorganized mass of callus which could be manipulated by transfer to a suitable regenerating medium to differentiate into plantlets.

Factors Affecting Anther Culture

The factors that affect androgenesis are as follows.

Genotype of the Donor plant: The genotype of the donor plant plays a significant role in determining the frequency of androgenesis. In wheat the induction of pollen callus and green plantlet formation is controlled by many genes. In *Hordeum* each genotype differs with respect to androgenic response in anther culture. Hence the genotype that responds well in anther culture should be selected. It has been possible to improve androgenic response of poorly responding tobacco and barley by following breeding approach.

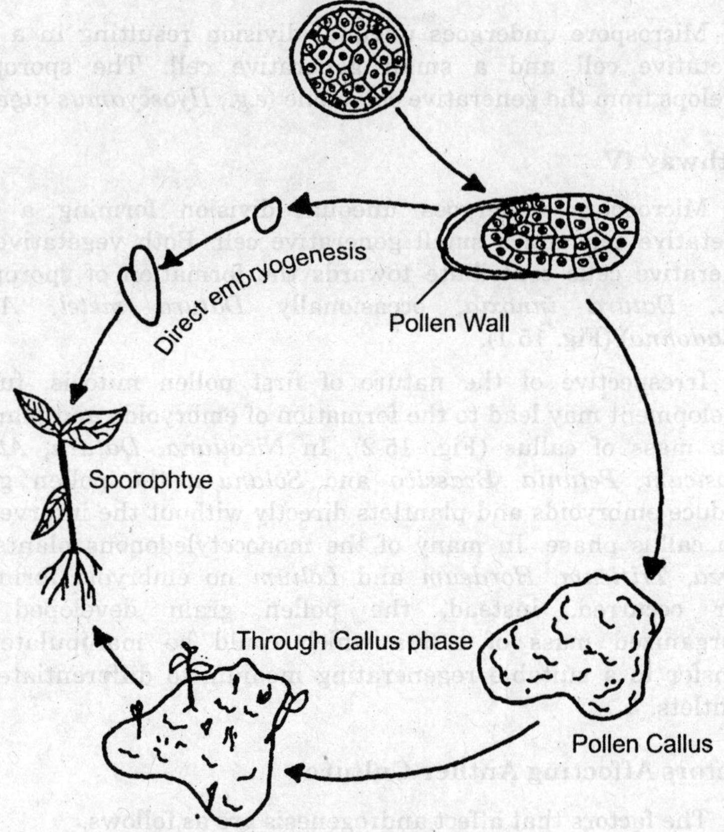

Fig. 15.2. Diagram showing origin of sporophyte from pollen grain (after Bhojwani and Razdan, 1983)

Anther wall factor(s): The effect of anther tissues on the ontogeny of embryoids is not fully understood. Histological studies support the role of anther wall factor(s) in pollen embryo development. The fact that in henbane the embryogenic pollen are restricted to the periphery of the anther locule suggests that a gradient of critical substances of tapetal origin is instrumental in the induction of embryogenic division in the pollen grains. Glutamine, L-serine and myinositol could replace anther wall factor(s) for isolated pollen cultures.

Culture medium: Culture medium plays an important role in inducing divisions in microspores. Basal media of Murashige and

Skoog (1962) and Gamborg *et al.* (1968) are especially advantageous for many species. Although usually sucrose 2% is added to the medium, anthers of barley and *Brassica* require a higher concentration of sucrose (10%). Similarly, in anther cultures of wheat 6 percent sucrose promoted pollen callusing. However, continuous culture in a high sucrose medium proved deleterious to the normal differentiation of embryoids into plantlets unless the embryoids are transferred to a low sucrose medium.

Iron in the medium is reported to be indispensable for the production of plantlets from microspores. Activated charcoal in the medium has been reported to have stimulatory effect on embryoid formation and plantlet development in tobacco. Charcoal possibly absorbs the inhibitory substances present in the anther wall.

Most Solanaceous species exhibit androgenesis on a medium containing mineral salts, vitamins and sucrose. In majority of non-solanaceous plant species known to exhibit androgenesis, pollen embryo formation occurs only when the medium is fortified with growth adjuvants like yeast extract, casein hydrolysate, coconut milk etc. For some others addition of auxin (2,4-D, IAA, NAA) or cytokinin (KN) promoted androgenesis.

Developmental stage of anther: The developmental stage of microspores at the time of culture is a critical factor for the induction of division. Anthers of some species give the best response if pollen is cultured at first mitosis or later stages (post-mitotic – *Datura innoxia, Nicotiana tabacum, Petunia hybrida*), whereas most others anthers are more productive when cultured at uninucleate microspore stage (pre-mitotic stage). The early bicellular stage is best for *Nicotiana sylvestris* and *Atropa belladonna*, but absolutely essential for *Nicotiana knightiana.* Young anthers showing meiotic stages in pollen mother cells and older ones with starch-filled grains failed to produce embryoids. However, there are certain exceptions to this rule. The ideal stage for androgenesis in *Arabidopsis thaliana* and *Lycopersicon esculentum* is when microspore mother cells are in meiosis, whereas *Brassica* species show better response when mature anthers or pollen, isolated from them, are cultured. External markers like the length of the stamen can be used for knowing the stage of anther.

Physiology of Donor plant: Age of the plant influences the androgenic response in anther culture. Buds developed towards the beginning and peak of flowering season responded better than those produced later in the season. Anthers, taken from plants grown under short-day (8 hr day) conditions, show relatively better response than long day (16 hr day) donor plants. Growth conditions of the donor plants also affect the embryogenic potential of the cultured anthers. By subjecting the anther of donor plants to nitrogen starvation, an increase in anther response and embryo yield could be obtained in tobacco.

Effect of temperature and light: A cold treatment given to flower buds before excising the anthers for culture enhances the induction frequency of microspore androgenesis. In tobacco and *Datura*, if the buds are treated at 5^{0}C for 72 hours induced 58 percent of anthers to yield pollen embryoids, while buds treated at 22^{0}C for the same period the response was only 21 percent. In rye, maize and *Pennisetum*, the induction of androgenesis has been found to be very effective if they are stored at low temperatures prior to anther culture.

In contrast to this high temperature treatments increased the response of cultured anthers in *Brassica napus*, whereas in *B. juncea* a combination of cold pre-treatment and high temperature shock was essential for inducing embryogenesis.

POLLEN CULTURE

Anther culture has some disadvantages. The plants obtained from anther culture may originate not only from pollen grain but also from various parts of the anther, resulting in mixed ploidy. Anther wall may produce inhibitory substances which are not conducive to pollen embryogenesis. The pollen population within an anther is generally heterogenous. Therefore, the plants arising from an anther is actually a chimera because numerous pollen in the anther contribute to the callus. These problems can be overcome by isolated pollen culture. In addition, through isolated pollen culture the problem of competition between a large number of pollen grains for growth in the limited space available within the anther, can also be solved.

The first report on the tissue formation from isolated pollen of an angiosperm was from Kameya and Hinata (1970). They used hanging drops to culture mature pollen isolated from *Brassica oleracea* and the hybrid *B. alboglabra* X *B. oleracea* at low temperature.

Nitsch (1974) described the method of raising haploid plants from isolated pollen grains of tobacco. Unopened flower buds are surface sterilized. About 50 anthers are placed in a beaker containing about 20 ml of the liquid medium. Pollen grains are squeezed out by pressing the anther with the help of a glass rod or a syringe piston. The whole solution is filtered through a nylon sieve. The suspension thus obtained is centrifuged at 500-800 rpm for 5 minutes. The pellet containing pollen is resuspended in fresh medium and the process is repeated twice. Finally the pollen is mixed with a suitable culture medium and the suspension is pipetted into petri-plates. The cultures are stored at around 25°C in diffuse light (500 lux).

In the following years nurse culture technique was found to be suitable for raising haploid tissue clones from isolated pollen grains. Intact anthers are obtained from a flower bud and placed horizontally on the surface of a semisolid medium. A filter paper disc is placed over the anther. Separately, a suspension of pollen from anthers of another bud was prepared in liquid medium and about 10 pollen grains are transferred to each filter paper disc. Within a month numerous clones were formed.

Callus derived from floral parts can replace the nurse tissue. The extract of cultured anthers has equal nursing effects on the growth and development of microspores. A purely synthetic medium developed for raising haploid plants from isolated pollen grains of tobacco and *Datura* may also be suitable. The crucial ingredients of this medium are glutamine, L-serine and inositol. Haploids isolated from pollen culture have been reported in tobacco, maize, *Petunia hybrida* and maple.

GYNOGENESIS

Production of haploid individual by the development of an unfertilized egg cell is known as Gynogenesis. This can be achieved through (a) delayed pollination, (b) through the use of abortive pollen, (c) pre-exposed to ionising radiation, and (d) by using alien pollen.

Efforts to raise parthenogenetic haploid plants using *in vivo* techniques yielded little success. San Noeum in 1976 was the first to demonstrate that gynogenesis, an essentially *in vivo* phenomenon, can be induced under *in vitro* conditions. Gynogenic haploids have been obtained in *Hordeum vulgare* using ovary culture (San Noeum, 1976). Zhu and Wu (1979) obtained haploid plants from cultured unpollinated ovaries of *Triticum aestivum* and *Nicotiana tabacum*. Gynogenic haploids have also been obtained from cultured pollinated ovules in rice, maize, sunflower, potato, *Beta vulgaris*, *Gerbera jamesonii*, *Hevea braziliensis*, *Hordeum vulgare*, *Nicotiana rustica*, *Allium tuberosum* and *Petunia axillaris*.

Application of Haploids in Plant Breeding

1. *In vitro* production of haploids can solve some problems in genetic studies since gene actions are readily manifested due to a single allelic dose present in chromosomes of an entire genome.
2. Free cells of the haploid in suspension could be used as an experimental material for mutagenic studies, and to evolve biochemical mutants resistant to pathogens, heat, cold, salinity and drought conditions. Some examples of using anther culture technique in mutant selection successfully are tobacco mutants resistant to blackshank disease and wheat lines resistant to scab (*Fusarium graminearum*).
3. Anther derived haploids have been extensively used in protoplast culture and somatic hybridization.
4. By haploid induction followed by doubling of chromosomes it is possible to obtain exclusively male plants in dioecious species. *Asparagus officinalis* is an example of this, in which androgenic haploids produced from male asparagus plants are either X or Y. Chromosome doubling of Y results in super males (YY) which can be vegetatively propagated. If XX plants are crossed with YY plants the progeny will be all male. Thevenin and Dore (1976) using this method, established high yielding male lines of *Asparagus*.

5. Transfer of desired alien genes: Chromosomal instability in haploids makes them potential tools for introduction of alien genes or chromosomes during wider crossing programmes. In rice the breeding or high yielding varieties for resistance to blast is conventionally achieved through back-cross. This is time-consuming process and may take more than 12 years to develop a cultivar with desired quality. Through hybridization and anther culture, a standard pollen-plant breeding procedure can be developed in rice to introduce genes for high yield and resistance to blast in a short span of 2 years.
6. For using hybrid vigour as a method of crop improvement, homozygous, true breeding cultivars are extremely important. Obtaining such cultivars through conventional method of plant breeding, especially in outbreeding crops requires many years of patient work. It involves selfing for many generations. Moreover, this technique can not be applied to self-incompatible plants. In this regard, haploids can be easily treated with colchicine to give homozygous diploids in a single generation.
7. Developing asexual lines of trees and perennial species: Chinese workers obtained a pollen-derived rubber tree taller by six metres which could then be multiplied by asexual propagation to raise several clones. Similarly, in poplar, the haploids desired genotypes are selected and then doubled after 7-8 years. This coincides with their flowering, enabling them to be propagated both asexually and sexually. Pollen-derived haploid plantlets have also been obtained in *Citrus microcarpa*, *Vitis vinifera*, *Litchi chinensis*, *Camellia sinensis*, etc.

NUCELLUS CULTURE

In some plants like *Citrus* and mango, embryos arise adventitiously from the cells of the nucellus. The nucellar embryos are of considerable importance to horticulturist, as they are genetically uniform and reproduce the maternal genotype. Another advantage of nucellar embryos is that, in contrast to clones established from cuttings or by conventional methods of propagation, plants derived from nucellar embryos are free from pathogens. Further, many of the desirable plant vigour and fruiting characteristics associated with juvenility are restored in trees established from nucellar seedlings.

Rangaswamy (1961) excised nucellus from pollinated ovules of *Citrus microcarpa* (a polyembryonate species) and cultured it on modified White's medium supplemented with casein hydrolysate. The nucellus proliferated into a callus and differentiated into numerous tumeroid outgrowths, called pseudobulbils which eventually developed into plantlets. Sabharwal (1963) reported similar results in *Citrus reticulata*, but opined that the pseudobulbils devlop from nucellar embryos.

Rangan *et al.* (1968, 1969) has shown that it is possible to induce nucellar embryogenesis in monoembryonate *Citrus* species (*C. grandis*, *C.limon*, *C.reticulata* and *C. sinensis*). In this investigation the nucellus was free from zygotic embryo, and cultured on a medium supplemented with malt extract.

Pollination and fertilization are generally considered as essential prerequisites for the induction of nucellar embryogenesis. However, Button and Bornman (1971) have demonstrated that nucellus from unfertilized ovules can also form embryoids. The nucelli from pre-fertilized ovules exhibit embryogenesis only if the medium is supplemented with malt extract and adenine.

Mitra and Chaturvedi (1972) induced embryogenesis in nucelli obtained from unfertilized ovules of *Citrus sinensis* and *C. aurantifolia*. They reported that embryos developed directly or from callus which proliferated from the nucellus.

In vitro nucellar embryogenesis has also been reported in *Aegle marmelos*, *Agrostemma githago*, *Malus domestica* and *Poncirus trifoliata*.

Studies on nucellus culture have been extended to species of other genera also. Mullins and Srinivasan (1976) succeeded in inducing nucellar embryony in ovule cultures of *Vitis vinifera*, a normally monoembryonate species. Unfertilized ovules cultured on Nitsch medium supplemented with Benzyl adenine and B-NOA formed a nucellar callus which subsequently produced embryoids. These embryoids when transferred to a semisolid medium containing Gibberellic acid and 2-isopentyladenine produced plantlets.

Rangaswamy (1982) cultured nucellus of several cucurbits – *Coccinea, Cucumis, Lagenaria, Luffa* and *Trichosanthes.* The nucelli cultured produced profuse callus, but embryogenesis was completely absent. Rangaswamy concludes that lack of embryogenic capacity in nucellar cultures of Cucurbitaceae are due to a chemical block cucurbitacins.

In unpollinated ovary cultures of *Cynanchum vincetoxicum*, Haccius and Hausner (1976) observed the development of green plantlets through the ovary wall. Histological studies proved them to be adventive embryos originating from the nucellar cells of unpollinated ovules. Some of these *in vitro* obtained nucellar plantlets were planted in soil where they developed into flowering plants.

OVULE CULTURE

White (1932), for the first time, cultured the ovules of *Antirrhinum majus* and reported callus formation from integument. La Rue (1942) cultured ovules of *Erythronium americanum* on White's medium fortified with IAA, although there was increase in size, the seeds failed to mature.

N. Maheshwari and Lal (1958, 1961) reared the ovules of *Papaver somniferum* after fertilization with zygote or 2-celled embryo and a few endosperm nuclei on White's medium fortified with kinetin. When the ovules were maintained in culture for 50 days they germinated. Sachar and Kapoor (1959) cultured the ovules of *Zephyranthes*, when they contained zygote and primary endosperm nucleus, on Nitsch's medium supplemented with coconut milk. The ovules showed mature embryo within 12 days and the seeds germinated *in situ* within 20 days.

Poddubnaja-Arnoldi (1959, 1960) cultured the ovules from pollinated ovaries of several orchids (*Calanthe veitchii, Cyprepedium insigne, Dendrobium nobile* and *Phalaenopsis schilleriana*) on simple sucrose 10% solution. Entry of the pollen tube into the ovule and development of embryo has been observed in these cultures. Rangaswamy (1961) cultured the ovules of *Citrus microcarpa* and observed that the nucellar embryos showed development and differentiation on White's medium (WM) containing 5 percent sucrose. However, the embryos dissected from such ovules showed normal growth only on addition of casein hydrolysate (400mg/l) to the medium.

Bajaj (1964) cultured the ovules of *Abelmoschus esculentus* 2 and 4 days after pollination (containing 4-celled proembryo with free nuclear endosperm and globular proembryo with cellular endosperm, respectively) and observed that ovules developed satisfactorily on medium containing casein hydrolysate and coconut milk. The ovules matured in 6 weeks and also germinated.

Guha and Johri (1966) cultured the ovules of *Allium cepa* before and after pollination on Nitsch medium + IAA + GA + 2% sucrose and observed that ovules failed to develop. They further observed that ovules cultured 13 days after pollination on Nitsch medium developed normally and germinated *in situ* after 20 days.

Wakizuka and Nakajima (1974) studied the *in vitro* development of ovules of *Petunia hybrida*. The ovules with early globular proembryo developed into mature seeds, on a medium supplemented with cucumber fruit-juice, adjusted to favourable osmotic value. They inferred that osmotic value of the medium influenced young ovules, which developed to maturity.

Chopra and Sabharwal (1963) cultured the pollinated ovules of *Gynandropsis*. They observed that the placental tissue had a beneficial effect on the growth and maturation of seeds.

The effect of plant hormones on fibre and seed development in cotton has been studied by Beasly (1971) and Beasly *et al.* (1974). Fertilized ovules excised 2 days after anthesis grew and developed fibres only when floated on the liquid medium. Submerged ovules or ovules placed on identical but agar medium callused or failed to grow. If cultured one day after anthesis, only a few ovules produced fibres, and none when cultured on the day of anthesis. Beasly and Ting (1973, 1974) studied the effect of growth substances on *in vitro* fibre development of cotton ovules. Both KN and ABA inhibited the development of fibres. The immature ovules enlarged and produced fibres if the medium contained IAA or Gibberellic acid or both.

In many interspecific and intergeneric crosses, the F_1 hybrid embryo frequently aborts in the developing seed, or the seeds produced are incapable of supporting the development of embryo. In their attempts on interspecific hybridization in *Abelmoschus*,

Gadwal *et al.* (1968) observed that the hybrid embryo failed to develop beyond the heart or torpedo-shaped stage. By employing both ovule and embryo culture, they successfully obtained viable hybrids in three out of five crosses attempted, namely *Abelmoschus esculentus* X *A. ficulneus*, *A.esculentus* X *A. moschatus*, *A. tuberculatus* X *A. moschatus*.

Ovule culture assumes great significance in plant breeding. The hybrids which usually fail to develop either due to abortion of embryo at an early stage or premature abscission of fruits could be obtained by ovule culture.

OVARY CULTURE

La Rue (1942) was the first to culture ovaries of a few angiosperms and obtained root formation from the pedicels. Jansen and Bonner (1949) cultured the ovaries of *Lycopersicon pimpinellifolium* on a medium supplemented with casein hydrolysate, IAA and a mixture of B vitamins. Nitsch (1949, 1951) successfully cultured the ovaries of *Lycopersicon esculentum. Nicotiana tabacum*, *Phaseolus vulgaris* and *Fragaria* sp.

Chopra (1958, 1962) studied the effect of IAA, IBA, KN and GA on both unpollinated and pollinated ovaries of *Althaea rosea.* On Nitsch medium supplemented with IBA the unpollinated ovaries developed into parthenocarpic fruits comparable in size with those which developed *in vivo.* Although the addition of KN or IAA alone had no pronounced effect, KN and IAA used in conjunction acted synergistically in inducing parthenocarpy. In culture of pollinated ovaries maximal growth of the fruit occurred on Nitsch medium + IBA in darkness; 50 percent of the fruits matured in 25-30 days. Chopra (1958, 1962) also observed that the growth of ovaries was considerably influenced by the presence of calyx. If the calyx was removed prior to culture, the fruits grew to a diameter of only 12 mm, the endosperm remained free nuclear and the embryo reached only heart-shaped stage. If the calyx was left intact, however, the fruits enlarged in diameter of 19 mm, the endosperm became cellular and the embryo showed normal development.

Similarly in *Iberis amara* (N. Maheshwari and Lal, 1961) flowers cultured 1 day after pollination produced normal fruits only if calyx was not removed before culture. If the calyx was removed, addition of sucrose (5%) was necessary to obtain satisfactory growth.

Puri (1963) studied the effect of growth substances such as CH, IAA, coconut milk and yeast extract on the growth of embryo in the culture of ovaries, flowers and portion of spikes of *Aerva tomentosa*, an obligate apomict. Whereas, in ovary culture only 7 percent of the ovules developed into seeds with poorly differentiated embryos, in flower cultures nearly 15 percent of the seeds bore mature embryos. The addition of yeast extract, casein hydrolysate or coconut milk increased the percentage of seed set only slightly. The best response was, however, elicited when portions of spike were cultured. In the presence of IAA and yeast extract the seed-set was comparable to that in nature, and the growth of the embryo and endosperm was quite normal.

Guha and Johri (1966) studied the minimum requirement of developing fruits of *Allium cepa*. The maximum growth of fruit occurred on Nitsch medium supplemented with GA, IAA and KN while only a few viable seeds were produced on the basal medium. The addition of IAA or Tryptophan increased seed-set to 20 and 30 percent, respectively.

Ovary culture has been successfully employed to obtain hybrids of diploid *Brassica chinensis* and autotetraploid *B. pekinensis* (Inomata, 1968), which are normally cross-incompatible. The ovaries excised 4 days after pollination were cultured on Nitsch medium supplemented with various growth substances, such as casein hydrolysate, Gibberellic acid, IAA, kinetin, yeast extract, tomato juice and coconut milk. Hybrid seeds were produced on a medium supplemented with yeast extract and casein hydrolysate, The triploid hybrid plants thus obtained showed intermediate characters of their parents.

FLOWER CULTURE

The technique of culturing young excised floral primordia affords a potentially useful tool for any study concerned with the control of flower morphogenesis. It permits assessment of the influence of growth regulators and nutrients on buds, without presence of intervening vegetative tissues. Secondly, one can test whether the floral apex behaves autonomously *i.e.,* whether the sequence of organ initiation and subsequent growth is regulated from within the isolated immature bud. Lastly, it facilitates easy handling and allows a variety of surgical manipulations of the excised buds in culture.

The technique of floral bud culture was first initiated by La Rue (1942) who cultured pollinated flowers of *Caltha*, *Kalanchoe* and *Lycopersicon*. Galun *et al.* (1962, 1963) achieved success in culturing detached floral buds of *Cucumis sativus*. All the floral buds, if left *in situ* would have developed into male flowers, depending on the nodes on which they were borne. However, the excised buds were induced into female flowers on modified White's medium supplemented with IAA. Gibberellic acid counteracted this effect, but had no stimulatory effect on its own.

Topfer *et al.* (1963, 1966) grew the floral buds of *Aquilegia formosa* to maturity. Blake (1966) first reported complete development of the excised buds into flowers in *Viscaria candida* and *V. cardinalis*. In these species excised floral apices were raised on MS medium supplemented with coconut milk, casein hydrolysate, inositol, amino acids, GA and IAA.

In *Begonia francois*, a monoecious species, when young floral primordia were cultured on a medium containing BA, normal flower buds developed. Gibberellins promoted the initiation of floral organs.

In *Ranunculus scleratus* (Konar and Nataraja, 1964, 1969) the buds comprising the primordia of various organs failed to complete normal development on any of the nutrient media tested, but they all formed callus which subsequently differentiated roots, shoots and embryoids leading to the formation of plantlets. The cut end of undifferentiated flower buds proliferated rapidly resulting in a mass of white, friable callus. Optimal callus formation was observed on White's medium containing sucrose 2%, coconut milk 10%, 500 ppm casein hydrolysate and 1 ppm 2,4-D.

EMBRYO CULTURE

The first systematic attempt to culture embryos of angiosperms was made by Hannig (1904). He cultured mature embryos of *Raphanus caudatus*, *R. landra*, *R. sativus* and *Cochlearia danica* (Family: Brassicaceae) in mineral salts medium supplemented with sugar, and obtained transplantable seedlings. While working with embryos from a wide range of families and genera Dietrich (1924) made two generalizations (a) the embryos, grown *in vitro*, usually skipped a rest period and germinated, (b) the embryos taken into culture at less than one third of mature size were unable to produce viable plants capable of reaching maturity. Embryos from immature seeds tended to form malformed seedlings. Later on many workers obtained plants by culturing excised embryos. Laibach (1925, 1929) cultured excised embryos from seeds of an interspecific cross *Linum perenne* X *L. austrianum* and succeeded in raising hybrid plants. The seeds formed after crossing shrivelled, became very light, incapable of germination. Laibach excised embryos from such seeds and cultured them on a moist filter paper or on cotton wadding containing glucose or sucrose and was able to raise the hybrids.

Embryo culture can be categorised into two types – culture of immature embryos and culture of mature embryos. Immature embryos originating from unripe seeds and hybrid seeds fail to germinate. Excising such embryos is difficult and they require a complex nutrient medium. Mature embryos are excised and cultured to avoid inhibition in the seed for germination and they require simple nutrient medium.

Generally, the ovule or ovaries are surface sterilized and the embryo is excised under aseptic environment. Media constituents of *in vitro* growth of young or immature embryos differ from those of mature embryos. This, often necessitates, the transfer of embryos from one medium to another in order to achieve their full development.

While carrying out *in vitro* culture studies of embryos in *Datura* van Overbeek *et al.* (1941, 1942) has made the following observations: (a) mature embryos form normal seedlings on mineral salts and dextrose, (b) young torpedo-shaped embryos (up to 0.5 mm) require a medium containing minerals, dextrose, glycine, thaiamine, ascorbic acid, nicotinic acid, vitamin B_{12}, adenine, succinic acid and pantothenic acid, and (c) embryos smaller than 0.5mm in length could be cultured successfully only upon the addition of coconut milk to the medium required by torpedo-shaped embryos.

Addition of coconut milk in the medium has proved to be beneficial for the development of embryo. In barley, undifferentiated embryos were successfully cultured by supplementing the medium with coconut milk 20%.

Osmotic value of the nutrient medium seems to be an important factor for the proper growth of embryo. Isolated mature embryos grow fairly well with sucrose 2% in the culture medium, but younger embryos require higher levels of carbohydrate for normal embryo development. Raghavan and Torrey (1969) had noted a stimulatory effect of high concentration (12-18%) of sucrose on *in vitro* development of excised globular embryos of *Capsella bursa-pastoris*.

Embryo-Nurse Endosperm Transplant (Embryo Rescue)

One of the main problems of hybridization is rescuing the hybrid embryo which normally aborts at a very early stage of development. Kreuse (1974) implanted immature embryos arising in a cross *Hordeum* X *Secale* on cultured barley endosperm and recorded significant improvement.

De Latour *et al.* (1978), Williams and de Latour (1980) described embryo rescue method (Fig. 15.3). They inserted an excised hybrid embryo into cellular endosperm dissected from a normally developing ovule of one of the parents or a third species and cultured the nurse endosperm with the transplanted embryo on the nutrient medium. They obtained interspecific and intergeneric hybrid plants of forage legumes – *Trifolium*, *Lotus* and *Ornithopus* – which could not be reared by culturing embryos directly on the medium.

Practical Applications

Obtaining hybrids

The culture of excised embryos can be advantageously exploited by the plant breeder. Very often interspecific crosses fail due to arrested development of embryo, either due to abortion of embryo at an early stage, or due to degradation of endosperm. A number of hybrids have been successfully raised through embryo culture. These include *Abelmoschus esculentus* X *A. manihot*, *A. esculentus* X *A. moschatus*, *A. tuberculatus* X *A. moschatus* (Gadwal *et al.*, 1968), *Hordeum vulgare* X *Secale cereale* (Kreuse, 1974), *Hordeum vulgare* X *Agropyron repens*, *Hordeum vulgare* X *Triticum* sp., *Lycopersicon esculentum* X *L. peruvianum*, *Brassica chinensis* X *B. pekinensis*.

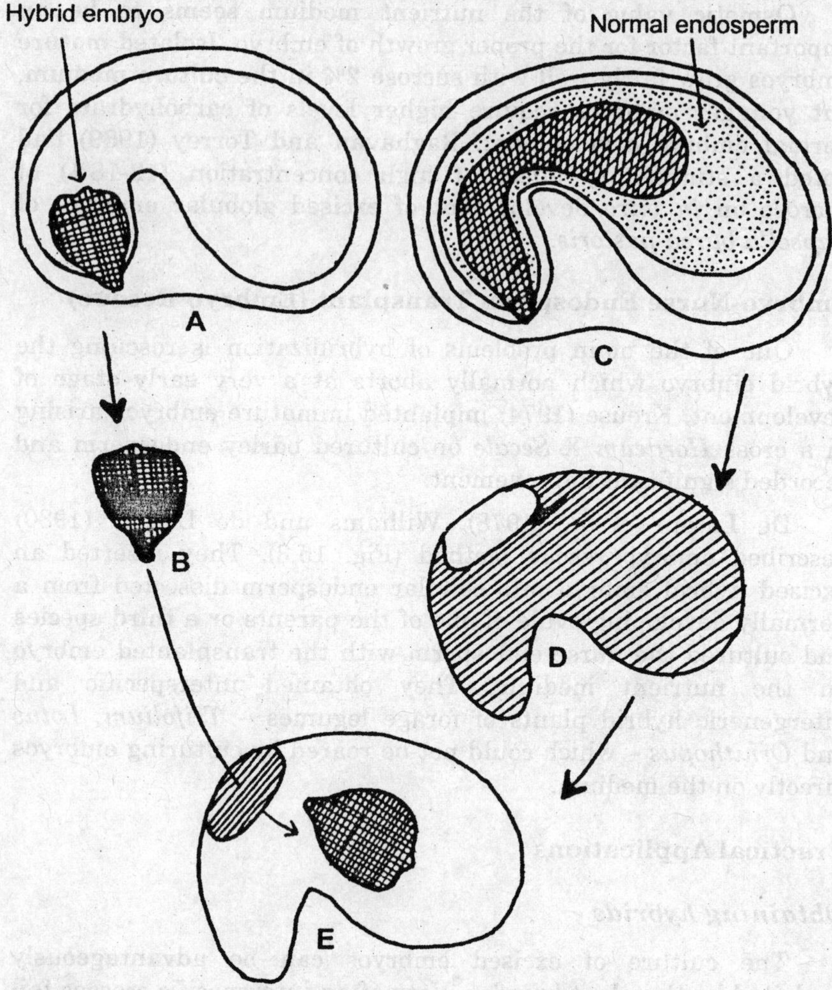

Fig. 15.3. Endosperm transplant for culture of hybrid embryos (after Williams and De Latour, 1989)

Overcoming Dormancy and Shortening Breeding Cycle

In horticultural crops breeding work is delayed due to long periods of dormancy in seeds. In these plants breeding cycle can be shortened using embryo culture technique. For example, in *Iris* Randolph and Cox (1943) could shorten the life cycle from 2-3 years to less than one year by using this technique. Similarly, in

Rosa it is possible to obtain two generations of flowering by culturing embryos. Overcoming seed dormancy and shortening of their breeding cycles have also been reported in weeping crabapple (*Malus* sp.), hollies (*Ilex*), American basswood (*Tilia americana*) etc.

Overcoming Seed Sterility

In early ripening fruit cultivar, seeds do not germinate because their embryos are still immature. Using the embryo culture method it is possible to raise seedlings from sterile seeds of early ripening stone fruits (sweet cherry), peach, apricot and plum.

"Makapuno" coconuts develop soft, solid, fatty endosperm in place of liquid endosperm. Such nuts are very rare and expensive. Under normal conditions such coconut seeds fail to germinate. DeGuzman *et al.* (1971) excised such embryos and cultured them and raised plantlets. 85 percent nuts borne by these plants were Makapuno.

Similar procedure can be applied for *Musa balbisiana, Maranta leuconeura, Colocasia esculenta* and *Pinus armandii X P. koriensis*.

Production of Monoploids

Embryo culture has been used in the production of monoploids (haploids) of barley. With the cross *Hordeum vulgare* X *H. bulbosum* fertilization occurs normally but thereafter chromosomes of *H. bulbosum* are eliminated resulting in a monoploid *H.vulgare* embryo. Such monoploid embryo can be rescued through embryo culture (Kashy and Kao, 1970). Following the bulbosum technique, monoploids were obtained from a cross *Agropyron tsukushiense* X *Hordeum vulgare* (see Kott and Kasha, 1985).

Rapid seed viability Test

The possibility of breaking seed dormancy by embryo culture also allows the use of this technique as a rapid test for seed viability. Barton (1961) has suggested that germination of excised embryos is reliable and capable of more exact interpretation than the commonly used staining tests for seed viability.

Clonal Micropropagation

The regenerative potential is an essential prerequisite in non-conventional methods of plant genetic manipulation. Due to their juvenile nature, embryos have a high potential for regeneration and hence may be used for *in vitro* clonal propag-ation. This is essentially true of grasses. Both organogenesis and somatic embryogenesis have been induced in major cereals and forage grasses from embryonic tissues. Generally callus derived from immature embryos of cereals has the desired morphological potential for regeneration and clonal propagation.

ENDOSPERM CULTURE

Lampe and Mills (1933) made the first attempt to culture endosperm. They grew the immature endosperm tissue of maize on a medium containing extract of potato or young corn kernels and observed slight proliferation of endosperm layers around the embryo. La Rue (1949), for the first time, succeeded in establishing a continuously growing tissue from immature endosperm of maize. Subsequently, tissue cultures were established from the immature endosperm of *Asimina triloba* (Lampton, 1952), *Lolium perenne* (Norstog, 1956) and *Cucumis sativus* (Nakajima, 1962). In all these cases endosperm grows in cultures only when excised 7-10 days after pollination. There are no reports on proliferation of free nuclear endosperm. Nakano *et al.* (1975) pointed out that in rice only the cellular endosperm proliferated. In maize, the endosperm becomes cellular 3-5 days after pollination and for 8 days the cells remain meristematic. Later, the mitoses are restricted to only the outermost layer.

In most cereals, the mature endosperm does not respond to cultural conditions. Even endosperm of some dicots (*Cucumis sativus*) exhibits similar behaviour.

In contrast to this mature endosperm of Euphorbiaceae, Santalaceae and Loranthaceae responds favourably to culture treatments. Association of embryo tissue in the initial stages seems essential for inducing proliferation of mature endosperm in members of Euphorbiaceae and Santalaceae. The embryo can be removed as soon as proliferation of endosperm has started. Presumably the embryo releases some factor(s) required to activate the cells to divide. In some species such as *Croton bonplandianum* and

Putranjiva roxburghii the embryo factor(s) may be replaced by Gibberellic acid.

The basal medium of White or MS is generally used in endosperm culture. To induce callus from immature endosperm the basal medium has been supplemented with tomato juice, yeast extract, grape juice and maize grain juice. For *Lolium* endosperm only yeast extract was effective.

The maiden report of cellular totipotency of endosperm tissue was that of Johri and Bhojwani (1965) in *Exocarpus cupressiformis*. They observed the development of eight shoot buds in a single explant (seed) cultured on White's medium supplemented with IAA, Kinetin and casein hydrolysate. These buds, which are triploid, originated from the peripheral cells of the endosperm and possessed distinct shoot apices and well differentiated vasculature. Since then organogenesis has been induced in the endosperm cultures of apple (Mu *et al.*, 1971), *Citrus grandis* (Wang and Chang, 1978), sandalwood (Lakshmi Sita *et al.*, 1980), rice (Nakano *et al.*, 1975), *Putranjiva roxburghii* (Srivastava, 1973), *Codiaeum variegatum* (Chikkannaiah and Gayathri, 1974), *Annona squamosa* (Nair *et al.*, 1986), Maize (Zhu *et al.*, 1988) and *Hordeum vulgare* (Sun and Chu, 1981).

Plantlet formation from endosperm callus may follow the embryogenic mode of development. Embryo differentiation occurs when the proliferated tissue is transferred from a callusing medium to a basal medium with or without Gibberellin.

Application of Triploids in Plant improvement

The technique of endosperm culture enabled the production of triploid plants. Triploid plants are self-sterile and usually seedless. This character increases the edibility of fruits and is desirable in such plants as apple, banana, grape, mango, watermelon etc., for their edible fruits. Seed sterility in timber, fuel or latex yielding plants is not a serious setback because their vegetative parts are important and these plants can be multiplied vegetatively. In such plants, triploids show better performance than their relative diploids or tetraploids. For example, the triploids of *Populus tremuloides* have better quality pulp wood which is important to the forest industry.

PROTOPLAST CULTURE AND SOMATIC HYBRIDIZATION

Plant cells without cell walls are referred to as protoplasts. The non-conventional genetic procedure involving fusion between isolated somatic protoplasts under *in vitro* conditions and subsequent development of their product to a hybrid plant is known as *Somatic hybridization* or *parasexual hybridization.*

Isolation of Protoplasts

The somatic cells of higher plants are bound by a rigid cellulose cell wall, and the adjacent cells are cemented together by a pectin rich matrix. Protoplasts from such cells can be isolated by enzymatic method. Pectinases and cellulases are used for large-scale isolation of protoplasts. These enzymes can be used: (1) sequentially (two-step) or (2) simultaneously (one-step). The concentration of the enzymes and the duration varies with the tissue. Cocking (1960) for the first time demonstrated the large scale isolation of protoplasts from root tips of tomato (*Lycopersicon esculentum*) by using concentrated solution of cellulase enzyme (prepared from the cultures of the fungus *Myrothecium verrucaria*). Further progress in enzymatic isolation of protoplasts was achieved as soon as cellulase and macerozyme (pectinase) enzymes became available commercially in 1968. Takebe *et al.* (1968) employed sequential or two-step procedure for isolating mesophyll protoplasts in tobacco using commercial preparation of enzymes.

Essentially the technique involves the following steps (1) surface sterilization of leaves, (2) peeling of lower epidermis, (3) pre-enzymatic treatment, (4) incubation of stripped leaf tissue in an enzymatic solution and (5) isolation of protoplasts.

Fully expanded leaves are excised from 7-8 week old plants growing in a green house. The leaves are surface-sterilized by first immersing in ethanol (70%) for 30 seconds followed by rinsing in sodium hypochlorite solution 0.4-0.5% for about 30 minutes. The leaves are then repeatedly washed in sterile distilled water. The lower epidermis of the sterilized leaves is carefully peeled with the help of fine foreceps and the stripped leaves are cut into small pieces of 4cm^2. Peeling the epidermis would be relatively easy if the sterilized leaves were allowed to become flaccid, or if the water supply to the plants is limited before the leaves are excised. From the peeled leaf tissues protoplasts are isolated by two-step enzymatic treatment. First stripped leaf pieces are placed in 20 ml enzyme medium (0.5% macerozyme, 0.8 M D-mannitol and 0.3% of potassium dextran sulphate at pH 5.8) in 100 ml flasks and gently shaken in a water bath at 25°C by a reciprocal shaker. Fresh medium is replenished at 30 min intervals for 2 hours. The mesophyll cells thus isolated by maceration are washed repeatedly with 0.8 M mannitol containing 0.1M $CaCl_2$ and suspended in 2 percent solution of cellulase (prepared from fungus *Trichoderma viride*) in 0.8 M mannitol at pH 5.4. The suspension is incubated at 36°C for 2-3 hours with occasional shaking. The cellulase digests the cell walls resulting in the release of protoplasts.

Power and Cocking (1968) mixed two enzymes together (simultaneous procedure) and isolated protoplasts in one-step. The peeled leaf tissues were placed directly in an enzymatic mixture containing macerozyme (0.5%) and cellulase (2%) in 0.7M sorbitol or mannitol solution. About 1 g of leaf material was added to 10 ml of the enzyme mixture (pH 5.4) incubated at 25°C for 4-6 hours in darkness. The leaf pieces are gently squeezed with a pasteure pipette to liberate protoplasts. The leaf debris is removed by passing through a 60-80 μM mesh sieve. The filtrate is transferred to tubes and centrifuged at least thrice at 100 g for 1 min. Each time the protoplasts sink to the bottom, the supernatant enzyme and the debris is decanted. The protoplasts are washed with 0.7 M sorbitol and finally sorbitol is replaced by sucrose solution (20%) and centrifuged at 100g for 10 min. Green protoplasts form a band at the top of sucrose solution. They are transferred with a pipette to another tube. Protoplast culture medium is added and centrifuged at 100 g for 3 min. Such washings are repeated 3 times. The cleaned protoplasts float, they are gently pipetted out with Pasteur pipettes and resuspended in culture medium.

Glusulase, a commercially available snail enzyme preparation, has also been used for the isolation of protoplasts from a number of leaf and root tissues (Pinto de Silva, 1969).

Culture of Protoplasts

Isolated protoplasts are cultured either in liquid medium or agar medium. The droplets of protoplasts suspension are placed in a liquid medium in petri dishes. It in turn is covered by another petri plate and sealed with paraffin. The culture dish is then maintained at 25-28°C.

For culturing the protoplasts on agar medium Bergman's cell plating technique may be followed. 2 ml aliquot of protoplasts is suspended in liquid culture medium and mixed with an equal volume of same medium containing agar (1%), the temperature of which should not exceed 45°C. On solidification of agar, the culture plates are sealed and maintained in an inverted position at 25-28°C. With this method, individual protoplasts or heterokaryons can be conveniently observed under microscope and plating efficiency readily determined. Protoplast concentration is critical for division. In liquid medium the minimum concentration neces-sary is between 10^4 and 10^5 protoplasts per ml, whereas for agar plating it is between 10^3 and 10^4 per ml.

The basic nutritional requirements for culture of protoplasts are more or less similar to those of cell culture. Mostly the salts of MS and B_5 media with some modifications have been found suitable. Nagata and Takebe (1971), for the first time, regenerated entire plantlets from mesophyll protoplasts of *Nicotiana tabacum*. Cell wall formation in cultured protoplasts usually takes place once the enzymes are removed. The cell undergoes divisions forming callus. The callus may be induced to undergo organogenic differentiation or whole plant regeneration.

Since the first report of plant regeneration from protoplasts, the list of species exhibiting this potentiality has increased. Initially, the majority of taxa which yielded protoplasts capable of regenerating callus or whole plants belonged to Solanaceae, but subsequent advances in protoplast technology resulted in differen-tiation of shoots of plants from isolated protoplasts of several other families including forage legumes, fibre or pulp crops, various tree species and even cereals, such as *Pennisetum*, rice and wheat.

Protoplast fusion and Somatic hybridization

The most promising application of protoplast culture is protoplast fusion. Power *et al.* (1970) was the first to induce fusion under the influence of sodium nitrate. Compared to low frequency fusion by $NaNO_3$, the polyethylene glycol (PEG) induced fusion gives better results. Induced fusion is influenced by an alkaline pH and high calcium concentration. The process of fusion is considerably enhanced if the protoplasts suspension is pre-treated with a membrane activating agent such as lysozyme.

Fusion product of protoplasts of same species is known as homokaryon, while the product of fusion of protoplasts of different species is known as heterokaryon.

The first report of successful production of a mature interspecific hybrid plant by fusion of protoplasts, is by Carlson *et al.* (1972) between *Nicotiana glauca* + *N. langsdorfii*. Power *et al.* (1976) obtained interspecific somatic hybrids of *Petunia hybrida* + *P. parodii*.

An outstanding example of intergeneric fusion of protoplasts is the work of Melchers *et al.* (1978). They reported regeneration of somatic hybrid plants of potato and tomato. Mesophyll protoplasts of *Lycoperson esculentum* and protoplasts obtained from liquid callus culture of *Solanum tuberosum* were fused in the presence of PEG with a high calcium ion concentration. The fused protoplasts divided and developed into callus masses some of which showed regeneration of green shoots. The shoots were either transferred to soil or grafted on to a tomato stock.

Sexual incompatibility barriers are the impediments in plant breeding programmes. These barriers may be overcome by somatic hybridization. Using protoplast fusion several interspecific and intergeneric hybrids have been obtained.

Further Readings

Bhojwani, S.S. and M.K. Razdan 1983. *Plant tissue Culture: Theory and Practice*. Elsevier Scientific Publication. Amsterdam.

Johri, B.M. 1982. *Experimental Embryology of Vascular Plants*. Springer-Verlag.

Pullaiah, T., M.V. Subba Ro and E. Sreedevi. 2017. *Plant Tissue Culture*. Scieentific Publishers, Jodhpur.

Raghavan, V. 1976. *Experimental embryogenesis in Vascular plants*. Academic Press, London.

Razdan, M.K. 2003. *An Introduction to Plant Tissue Culture*. Oxford & IBH, New Delhi.

Chapter – XVI

PRACTICALS

Practical 1: Preparation of Pollen grains by Acetolysis method

For studying their external morphology, the pollen grains are acetolyzed. Erdtman's (1960) acetolysis technique is the most popular method, which is followed almost throughout the world. The different steps involved in this technique are given below:

(1) Collection and preservation of pollen material: It involves collection of anthers from mature flower buds just before opening and anthesis of flowers by using clean forceps and preserving the anthers or sometimes anthers with a portion of filament of stamens in clean glass vials with glacial 70% acetic acid. Some palynologists recommend ethyl 70% alcohol for preserving the anthers in place of glacial acetic acid. In this condition, the anthers packed with pollen can be stored indefinitely.

(2) Anthers are transferred to a centrifuge tube and washed thoroughly with distilled water. The anthers are crushed with a glass rod and the mixture is filtered through cheesecloth or a sieve (c. 180 μm mesh). The lower fraction is recollected in the centrifuge tube.

(3) Freshly prepared acetolysis mixture consisting of acetic anhydride and concentrated H_2SO_4 (sulphuric acid) in the ratio of 9:1 was poured in the centrifuge tube containing washed pollen. Care has to be taken to add conc. H_2SO_4 to acetic anhydride drop by drop to avoid excess heating. The acetolysis solution reacts violently with water. Glacial acetic acid washes are necessary to replace water with acetic acid, which does not react violently with the acetolysis solution. Although *acetolysis*

procedure varies from palynologist to palynologist, the following procedure seems to be most suitable. From the first to last acetic acid wash, caution has to be taken so that no water comes in contact with the solution. Only distilled water has to be used for washes to avoid any contamination.

(4) The centrifuge tube containing pollen in suspension of an acetolysis mixture is placed in a water bath and heated at 70°C for 5-10 minutes, stirring the contents with a glass rod intermittently. The time required for acetylation varies from material to material. Pollen after completion of acetolysis process turns light to golden brown in colour.

(5) Acetolyzed pollen suspension is centrifuged and supernatant liquid is decanted.

(6) Pollen sediment in a centrifuge tube is immersed in glacial acetic acid for a few minutes and the superfluous liquid is decanted after 5 minutes centrifugation (c. × 2000 rpm). This is washed with distilled water 2-3 times.

(7) The pollen material is ready for mounting on the slides for microscopic examination. However, it can be stored in this condition in vials containing glycerine (50%).

(8) Acetolyzed pollen material is mounted in either glycerine jelly, polyvinyl alcohol, lactophenol, Canada balsam, corn syrup, or synthetic resin: piccolyte. The choice of the mounting medium depends on the individual palynologist. It is proved beyond doubt that molten glycerine jelly prepared by using the following method is the most ideal one. Pollen mounted in glycerine jelly remains unaffected even after 4-5 decades. Water solubility and viscosity of glycerine jelly makes it an ideal mounting material for pollen.

(9) After mounting the cover glass on acetolyzed pollen in glycerine jelly, edges of the cover glass are sealed with paraffin wax or nail polish to avoid desiccation.

As mentioned earlier, the assets of acetolysis of pollen grains include bringing transparency to and expansion of pollen to facilitate examination of pollen morphological characters.

However, there are some disadvantages of this technique. The pollen grains sometimes get over expanded and become unfit for original size measurements. It has been reported that fossil pollen and modern pollen when subjected to acetolysis expand to different degrees. Hence, it is advisable to use unacetolyzed pollen or spores when size determinations are critical.

Sometimes discolouration of pollen is required, which is achieved by chlorination of sample as follows:

Add 5 cc of glacial acetic acid to a centrifuge tube containing pollen material and stir the contents. Then add 1-3 drops of saturated solution of sodium chlorate with a 0.5 cc pipette. Follow this with 1-2 drops of conc. HCl. The solution changes colour in 1-2 minutes as a result of liberation of chlorine gas. Centrifuge sample, decant chlorination mixture, add water, transfer to the acetolysed fraction, stir and follow the other regular actolysis procedure.

Prepration of Glycerine Jelly

Kisser's method of preparation of glycerine jelly is usually followed by palynologists. The following ingredients are required for glycerine jelly preparation:

Gelatin 50 gms

Distilled water 175 ml

Glycerine 150 ml

Phenol crystals 7 gms

Gelatin is first soaked in 175 ml cold water for 2-3 hours. Superfluous water is discarded and gelatine is heated till it melts. 150 ml of glycerine is added to molten gelatin and while still hot it is filtered through spun glass pressed into the lower part of the heated funnel. Glycerine jelly has its natural light golden yellow to brown colour. It is stored in Petri dishes or a suitable glass container in a refrigerator.

Practical 2: Methods for Pollen Germination

A large numbers of pollens have been successfully germinated under laboratory conditions on relatively simple media. The composition of a germination medium to obtain optimal responses, has to empirically formulate for each species. Brewbaker and Kwack (1963) medium found to be suitable for most species.

Ingredients	Brewbaker-Kwack Medium	Roberts Medium	Hodgkin and Lyon's Medium
Sucrose (5 - 20%)	Adjust according to species		
Boric Acid	100 mg/l	10 mg/l	100 mg/l
Calcium Nitrate	300 mg/l	362 mg/l	400 mg/l
Potassium nitrate	100 mg/l	100 mg/l	100 mg/l
Tris	–	60-130 mg/l	–
Manganese sulfate	–	–	4.86 g/l
TAPS	–	–	200 mg/l

One must keep in mind that time and method of collection and storage age of pollen affect viability (Shivanna and Johri 1989, Shivanna 1993). Pre-germination relative humidity probably affects the internal solute potential and wall properties of the pollen grain and has a subsequent effect on pollen germination (Shi and Yang 2010). Any work requiring maximally viable pollen should be conducted shortly after anther dehiscence. Germination is most successful immediately after anthesis and viability deteriorates rapidly in most species. Several methods of *in vitro* raising pollen cultures are available. The selection of a method depends on requirement of the study and facility. The methods are as follows (Shivanna and Rangaswamy 1992).

Hanging Drop Culture

The hanging drop culture method has been most commonly used in the past. Essentially, it involves suspending the pollen grains in a drop of nutrient medium (on a cover glass) hanging over a shallow depression. To prevent evaporation of the culture medium, the hanging drop culture is sealed suitably.

Sitting Drop Culture

The sitting drop culture method is simpler than the hanging drop method. It involves culturing of pollen grains in a drop of culture medium placed on a microslide. The culture is then maintained in a humid chamber to prevent evaporation.

Suspension Culture

In large samples of pollen grains in 2 to 10-ml culture medium in suitable vials incubated on a shaker and observed for germination by placing drop in slide.

Practical 3: Clearing technique for the study of Embryo sacs

Herr (1971) described a method of clearing technique to study the embryo sac. The simple, efficient method described here for the study of ovule and megagametophyte development in angiosperms provides for the extension of investigation beyond the limits imposed by the traditional but arduous section technique. Excised pistils previously fixed in FPA50 and stored in ethanol 70% are placed in a clearing fluid composed of lactic acid (85%), chloral hydrate, phenol, clove oil, and xylene (2:2:2:2:1, by weight). After 24 hr, ovules dissected from the ovlllaries are transferred with some of the fluid to a slide, covered so that the cover glass is supported laterally by two permanently affixed covers, and examined with phase contrast optics. The unique action of the clearing fluid permits the study of cellular structure with the phase off objective focused at any focal plane within the ovule. Downward focusing thus reveals a series of optical sections in the sagittal, frontal, or transverse plane depending on the orientation of the ovule. Orientation can be altered by a slight shifting of the cover glass on the lateral support mounts. The ovules become quite fragile in the clearing fluid. Pressure applied to the cover glass gradually breaks the ovule apart without disrupting the structural integrity of individual cells. This squash procedure provides for extending observations to cytological features of megasporocytes, megaspores, and megagametophytes previously identified in intact ovules. For best results, the ovules of *Ludwigia* must be pretreated in lactic acid (85%) for 24 hr prior to application of the clearing fluid. Other methods for pretreatment likely will be required as the technique is applied to a wider range of flowering plant species.

Young *et al.* (1979) described a protocol for the study of embryo sacs. The ovules are fixed in FAA (95% ethanol: water : 40% formalin : glacial acetic acid, 40:14:3:3 v/v) for 24 hours and stored in ethanol 50%. For clearing, the fixed ovules are passed through a dehydration and clearing series in the following sequence: ethanol 70%, ethanol 85%, ethanol 100% (three changes), 1:1 ethanol-methyl salicylate (MESA), 1:3 ethanol-MESA and MESA 100% (two changes). The duration of each treatment may vary with the material and the number of ovules. For individual ovules of *Cenchrus ciliaris,* 1 ml of liquid and 30 min at each step were found suitable, but for processing 10-20 ovules together 2 ml of liquid and 2 hours of treatment at each step are recommended. The cleared material is stored in MESA 100% of mounted in it and observed with phase contrast microscope.

Practical 4: Isolation of Embryo sac.

Isolation of live egg and other cells of the female gametophyte involves enzymatic treatment of ovules followed by micro-dissection (Kranz-Lorz, 1993). The ovules are excised from the placenta and immediately placed in the enzymatic solution containing pectinase (0.75%), pectolyase 423 (0.25%), hemi-cellulase (0.5%) and cellulase onozuka RS (0.5%) with osmolarity set at 540 mosmol kg^{-1} using mannitol and pH 5. After 40-60 minutes of incubation at 24 ± 0.5°C the ovules are treated with micro-glass needles to liberate the gametic cells. The cells are manually lifted with microcapillary (200 μm opening) and transferred to clean mannitol solution of the same osmolarity as the enzyme solution and used to study their structure or for *in vitro* fertilization. Kranz *et al.* (1991) could isolate 5 intact egg eclls from 20 ovules and up to 40 eggs in 2-3 hours. The freshly isolated egg and synergid cells are like sperm cells, protoplasts.

SUBJECT AND PLANT INDEX

Aa 56
Abelmoschus 292, 297
Abiotic Pollination 118
Acacia 56, 126
Acalypha 69, 70, 71, 90, 91
Acanthaceae 11, 60, 70, 127, 191, 237
Acanthus 129, 191, 237
Adhatoda 130, 191
Adoxa 89, 96, 97
Adoxaceae 9
Aegle 70, 72, 230, 290
Aegopodium 178
Aeration 276, 277
Aerva 294
Agrostemma 175, 290
Aizoaceae 60, 62, 73, 96, 206, 226
Albizzia 56
Alchemilla 3, 244, 253
Alectra 14
Aleurone Tissue 201
Alisma 3, 53, 120
Alismataceae 15, 54, 59, 87, 88
Allium 86, 87, 88, 96, 97, 100, 108, 198, 200, 288, 292, 294
Alnus 45, 49, 252
Alpinia 133
Althaea 136, 293
Amaranthaceae 15, 59, 60, 73, 206, 226
Amaranthus 50
Amaryllidaceae 54, 62, 68, 70, 74, 87, 218
Amaryllis 134
Amentiferae 3, 51, 118, 139
Ammobium 87
Amoeboid tapetum 17
Amphitropous ovule 59
Anacardiaceae 10, 74, 198, 230
Androgenesis 253, 279, 281
Andrographis 191, 196
Anemone 72 182, 207
Anemophily 118
Annona 159, 197, 301
Annonaceae 11, 22, 56, 72, 196, 206
Antennaria 3, 246, 248, 252
Anthephora 246
Anther culture 6, 283, 286
 tapetum 10, 11, 12
 wall 6, 10, 12, 13
Antipodal haustoria 105, 107
Antipodals 104, 105, 245
Antirrhinum 7, 127, 291
Ants 124
Apertural patterns 52
Aphanes 244

Apiaceae 11, 68
Apomixis 242
causes 254
Apospory 243, 261
Aquilegia 104, 116, 295
Arabidopsis 140, 141, 142, 143, 158, 285
Arabis 235, 248, 251
Araceae 9, 68, 123
Arceuthobium 9
Ardisia 63
Areca 187
Arecaceae 186
Argemone 50, 108, 174, 198, 200, 235
Aril 64
Arillode 64
Aristolochia 24, 75, 124, 159, 234
Aristolochiaceae 70, 124, 196
Arnica 21, 247
Artemisia 106, 243
Arum 14, 17, 18, 201
Asclepiadaceae 9, 11, 24, 26, 52, 56, 69, 72, 124
Asclepias 69
Aseptic conditions 276, 277
Asimina 300
Asparagus 239, 288
Asphodelus 194, 195, 198, 200
Aster 105
Asteraceae 9, 11, 15, 17, 28, 52, 54, 59, 60, 65, 72, 105, 108, 122, 150, 163, 167, 204, 206, 218, 243, 244, 247, 248
Asterad type 206, 209, 275
Asystasia 191
Atkinsonia 113
Atraphaxis 237
Atropa 178, 283, 285
Axenic 279
Azadirachta 50, 234, 235
Balanites 65
Balanophora 248
Balsaminaceae 22, 206
Barleria 191
Basal medium 276, 279
Basella 9, 21, 210, 211, 272
Basellaceae 11, 206, 271, 272
Bassia 63
Bats 130
Bauhinia 54, 130
Bees 126
Begonia 295
Begoniaceae 62, 68
Berberis 34
Beta 27, 163, 244, 257, 288
Betula 3
Bidens 106, 198, 200
Bignoniaceae 11
Biological Significance 171
Biotic pollination 122
Birds 129
Bisporic embryo sacs 86
Blumea 9
Boerhavia 9, 21
Borassus 49
Bothriochloa 246, 253, 255
Bouteloua 246
Brachiaria 24
Brachycome 247
Brachysiphon 91
Brassica 29, 30, 49, 136, 164, 173, 175, 176, 177, 198, 200, 283, 285, 286, 287, 294, 297

Brassicaceae 11, 15, 28, 52, 163, 167, 169, 170, 182, 206, 218, 222, 295
Breeding behaviour 260
Bromhedia 56
Bromus 258
Brunella 14
Bryonia 222
Burmannia 57, 247, 252
Butomopsis 53
Butomus 53

Cactaceae 59, 60, 63, 76, 135, 230, 232
Cadaba 128
Caesalpinia 128
Calamogrostis 254
Calanthe 56, 115, 291
Calliandra 56, 186
Callus 279, 287
Caltha 53, 152, 295
Campanulaceae 206
Campylotropous ovule 59
Cannaceae 54, 226
Canscora 13, 275
Cantherophily 122
Capillipedium 237, 246, 253
Capparaceae 59, 60
Capparis 13, 127
Capsella 3, 99, 104, 152, 188, 208, 219, 220, 227, 228, 297
Capsicum 96, 239, 282, 283
Caralluma 69
Cardiospermum 64
Carex 40
Carica 160, 188
Carthamus 234, 235
Caruncle 68, 69
Caryophyllaceae 175, 206
Caryophyllad type 206, 211
Cassia 130, 186
Casuarina 3, 50, 51, 76, 119, 139, 237
Celastraceae 9, 68
Cenchrus 233, 237, 238, 255, 311
Central cell 110
Cephalanthera 56
Ceratocephalus 70
Ceratostigma 92
Cercidiphyllum 188
Chaenostoma 237
Chalazogamy 139, 140
Chalazosperm 227
Chenopodiaceae 11, 59, 60, 73, 206, 272
Chenopodiad type 206, 210
Chiropterophily 130
Chloris 181
Chondrilla 248, 253
Chrozophora 70, 73, 123
Chrysanthemum 95, 96
Cicer 54
Circinotropous ovule 59, 61
Cistaceae 60, 63
Citrullus 183
Citrus 3, 7, 49, 229, 230, 237, 240, 289, 290, 291, 301
Clerodendron 128
Clintonia 93
Clonal propagation 279
Coccinia 184
Coccoloba 196, 197
Cochlearia 6, 295
Cochlospermum 126, 130

Cocos 73, 127, 186, 225, 237, 238
Codiaeum 301
Coffea 182
Coix, 251
Coleus 14
Coma 69
Composite endosperm 192
Convolvulaceae 22, 72, 198
Cooperia 247, 251
Coreopsis 106, 243
Cornus 93
Cortaderia 100, 116, 246
Cosmos 34, 106, 175
Cotula 100
Crassinucellate 58, 60, 72, 73, 230, 266
Cremophyllum 71
Crepis 165, 234, 243, 253
Crinum 70, 201
Crossandra 191
Crossosoma 64
Crotalaria 183, 186
Crucianella 92
Cruciferae 206
Cryopreservation 154
Cucumis 184, 185, 291, 295, 300
Cucurbita 20, 139, 160, 184
Cucurbitaceae 15, 46, 60, 62, 72, 73, 182, 183, 185, 291
Cuscuta 71
Cyamopsis 186
Cybrid 279
Cymbidium 222
Cynanchum 291
Cyperaceae 5, 11, 39, 40, 74, 118, 186, 264, 268
Cyprepedium 56, 115, 222, 291
cytochemistry 115, 218
Cytodifferentiation 279
Cytokinesis 23, 265
Cytological tests 261
Cytomixis 23

Dactylis 177
Datura 6, 65, 66, 127, 135, 139, 175, 178, 227, 240, 281, 282, 283, 285, 286, 287, 296
Daucus 178
Dedifferentiation 279
Delosperma 96
Dendrobium 115, 222, 291
Dendrophthoe 78, 130, 201
Deschampsia 254
Desmanthus 56
Dianthus 174
Dicanthium 237, 238, 246
Dichogamy 159
Dichrostachys 186
Dicraea 73, 222, 267, 268
Dicranostigma 174
Dioecy 160
Dioscorea 235
Diplospory 246, 250, 262
Dipteracanthus 191, 237
Dipterocarpaceae 13
Distyly 161
Double fertilization 149, 175
Dracaena 133
Drimys 26, 27, 123
Droseraceae 56, 68
Drusa 89, 92, 96

Echinocystis 184, 225
EDP system 44

Egg 102
Eichhornia 161
Elaiosome 70
Elatinaceae 68
Elatostemma 244, 247, 253
Elephantopus 9, 21, 100, 106, 107
Elodea 120
Elytraria 191, 196, 197
Embryo 213
Embryo culture 6, 296, 299
Embryo rescue 280
Embryo sac haustoria 110
Embryogeny 206, 266, 270, 275
Embryoid 279
Enaulus 9
Endosperm 179
    Cellular 188
    Helobial 193
    Nuclear 181
Endosperm Culture 300
Endosperm development 179
Endosperm nodules 188
Endothecium 13
Endothelium 65, 275
Endymion 86, 87, 96, 97
Epidendrum 98, 102, 152, 222
Epigonium 56
Epistase 73
Eragrostiella 233
Eragrostis 246
Eranthis 218
Eremopogon 246
Ericaceae 13, 24, 63, 122
Erigeron 96, 247, 248, 249, 253
Erycibe 198
Erythronium 89, 236, 291
Eucalyptus 50
Eugenia 49
Eulophia 222, 236
Eupatorium 19, 20, 25, 247, 248, 253
Euphorbia 28, 70, 73, 96, 123, 124, 125, 222
Exine 54
Exocarpus 111, 113, 190, 237, 301
Experimental embryology 1, 178

Fabaceae 15, 54, 60, 163, 182, 185, 222
Female gametophyte 1, 78, 81, 82, 88, 89, 96, 115, 116, 132, 140, 200, 229, 280, 311
Fertilization 174, 178, 265
Festuca 258
Filiform apparatus 98, 101, 103, 104, 140, 141, 143
Flies 123
Flower culture 294
Foeniculum 188
Forage Grasses 258
Fragaria 160, 234, 293
Frankenia 182
Fritillaria 3, 89, 92, 93, 94, 97, 132, 179

Gaillardia 93, 94
Galinsoga 182
Gametophytic self-incompatibility 169
Generative cell 27
Gentianaceae 22, 217, 275
Geodorum 236
Gerbera 288
Gladiolus 63, 134

Gloriosa 76, 160
Gmelina 130
Gossypium 63, 69, 99, 102, 135, 205
Grevillea 182, 183
Grindelia 105, 108
Gynandropsis 292
Gynogenesis 280, 287

Habenaria 123, 222, 236
Haemanthus 27, 198, 200
Haplanthus 191
Haplopappus 108
Hartmannia 86
Helianthus 18, 29, 98, 99, 102, 116, 159, 198, 200
Helicanthes 78, 79
Heliconia 54
Helicteres 130
Helixanthera 78, 79, 130
Helleborus 15, 33, 159
Hemigamy 218, 251
Herkogamy 160
Hernandiaceae 198
Heteromorphic Incompatibility 160
Hevea 160, 288
Hibiscus 152, 160
Hieracium 243, 244, 247, 253
Holoptelia 49
Homomorphic Incompatibility 162
Hordeum 133, 152, 177, 201, 219, 282, 283, 288, 297, 299, 301
Hosta 3
Hydrilla 120
Hydrocharitaceae 11, 13, 120, 213
Hydrophily 120
Hymenantherum 188, 189
Hymenocallis 63
Hypericum 244, 251
Hyphaene 25
Hypostase 73

Iberis 7, 198, 200, 293
Identifying apomixis 260
Impatiens 108, 159, 189, 190
In vitro fertilization 141, 311
Incompatibility 159
Indotristicha 73, 88, 268
Integumentary embryos 68
Integumentary polyembryony 234
Iodina 190
Ipomoea 57, 223
Iridaceae 54, 213
Iris 298
Irradiation 172
Ixeris 246, 249, 250, 253

Jaculator 70
Jasione 98, 101, 116
Juglans 3, 51, 160, 237
Juncaceae 24, 56, 118
Juncus 193

Kalanchoe 295
Kingiodendron 198

Lagascea 65, 66
Lagenaria 100, 104, 291
Lamiaceae 14, 52, 57, 59, 60, 127
Lathyrus 173
Lauraceae 9
Leguminosae 3, 183, 185
Leitneriaceae 60
Lemna 68, 69

Lemnaceae 9, 12, 73
Leontodon 243
Leptomeria 109, 110
Leucaena 56, 186
Liliaceae 11, 62, 68, 70, 74, 87, 133, 150, 243
Lilium 3, 34, 63, 93, 108, 132, 134, 135, 168, 176, 179, 235
Limnanthes 181
Linaceae 161, 206
Linaria 115, 116
Linum 6, 105, 133, 219, 238, 240, 296
Liparis 56
Lobelia 57, 120
Lolium 177, 259, 283, 300, 301
Lomatia 182, 184
Loranthaceae 75, 78, 79, 115, 129, 192, 222, 264, 268, 300
Lotus 297
Luffa 185
Lusia 222
Lycopersicon 66, 127, 164, 168, 172, 176, 178, 285, 293, 295, 297, 302
Lycopsis 222
Lysiana 78, 79
Lythrum 161

Macadamia 62, 182
Magnolia 63, 123, 189, 190
Magnoliaceae 62, 64, 206, 270
Mahonia 14, 18
Maize 38, 114, 125, 159, 201, 257, 259, 301
Male gametophyte 1, 4, 9, 21, 28, 36, 37, 164, 200, 229, 279
Male sterility 20, 38
Mallotus 92
Malpighiaceae 59, 60, 91
Malus 244, 251, 290, 299
Malva 136
Mamelon 78
Mammillaria 232
Mangifera 3, 127, 160, 231, 232, 240
Maranta 299
Mathiola 201
Megaspore Haustoria 76
Megasporogenesis 77, 84, 265
Melampodium 67, 234, 245
Melampyrum 191, 201
Melandrium 175
Melittopalynology 49
Melittophily 126
Melothria 185
Mentor pollen 175
Menyanthaceae 275
Mercurialis 1, 119
Mesogamy 139, 140
Michelia 63
Microsporangium 9
Microspore tetrads 24, 266
Microsporogenesis 9
Middle layers 10, 13, 18, 68
Mimosa 56, 186
Mimosaceae 22, 26, 56
Molluginaceae 206
Momordica 184
Monimiaceae 24, 56
Monoecy 160
Monosporic embryo sac 77, 238
Monotropa 3
Moringa 63, 130
Moringaceae 9
Morus 51

Multiple polyembryony 237
Musa 76, 160, 299
Myophily 123
Myriocarpa 70
Myristica 64, 123, 196, 197
Myrmecophily 124

Najadaceae 12
Najas 120, 214, 234
Narcissus 175
Nardus 247, 254
Nelumbo 53
Nelumbonaceae 53
Neptunia 56, 120
Nicotiana 66, 99, 109, 163, 168, 173, 174, 175, 178, 212, 213, 281, 282, 283, 285, 288, 293, 304, 305
Nigella 16
Nothoscordum 3, 200
NPC system 43, 52
Nucellus 72
Nucellus culture 7
Nutrient media 276
Nuytsia 78, 79, 111, 113
Nyctaginaceae 46, 226
Nymphaeaceae 53, 75

Obturator 70
Oenothera 78, 82, 85, 86, 88, 168, 172, 176, 200, 239, 267, 269, 271
Ombrophily 121
Onagraceae 5, 68, 75, 76, 86, 163, 206, 267, 271
Onagrad type 206, 208, 271
Operculum 68
Ophrys 222
Opuntia 61, 232
Orbicules 16
Orchidaceae 11, 76, 124, 133, 179, 203, 217, 218, 222, 225, 230, 233
Orchis 2, 76, 235
Oreorchis 96
Ornithophily 129
Orobanchaceae 62, 218, 225
Orobanche 218
Orthotropous ovule 59
Oryza 38, 135, 238, 283
Ovary culture 7, 294
Ovule 7
Ovule culture 7, 293
Oxalis 161
Oxyspora 181
Oxystelma 14, 69

Paeonia 4, 216, 217, 228, 269, 270, 271
Palynology 41, 46, 47, 57
Panicum 245, 246, 251, 254, 255
Papaver 7, 163, 169, 171, 174, 198, 200, 291
Papaveraceae 70, 169, 170, 171, 174
Paphiopedilium 56
Papilionaceae 68
Parrotiopsis 188
Parthenium 51, 165, 247, 251
Parthenogenesis 252
Paspalum 99, 101, 153, 237, 246, 248, 251, 253, 255
Passiflora 188, 196, 197
Pedicularis 65, 201, 222
Peltiphyllum 234
Peltophorum 130
Penaea 89, 91, 222
Penaeaceae 91

Pennisetum 153, 159, 188, 245
Peperomia 89, 90, 91, 200
Periplasmodial tapetum 16, 18
Perisperm 226
Peristylus 56
Persea 225
Petunia 99, 164, 168, 283, 292, 305
Phalaenophily 127
Phalaenopsis 291
Phalaris 163
Phaseolus 293
Philydraceae 24, 213
Phlox, 159
Phylidrum 193
Physalis 163
Phytolacca 53
Phytolaccaceae 53
Pinus 50, 299
Piper 73, 93, 226
Piperaceae 59, 60, 205, 206, 226
Pistacia 139
Pisum 223
Platanthera 56
Plectranthus 14
Plumbagella 89, 94
Plumbaginaceae 22, 59, 92
Plumbago 31, 89, 103, 146
Poa 76, 234, 237, 244, 247, 253
Poaceae 11, 118, 150, 163, 166, 233
Podostemaceae 267, 268
Podostemon 74, 88
Polarity 108
Pollen association 52
pollen culture 286, 287
Pollen embryo sacs 31
Pollen germination 137, 155, 156
Pollen kitt 21, 35
Pollen nuclear number 52
Pollen storage 153
Pollen tube entry 143
growth 144
Pollen tube-ovule interaction 166
Pollen tube-style interaction 166
Pollen viability 155
Pollen-pistil interaction 148, 155
Pollen-stigma interaction 166
Pollination Mechanisms 117, 118
Pollinium 24, 56
Polygonaceae 59, 60, 267
Polygonum 84
Polyploidy 253
Polyspermy 157
Polyspory 25
Polystachya 56
Poncirus 230, 290
Pontederiaceae 60, 213
Populus 53, 176, 301
Portulaca 2
Portulacaceae 206, 226, 272
Potamogetonaceae 213
Potentilla 76, 244, 247, 253, 255
Primula 161
Primulaceae 59, 60
Prionopsis 108
Prosopis 50, 56, 159
Protoplasts 302, 303
Prunus 172
Pseudo embryo sac 226
Pseudogamy 251, 252
Psychophily 128
Putoria 76
Pyrus 225

Quercus 51, 114, 116, 152, 219
Quinchamalium 100, 106, 108, 109

Ranunculaceae 218, 269, 270, 271
Ranunculus 121, 152, 198, 200, 244, 251, 255, 295
Raphanus 6, 176, 201, 295
Redifferentiation 279
Rhamnaceae 10
Rheum 76
Rhizophora 237
Rhus 70
Ricinus 1, 50, 71
Rivinia 53, 70
Rosa 299
Rosaceae 68, 72, 76, 163, 169, 171, 243
Rosularia 76, 77
Rothia 186
Rubiaceae 11, 76, 161, 222
Rudbeckia 93, 218, 249, 251, 252
Ruellia 191
Rumex 76
Ruminate endosperm 196

Sagina 211, 212
Salix, 51, 53, 188
Salvia 14
Sambucus 89
Santalaceae 100, 108, 109, 115, 196, 237, 300
Santalum 109, 110, 111, 126
Sapindus 49
Sapotaceae 104
Sarcotesta 64
Sarracenia 188
Sasa 105
Satyrium 222
Saxifraga 159
Schisandra 86, 88
Scilla 87, 105
Scleria 186, 187
Scrophulariaceae 109, 134, 163, 169, 171, 206
Secale 201, 238, 239, 297
Sedum 75, 76, 77, 100
self incompatibility 163, 164, 165
Semigamy 218, 251, 252
Senecio 106, 209, 210
Sesamum 53, 176
Silene 34, 38, 175
Smilacina 92
Sobralia 222
Solanaceae 11, 65, 68, 134, 163, 169, 170, 171, 175, 206, 304
Solanad Type 206, 212, 267, 270, 271, 275
Solanum 127, 159, 164, 173, 176, 178, 238, 283, 305
Solid style 135
Somatic Hybridization 177
Sorghum 125, 133
Sparganiaceae 193, 213
Sparganium 24, 193
Sperms 146
Spigelia 197
Spinifex 237
Spiranthes 26, 27, 56, 233
Spirodela 68
Stackhousia 105, 188
Statice 93, 249
Stellaria 198, 220, 221, 223, 224
Stigma 133, 134, 163, 169
Stipa 98

Strophiole 68
Struthanthus 192
Stump Pollination 173
Style 133, 134, 168
Stylosanthes 185
Suspensor 223, 236, 237, 271
Suspensor polyembryony 236, 237
Swertia 13
Synergid haustoria 100, 107
Synergid polyembryony 234, 235
Synergids 92, 98, 99, 116, 145, 235
Syngamy 132, 149, 151, 204

Tagetus 176
Tamarix, 69, 73, 93
Tanacetum 92
Tapetum 14
    Amoeboid 15
    Dimorphic 14
    Function 20
Taraxacum 139, 246, 248, 249, 253
Tellima 234
Tenuinucellate 60, 266
Terniola 73, 88
Test tube Fertilization 174
Tetragonia 70
Tetrasporic Embryo Sacs 88
Theaceae 206
Themeda 246, 253
Theobroma 53, 169, 175
Thesium 109, 200
Thismiaceae 104
Thunbergia 25, 191
Tilia 222, 299
Tiliaceae 10
Tithonia 65, 66, 182
Toddalia 230
Tolypanthus 192
Torenia 98, 109, 110, 140, 141, 142
Totipotency 279
Tradescantia 2, 18, 26, 35, 176, 220
Trapa 73, 271
Trapaceae 179, 203, 222, 271
Trianthema 63
Tridax 87
Trifolium 134, 164, 172, 176, 237, 297
Tripsacum 257, 258, 259
Tristyly 161
Triticale 201, 282
Triticum 177, 199, 201, 215, 216, 239, 256, 283, 288, 297
Tropaeolum 221
Tulipa 89, 93, 237

Ubisch bodies 16
Ulmus 51, 89, 96
Ultrastructure 18, 116
Unisexuality 160
Urginea 9
Urochloa 246
Ursinia 100
Urticaceae 60, 230, 243

Vanda 219, 221, 222
Vanilla 56
Vectors 117, 118
Vegetative cell 27
Verbascum 197
Vernonia 21
Violaceae 206
Viscaceae 87
Viscaria 175, 295

Viscum 201
Vitaceae 10
Vitis 154, 160, 289, 290
Vogelia 92

Wasps 125
Wickstroemia 248
Winteraceae 10, 24, 56

X-bodies 146
Xeromphis 128
Zea 29, 38, 45, 102, 105, 116, 119, 133, 142, 201, 237, 238, 257, 258
Zephyranthes 7, 100, 114, 200, 247, 251, 291
Zingiberaceae 54, 75, 108, 226
Zinnia 9, 65, 66
Zostera 120